**Fundamentals of
numerical analysis**

Fundamentals of numerical analysis

STEPHEN G. KELLISON
Fellow of the Society of Actuaries
University of Nebraska–Lincoln

 1975

RICHARD D. IRWIN, INC. Homewood, Illinois 60430
Irwin-Dorsey International London, England WC2H 9NJ
Irwin-Dorsey Limited Georgetown, Ontario L7G 4B3

First Printing, January 1975

ISBN 0-256-01612-7
Library of Congress Catalog Card No. 74–78157
Printed in the United States of America

Foreword by the Casualty Actuarial Society and the Society of Actuaries

SINCE THE INTRODUCTION of numerical analysis as a subject on the examinations required for admission to membership in the Casualty Actuarial Society and the Society of Actuaries, their respective Education and Examination Committees have felt the need for a textbook which would fully meet the requirements of students preparing for the examinations. This textbook has been accepted by the Education and Examination Committees of the Casualty Actuarial Society and the Society of Actuaries as the reference on numerical analysis. It should also be useful to practicing actuaries and others who are concerned with this subject. The Societies are pleased that Richard D. Irwin, Inc. has undertaken to publish this text.

The Casualty Actuarial Society and the Society of Actuaries appreciate the great amount of work undertaken by Mr. Kellison in preparing this textbook and express their gratitude for this important contribution to the literature on numerical analysis.

PAUL S. LISCORD *President*
Casualty Actuarial Society

GEORGE D. MORISON
General Chairman
Education and
Examination Committee
Casualty Actuarial Society

EDWARD A. LEW *President*
Society of Actuaries

PETER W. PLUMLEY
General Chairman
Education and
Examination Committee
Society of Actuaries

Preface

THIS BOOK is an introductory treatment of the mathematical theory and practical applications of numerical analysis. It is designed to serve as a textbook for students preparing for the examination jointly administered by the Casualty Actuarial Society and the Society of Actuaries covering numerical analysis and as a reference for practicing actuaries. Its usefulness is not restricted to actuarial science, however, for it includes a wide range of applied disciplines in which the numerical solution of mathematical problems is required.

Numerical analysis had its origin in the mathematical theory of finite differences. This latter subject is of academic interest as the finite analogue of infinitesimal calculus. However, finite differences proved to be of more than theoretical interest, since certain results were found to be useful in developing numerical solutions to mathematical problems for which a direct, analytical solution was difficult or impossible to determine.

Over time, the scope and techniques of numerical analysis have expanded considerably from their finite-difference origin. A major impetus for this growth has been the development of high-speed digital computers which make many techniques feasible that formerly were too time-consuming or expensive to implement. Thus, numerical analysis is a dynamic subject reflecting the continuing development of new and more powerful techniques for the numerical solution of mathematical problems.

vii

The algorithms presented in this book are capable of being efficiently programmed for the computer. However, unlike so many other textbooks in numerical analysis, a knowledge of some specific programming language, such as FORTRAN, is not required. Although many of the various algorithms are presented with their suitability for programming clearly in mind, specific references to a particular programming language do not appear until Chapter 16, at which point the elementary language BASIC is introduced. This approach is chosen so that the book can be used successfully by students with a wide variety of backgrounds. It is the author's conviction that a knowledge of computer programming, although quite useful, is not essential for an understanding of the fundamentals of numerical analysis.

A knowledge of elementary calculus is required background for the entire book. Some exposure to differential equations is helpful, but not essential, for the numerical solution of differential equations contained in Chapter 10. Also some knowledge of linear algebra is useful for the discussion of systems of linear equations and linear programming presented in Chapters 13 and 14, respectively. A review of the basic properties of matrices and determinants is given in Chapter 12 for those requiring this background. Readers with some knowledge of linear algebra will find much of the material in Chapter 12 to be familiar to them.

The book is designed to be appropriate for both classroom use and for self-study. The amount of material is sufficient to constitute a two-semester sequence. Such a comprehensive treatment of the subject could be supplemented with additional treatment of programming applications. A one-semester course can also readily be structured by selecting those chapters and sections which an instructor wishes to emphasize. Although there are a certain number of topics for which the material is dependent upon previous development, there are many other topics which are relatively independent and which can be learned without extensive previous development. This independence was maintained to the extent possible to facilitate the omission of certain chapters and sections by those readers requiring less than a complete development of the subject.

Since a thorough working knowledge of numerical analysis can only be obtained by the solution of problems, each chapter (except for Chapter 16) includes a large number of examples and exercises. Throughout the book alternative approaches to problems are emphasized. The answers to all of the exercises are given at the end of the book. The exercises are intended to be nonrepetitive, with each exercise illustrating a slightly different point. Although a few exercises do entail a significant amount of computation, the great majority do not and can be worked without even the aid of a calculator. The author has attempted to design exercises which illustrate the theory and techniques without requiring inordinate amounts of computation. Readers with access to digital computers or programmable calculators are encouraged to apply the techniques developed to problems involving more extensive computation.

The reading material in the 16 chapters is supplemented with 8 appendixes. These appendixes cover topics which may be of interest to certain readers, but are not of sufficiently broad interest to warrant inclusion in the chapters. Each of the appendixes is referred to in the text that is most closely related to its contents.

The author is deeply indebted to Michael J. Cowell, F.S.A., who reviewed the entire manuscript and made numerous helpful suggestions increasing the clarity of exposition. His review was more thorough and painstaking than most and was invaluable in the preparation of the manuscript. Furthermore, Chapter 16 and Appendix H were adapted from a study note written by Mr. Cowell for the Society of Actuaries with only minor editorial changes, and the credit for these portions of the book should be attributed to him.

The author also wishes to acknowledge that the presentation of osculatory interpolation contained in Section 15.2 was largely adapted from a study note of the Society of Actuaries on graduation written by T.N.E. Greville, F.S.A. The statement concerning the optimum number of terms to use in the iteration method of matrix inversion contained in Section 12.6 is based on unpublished research of Professor Lebert Alley of the University of Nebraska–Lincoln.

The author is grateful to the University of Nebraska–Lincoln students who used a rough draft of the book under classroom conditions during the 1972–73 and 1973–74 academic years. As a result of this classroom experience numerous improvements were incorporated into the book. The long and arduous effort in typing the manuscript was diligently and cheerfully performed by Geralyn Dukich and Peggy Jaixen, and the author is appreciative of their dedicated effort. Finally, the author wishes to express his gratitude to the Education and Examination Committees of the Casualty Actuarial Society and the Society of Actuaries for their acceptance of this textbook for use on the actuarial examination covering numerical analysis.

December 1974   STEPHEN G. KELLISON

Contents

xi

1

Introduction to numerical analysis

1.1. INTRODUCTION

NUMERICAL ANALYSIS can be defined as the development and implementation of techniques to find numerical solutions to mathematical problems.

In classical mathematics, such as basic algebra and calculus, the approach taken is to present a body of theory which guarantees the existence or non-existence of solutions to various general types of problems. This theory is then usually applied to illustrative problems in which the solution proceeds along direct analytical lines to the final answer. A solution of this type is sometimes called a *closed-form* solution.

However, the student quickly discovers that classical mathematics can be capricious. Whereas one problem may possess a direct analytical solution, another quite similar problem may lack such a solution. Two examples of seemingly elementary, straightforward problems which lack analytical solutions are: (1) to find the roots of a nontrivial fifth-degree polynomial, and (2) to find the integral of e^{x^2}. The capricious nature of classical mathematics is well illustrated by example (2). Had the function been e^x, xe^x, or xe^{x^2}, a straightforward analytical solution would be available.

Even if an analytical solution to a problem does exist, it may be difficult to implement. In many such cases a numerical approach may be superior to the analytical approach. Two examples of problems for which cumbersome

1

analytical solutions exist but for which a numerical approach is superior are: (1) to find the roots of a nontrivial fourth-degree polynomial, and (2) to find the solution of 100 linear equations in 100 unknowns by the traditional method of evaluating determinants.

Finally, some problems arise in which classical mathematics cannot be applied at all. For example, data on mortality rates by age may be available at every age which is a multiple of five. If a rate is needed at age 23, some type of interpolation is required. A closed-form solution does not exist, since no underlying mathematical formula exists for data of this type other than on an approximate basis.

The techniques of numerical analysis are designed to find numerical answers for the types of examples mentioned above and for a variety of other problems as well. Whereas the emphasis in classical mathematics is upon the development of general theory for a large class of problems, the emphasis in numerical analysis is upon the development of accurate and efficient techniques to solve specific problems.

Thus, one disadvantage of numerical analysis is that the approaches taken are specific and relate to only one problem or to a limited number of problems. If the computations involved are quite extensive, as is often the case in practice, the volume of such computation may not be justified. However, the advent of high-speed digital computers has made feasible many computational techniques that formerly were too time-consuming and too expensive to implement. Modern computers are capable of doing a volume of computations in seconds that would take years by hand. In fact, it would not be an overstatement to say that the rapid growth in the use of the techniques of numerical analysis in recent years has directly resulted from the computer revolution.

1.2. ALGORITHMS

The basic approach used to solve problems in numerical analysis is the *algorithm*. An algorithm is a complete, well-defined procedure for obtaining a numerical answer to a given mathematical problem. Algorithms are expressed as a finite number of ordered computational steps. Algorithms are really nothing new to the student; many techniques of classical mathematics are expressible as algorithms.

Algorithms are essentially mathematical models which take input data from some source and process the data to produce a required output. Most of applied mathematics is concerned with the building of mathematical models. Much of the motivation for the discipline of numerical analysis arises from the need to be able to solve the mathematical formulas in these models. The concept of an algorithm is illustrated by the elementary flowchart in Figure 1.1.

In classical mathematics different solutions to a given problem are all equivalent in one sense. A mathematician may be able to say of two correct solutions to a problem that solution A is more "elegant" than solution B, but it really cannot be said that there is anything wrong with solution B.

Similarly, in numerical analysis different algorithms for the same problem often exist. However, it is usually quite important to be able to say whether one algorithm is "better" than another in a computational sense.

FIGURE 1.1

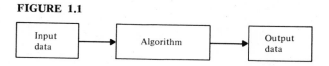

Two important criteria in judging the acceptability of an algorithm are *speed* (sometimes called *efficiency*) and *accuracy*. With many mathematical models involving numbers of computations in the millions, speed is clearly of vital concern. It is not uncommon for one algorithm to be many times faster than another analytically equivalent algorithm.

The accuracy of an algorithm is also crucial. The errors arising from the approximations used and the large volume of computations performed are an ever-present problem in numerical analysis. Sections 1.3, 1.4, and 1.5 will examine the problem of error more completely.

Fortunately, in many circumstances, there is no necessary trade-off between speed and accuracy, that is, one does not have to sacrifice speed to improve accuracy, and conversely. The challenge in numerical analysis is to find algorithms which are both "fast" and "accurate."

Thus, numerical analysis is seen to be an art as well as a science. Part of the work of the numerical analyst is to search for better algorithms to improve speed, accuracy, or, better yet, both. Numerical analysis is not a static discipline; algorithms may become obsolete and be replaced by more powerful algorithms as computer capability increases or as new techniques are developed.

Classical mathematics may be of only limited value in the search for better algorithms. One algorithm which is elegant analytically may be a poor approach to use in computation, whereas another analytically more clumsy algorithm may be quite powerful to use in computation. The situation is further complicated by the fact that an algorithm which works well on one set of input data may prove to be much less successful on another. Universal approaches which are fast and accurate for all problems of a given type are comparatively infrequent in numerical analysis.

Example 1.1.   Compute $(.987)^{-1}$ *and* $(.321)^{-1}$ *to three decimal places using the following algorithms:*
   (1)   Long division.
   (2)   Series expansion of $(1 - x)^{-1}$.

Computation of $(.987)^{-1}$:

(1)   Computing $(.987)^{-1}$ with long division gives

$$
\begin{array}{r}
1.013 \\
.987\,\overline{)\,1.000000} \\
987 \\
\hline
1300 \\
987 \\
\hline
3130 \\
2961 \\
\hline
169
\end{array}
$$

Thus, the answer is 1.013 to three decimal places.

(2).   The series expansion for $(1 - x)^{-1}$ is given by

$$\frac{1}{1 - x} = 1 + x + x^2 + x^3 + \cdots \qquad \text{for } -1 < x < 1$$

This series is derived by using the binomial theorem on the left-hand side of the equation or by using the sum of an infinite geometric progression on the right-hand side.

In this example $x = 1 - .987 = .013$, and the answer is immediately $1 + x = 1.013$, since x^2 and higher powers contribute nothing to the third decimal place.

Computation of $(.321)^{-1}$:

(1)   Computing $(.321)^{-1}$ with long division gives

$$
\begin{array}{r}
3.115 \\
.321\,\overline{)\,1.000000} \\
963 \\
\hline
370 \\
321 \\
\hline
490 \\
321 \\
\hline
1690 \\
1605 \\
\hline
85
\end{array}
$$

Thus, the answer is 3.115 to three decimal places.

(2)   It is evident that the series expansion will require considerable computation to produce three-decimal accuracy. Rather than to routinely perform all this computation, it is instructive to make the following analysis in determining how many terms of the series must be used to produce three-decimal accuracy.

The series expansion to $n + 1$ terms is

$$1 + x + x^2 + \cdots + x^n = \frac{1 - x^{n+1}}{1 - x}$$

To produce three-decimal accuracy this must differ from the true answer by less than .0005. Thus, we have

$$\left| \frac{1}{1 - x} - \frac{1 - x^{n+1}}{1 - x} \right| < .0005$$

$$\frac{x^{n+1}}{1 - x} < .0005$$

$$x^{n+1} < .0005(1 - x)$$

and letting $x = 1 - .321 = .679$, we have

$$(.679)^{n+1} < (.0005)(.321) = .0001605$$

If we solve this inequality, we discover that $n + 1 = 23$, that is, 23 terms are required to produce only three-decimal accuracy!

This elementary example shows that more than one algorithm may exist for even the simplest problem. It also shows that an algorithm which works well with one set of data may be much less successful with another set.

In considering speed, algorithm (2) would be preferred for $(.987)^{-1}$, while algorithm (1) would be preferred by a wide margin for $(.321)^{-1}$.

In considering accuracy, there is little difference in the two algorithms, since either is capable of producing three-decimal accuracy. It should be noted, however, that in computing $(.321)^{-1}$ with algorithm (2), roundoff error may become severe in view of the large volume of computations. In particular, it will be necessary to carry more than three decimal places in each term of the series to produce three-decimal accuracy in the answer. The problem of roundoff error is examined more completely in Section 1.5.

Finally, in algorithm (2) an infinite series is replaced with a finite series. The process of replacing one series by another with fewer terms is called *truncation*, and the error arising from it is examined in Section 1.4.

Example 1.2.   Find the two roots of the quadratic equation $x^2 - 20x + 1 = 0$
using three-digit arithmetic.

The classical algorithm for finding the roots of the quadratic equation
$ax^2 + bx + c = 0$ is

$$x = \frac{-b \pm \sqrt{b^2 - 4ac}}{2a}$$

Applying this formula to the example gives

$$x = \frac{20 \pm \sqrt{400 - 4}}{2} = 10 \pm \sqrt{99}$$

The student should note that "three-digit arithmetic" is not the same as
the "three-decimal arithmetic" used in Example 1.1. Three-digit arithmetic
uses the first three significant digits[1] regardless of where the decimal point
appears. This is typical of the operation of most modern computers and programming languages, although normally machine capacity for a number is
much larger than three digits. These concepts will be explored more fully in
Section 1.5.

The larger root presents no problem, since

$$10 + \sqrt{99} = 10.0 + 09.9 = 19.9$$

The student should note that 09.9 has only two-digit accuracy. However, the
leading zero is necessary to perform the addition with 10.0. This is typical of
the addition and subtraction processes with most computers.

The smaller root does present a problem, since

$$10 - \sqrt{99} = 10.0 - 09.9 = 00.1$$

The answer is accurate to only one significant digit. In this case, significant
digits are lost because 10.0 and 09.9 are so nearly equal. The loss of significant
digits in the subtraction of two nearly equal numbers is one of the more
frustrating types of roundoff error encountered in numerical analysis.

This severe roundoff error can be avoided by using an alternative algorithm.
We have

$$10 - \sqrt{99} = \frac{(10 - \sqrt{99})(10 + \sqrt{99})}{10 + \sqrt{99}} = \frac{1}{10 + \sqrt{99}}$$

$$= \frac{1}{10.0 + 09.9} = \frac{1}{19.9} = .0503$$

---

[1] "Significant digits" will be precisely defined in Section 1.5.

using three-digit arithmetic. Modern computers are generally capable of producing this level of accuracy in the multiplication and division processes. Note that no more than three significant digits are carried anywhere in the computations, although the computer is capable of properly positioning the decimal point.

1.3. TYPES OF ERROR

It is important in applying techniques of numerical analysis to be able to analyze the sources and magnitudes of various types of *error* and to choose algorithms which will minimize error. Figure 1.2 follows from Figure 1.1 and illustrates the presence of error. Our objective is to hold *output error* to a minimum. Output error can arise from both *input error* and *algorithm error*.

FIGURE 1.2

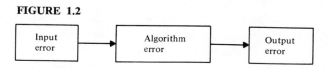

Input errors, other than mistakes or gross blunders, typically arise from imperfect measurement of real-world phenomena. Measurement of any continuous quantity, such as temperature or weight, will always contain some errors of this type. Also, input data may be based upon various types of approximations which introduce error. Finally, input data may be subject to random error in a statistical sense.

The control of input error is dependent upon the nature of the input and is largely beyond the scope of this book, although Section 2.7 describes briefly one technique which may be of some value in detecting error of this type. Input error arising from rounded measurements can be analyzed in the same manner as algorithm roundoff error.

Algorithm errors, other than mistakes or gross blunders, are of two types, *truncation error* and *roundoff error*. Truncation error refers to the error involved in using approximate processes and formulas, whereas roundoff error refers to the error involved in using approximate numbers. These two types of error are illustrated in Examples 1.1 and 1.2 and will be analyzed in more detail in Sections 1.4 and 1.5.

Numerical analysis is concerned with the measurement and minimization of algorithm error. Since truncation error and roundoff error can both be present in any problem, we have

$$\text{total algorithm error} = \text{truncation error} + \text{roundoff error} \qquad (1.1)$$

Truncation error and roundoff error can be either positive or negative. Thus, they may partially offset each other or reinforce each other, depending upon the situation. Total algorithm error is often quoted as a positive number, so that the absolute value of the sum of truncation error and roundoff error is used. The term *absolute* when attached to a descriptive phrase involving error refers to absolute value.

In general, one of our objectives is to find algorithms which minimize the growth of algorithm error for an increasing number of computations. As an intuitive aid, it can be said that error which grows proportionally, or slower, to an increasing number of computations is normal and often unavoidable. Error growth which is more rapid than the increasing number of computations is usually disastrous.

Care must be taken in using output data not to attribute to them a greater degree of accuracy than is warranted by the input and algorithm errors. Also, it is usually wasted effort to attempt to develop highly sophisticated algorithms to use on crude input data, since the accuracy in the algorithm will not overcome the inaccuracies in the input.

In considering error it is important to distinguish between *actual error* (or just *error*) and *relative error*:

$$\text{actual error} = \text{true answer} - \text{computed answer} \qquad (1.2)$$

$$\text{relative error} = \frac{\text{true answer} - \text{computed answer}}{\text{true answer}} \qquad (1.3)$$

Relative error is generally a better measure of the extent of error than is actual error. For example, an error of .5 in rounding numbers in the vicinity of 1000 is much less important than the same error for numbers in the vicinity of one. The student should note that relative error is undefined if the true answer is equal to zero.

Also, it is useful to distinguish between *errors* and *mistakes*, as we are using the terms. A *mistake* is an unintentional blunder which is correctable when found (although it may be quite difficult to find). A mistake may occur in either the input data or in the algorithm. An example of an input mistake would be to misread gauges in taking a reading for a physics problem. Examples of algorithm mistakes would be a formula error or a "bug" in a computer program. If the proper care is taken in preparing the input and in formulating the algorithm, mistakes should be held to a minimum.

We shall use the term *errors* to include mistakes and also the more subtle inaccuracies of other types described above. These other errors arise from such sources as measurement limitations in input data, use of approximation techniques, roundoffs, etc. These types of errors are often unavoidable by-products of the types of input data available and the numerical analysis techniques used. We shall primarily concern ourselves with errors of this latter type rather than with mistakes, although it may not always be clear-cut whether an error is a mistake or another type of error.

Example 1.3.   *Find the absolute actual and relative errors in Example 1.2.*

We cannot determine the true answers in a computational sense. However, we shall assume that six-digit (not six-decimal) accuracy is sufficiently close to "true" for purposes of this example.

The larger root to six digits is

$$10 + \sqrt{99} = 19.9499$$

Absolute actual error for the original algorithm is

$$|19.9499 - 19.9| = .0499$$

Absolute relative error for the original algorithm is

$$\left| \frac{19.9499 - 19.9}{19.9499} \right| = .0025 \quad \text{or } .25\%$$

The smaller root to six digits is

$$\frac{1}{10 + \sqrt{99}} = .0501256$$

Absolute actual error for the original algorithm is

$$|.0501256 - .1| = .0498744$$

Absolute relative error for the original algorithm is

$$\left| \frac{.0501256 - .1}{.0501256} \right| = .9950 \quad \text{or } 99.50\%$$

which is clearly unacceptable.

Absolute actual error for the alternative algorithm is

$$|.0501256 - .0503| = .0001744$$

Absolute relative error for the alternative algorithm is

$$\left| \frac{.0501256 - .0503}{.0501256} \right| = .0035 \quad \text{or } .35\%$$

which is only slightly poorer than the absolute relative error for the larger root from the original algorithm.

It is interesting to note that the absolute actual error in the original algorithm is approximately equal to .0499 for both the larger and smaller roots, whereas the absolute relative error is .25 % for the former and 99.50 % for the latter. This vividly illustrates the significance of relative error. Small actual errors can become large relative errors if the true answers are close to zero.

1.4. TRUNCATION ERROR

In Example 1.1 *truncation* is defined as the replacement of one series by another with fewer terms. *Truncation error* is the error arising from this approximation.

Actually we shall use a somewhat broader definition of truncation error than the above to include the error involved in using approximate processes and formulas in general. Examples of truncation errors which will be introduced in later sections of the book are the error arising from using polynomials to approximate nonpolynomials and the error involved in various approximate integration techniques.

In analyzing the errors arising from the truncation of series, several types of series expansions commonly occur. These include, but are not limited to, the following:

1.   Infinite geometric progression.
2.   Binomial expansion.
3.   Taylor series.

It is assumed that the student is familiar with these basic series expansions from previous courses in algebra and calculus.

Example 1.1 considered truncation error in a problem involving an infinite geometric progression. The following examples are instructive in illustrating some other types of series and approaches.

Example 1.4.   It is required to compute e^x for $-1 \leq x \leq 1$ by means of the standard series expansion.
  (1)   Find how many terms must be used to guarantee three-decimal accuracy for all x in the interval $-1 \leq x \leq 1$.
  (2)   Find an upper bound for absolute relative error.

In working this problem we assume that enough decimal places are carried so that roundoff error does not affect the answers.

(1)   The standard series expansion for e^x is

$$e^x = 1 + x + \frac{x^2}{2!} + \cdots + \frac{x^n}{n!} + \cdots$$

This is derived from the Taylor series

$$f(x) = f(0) + xf^{(1)}(0) + \frac{x^2}{2!} f^{(2)}(0) + \cdots + \frac{x^n}{n!} f^{(n)}(0) + R_{n+1}(x)$$

where

$$R_{n+1}(x) = \frac{x^{n+1}}{(n+1)!} f^{(n+1)}(\xi)$$

for some ξ between 0 and x. Since $f^{(i)}(x) = e^x$ for $i = 0, 1, 2, \ldots$, the remainder term gives

$$|R_{n+1}(x)| = \left| \frac{x^{n+1}}{(n+1)!} e^\xi \right| \leq \frac{e}{(n+1)!}$$

for $-1 \leq x \leq 1$ and $-1 \leq \xi \leq 1$.

To produce three-decimal accuracy, we set

$$\frac{e}{(n+1)!} < .0005$$

or

$$(n+1)! > \frac{e}{.0005} = \frac{2.7183}{.0005} = 5437$$

to the nearest integer. Now $7! = 5040$ and $8! = 40{,}320$, so that $n + 1 = 8$, and eight terms will be required (that is, through $n = 7$).

(2) To obtain an upper bound for absolute relative error, note that

$$|\text{relative error}| = \left| \frac{\text{actual error}}{\text{true answer}} \right|$$

so that we must maximize the numerator and minimize the denominator. Thus,

$$|\text{relative error}| < \frac{.0005}{e^{-1}}$$

$$= \frac{.0005}{.3679}$$

$$= .0014 \qquad \text{or about .14\%}$$

This upper bound is likely to be quite pessimistic. For example, if we compute eight terms of the series for e^{-1}, we obtain .36785714 to eight decimal places. The exact value of e^{-1} to eight decimal places is .36787944. This produces a relative error of

$$\left| \frac{.36787944 - .36785714}{.36787944} \right| = .000061$$

or only about .0061%.

Example 1.5.   Express $\log_e 2$ as a series expansion and find how many terms must be used to guarantee three-decimal accuracy.

Again we assume that enough decimal places are carried so that roundoff error does not affect the answer.

The Taylor series expansion for $\log_e (1 + x)$ is given by

$$\log_e (1 + x) = x - \frac{x^2}{2} + \frac{x^3}{3} - \cdots + (-1)^{n+1} \frac{x^n}{n} + \cdots$$

which converges for $-1 < x \le 1$. Thus, we have

$$\log_e 2 = 1 - \frac{1}{2} + \frac{1}{3} - \cdots + (-1)^{n+1} \frac{1}{n} + \cdots$$

In this example we utilize a result from calculus for infinite series with monotonic decreasing terms in absolute value and alternating signs. If S is the true sum of an infinite series

$$S = a_1 + a_2 + \cdots$$

and S_n is the sum to n terms

$$S_n = a_1 + a_2 + \cdots + a_n$$

then the result is

$$|S - S_n| < |a_{n+1}|$$

In other words, this result states that the error in truncating such a series after n terms is less than the first omitted term.

Thus, we have

$$|S - S_n| < |a_{n+1}|$$

$$= \frac{1}{n + 1}$$

$$< .0005$$

or

$$n + 1 > \frac{1}{.0005} = 2000$$

Thus, we need 2000 terms to guarantee only three-decimal accuracy! This is indeed a slowly converging series. An alternative algorithm which is much more efficient for this example will be developed in Exercise 8.

1.5. ROUNDOFF ERROR

Roundoff error is the computational error in a numerical problem arising from the use of approximate numbers. In a sense, roundoff error could be called a form of truncation error, since finite decimal expansions are used instead of infinite decimal expansions. However, it is useful for our purposes to let *truncation error* refer strictly to the error involved in the use of approximate processes and formulas other than in the computations.

As we are using the term, *roundoff error* is a form of algorithm error arising from the computations performed in the algorithm. However, the input data on which the algorithm is used are often rounded off and are not exact. This type of input error can be handled analytically as an additional roundoff error using the principles developed in this section.

The term *significant digits* is used in the preceding sections, and the student has undoubtedly encountered this term before. In considering roundoff error it is important to be precise in the usage of this term. The significant digits in a number are those which carry real information about the true magnitude of the number. To be significant, the last digit contained should be accurate within half a unit in the last decimal place. For example, if an approximate value of x is said to equal 3.72 and the true value of x lies in the interval $3.715 \leq x \leq 3.725$, then x is said to have three significant digits. However, if somehow x could equal 3.712 or 3.727, then the number 3.72 would have fewer than three significant digits.

The significant digits in a number are not dependent upon the position of the decimal point in that number. For example, 7328 and .7328 both contain four significant digits. However, the student should be careful to note which digits are significant when using leading or following zeros. In some cases the position of the decimal point may have an effect on the number of significant digits, if the numbers are written in conventional form.

In considering significant digits, the following rules are typically used for numbers written in conventional form:

1. Leading zeros are not significant.
2. Following zeros that appear after the decimal point are significant.
3. Following zeros that appear before the decimal point may or may not be significant. More information is required.

These rules can be illustrated by the following examples:

1. The number .006037 has four significant digits. The leading zeros are not significant.
2. The number .0060370 has five significant digits. The following zero should not be written unless it is significant.
3. The number 603,700 may have four, five, or six significant digits depending upon the situation. The conventional form of writing numbers is somewhat ambiguous in this instance.

It is important to distinguish between *rounding* and *chopping*. In eliminating ending digits with rounding, portions that are less than one half a unit in the last retained decimal place are rounded down, while portions greater than one half a unit are rounded up. With chopping, ending digits are always dropped. Thus .3877 rounded to two decimal places is .39, but it is .38 chopped to two decimal places. Some computers and programming languages round numbers and some chop them. Clearly, rounding is to be greatly preferred to chopping, all other things being equal. It should be noted that chopping numbers will result in the loss of significance of the final digit about half the time and will tend to double roundoff error. Furthermore there is much greater danger of a significant nonrandom accumulation of roundoff error when chopping.

One minor problem occurs in rounding numbers which differ by exactly one half a unit in the last rounded digit. One possibility is to always round down, while a second possibility is to always round up. A third possibility, which is commonly used, is to round so that the result is always even. For example, 2.735 and 2.745 would both be rounded to 2.74 while 2.755 and 2.765 would both be rounded to 2.76. This third approach would seem to be less biased than either of the first two approaches, since the round ups and round downs might tend to offset each other.

Examples 1.1 and 1.2 illustrate the difference between what is termed "three-decimal arithmetic" and "three-digit arithmetic." The former is an example of *fixed-point arithmetic*. This procedure, in which a fixed number of decimal places is carried, is typically used in computations by hand or with a desk calculator and is also used in certain programming languages.

The latter is an example of *floating-point arithmetic*. This type of computation is commonly used in scientific work and in certain programming languages. Numbers used in floating-point arithmetic are expressed in the form $a \times 10^b$. If $.1 < a \le 1$, then the number is said to be *normalized*. For example, $91.65 = .9165 \times 10^2$ and $.003217 = .3217 \times 10^{-2}$ are two normalized floating-point numbers. Multiplication and division are especially easy to handle with numbers in this form, since the coefficients, a, can be multiplied or divided and the exponents, b, can be added or subtracted. The student should note that the number of significant digits contained in any number expressed in normalized floating-point form is equal to the number of decimal places in the coefficient, a.

The above discussion of floating-point numbers assumes that the *decimal number system* (base 10) is being used. The decimal number system is typically used in computations performed by hand or with a desk calculator. However, almost all computers internally use the *binary number system* (base 2), since the storage in a computer is composed of millions of electrical connections, each of which may be on or off. The use of the binary number system will not be especially important in our work in numerical analysis, since the computer automatically converts number bases. The computer accepts decimal numbers as input, converts them to binary form for internal processing, and then converts the output back to decimal form.

Analyzing the amount of roundoff error is one of the more frustrating problems encountered in numerical analysis, and no ideal solution has yet been found. Roundoff errors which are independently distributed in a statistical sense are usually not serious; however, if they become correlated or accumulative, then the distortions can become severe.

One approach in analyzing roundoff error is to perform all the computations two times carrying twice as many significant digits the second time as the first. The second computation will then give an estimate of the roundoff error in the first. This is basically the approach used in Example 1.3. In computer terminology the first computation would be termed *single precision* and the second *double precision*. Computers use single precision unless double precision is specified. Double precision does lead to twice as many significant digits but typically requires more than a doubling of computer capacity and running time to implement. Thus, double precision may not be possible to implement in a computer with limited capacity, and it is expensive in any event even if it can be implemented.

A more mathematical approach is given by *interval arithmetic*. In this approach each rounded number is replaced with two numbers representing its maximum and minimum possible values. Then two computations are performed in which numbers are chosen at each step to maximize the answer in one case and to minimize it in the other. The true answer must lie somewhere in the interval for the answer. Unfortunately, if a significant volume of computations is performed, the interval for the answer can become too "wide" to produce satisfactory results. Moreover, it is often not safe to say that the true answer is about half way through the interval for the answer.

If we denote the true values of a variable by x_i and approximate (rounded) values by x_i^* and if we denote actual errors by E_i and relative errors by E_i^R, then formulas (1.2) and (1.3) become

$$E_i = x_i - x_i^* \tag{1.2}$$

$$E_i^R = \frac{x_i - x_i^*}{x_i} = \frac{E_i}{x_i} \tag{1.3}$$

Let us first consider the process of addition in which it is required to find $\sum_{i=1}^{n} x_i$. Since only the rounded numbers are available, we compute $\sum_{i=1}^{n} x_i^*$. From formula (1.2) this will entail a roundoff error in the answer of

$$\text{actual roundoff error} = \sum_{i=1}^{n} E_i \tag{1.4}$$

In analyzing roundoff error the absolute value of the error is of primary concern. Assume that $|E_i| \leq E$ for $i = 1, 2, \ldots, n$. Then maximum roundoff error can be obtained from formula (1.4) as

$$\left| \sum_{i=1}^{n} E_i \right| \leq \sum_{i=1}^{n} |E_i| \leq En$$

Thus, we have

$$\text{maximum roundoff error} = En \tag{1.5}$$

For example, in adding 100 numbers each of which has been rounded to three decimal places, E is equal to .0005 and the maximum roundoff error is equal to $(.0005)(100) = .05$.

It should be noted that the answer produced by formula (1.5) agrees with that produced by interval arithmetic using the maximum (or minimum) value in each term of the summation. It should also be noted that formula (1.5) is applicable to subtraction, since subtraction can be viewed as the addition of a corresponding negative quantity.

The situation for multiplication is somewhat more complex. In this case the use of the relative error formula (1.3) is quite useful. Consider the relative error of the product of two numbers x_1 and x_2. If we denote the relative error of the product by $E_{x_1 x_2}^R$, then we have

$$
\begin{aligned}
E_{x_1 x_2}^R &= \frac{x_1 x_2 - x_1^* x_2^*}{x_1 x_2} \\
&= \frac{(x_1^* + E_1)(x_2^* + E_2) - x_1^* x_2^*}{x_1 x_2} \\
&= \frac{x_2^* E_1 + x_1^* E_2 + E_1 E_2}{x_1 x_2}
\end{aligned}
$$

Now if E_1 and E_2 are "small" relative to x_1 and x_2 respectively, we have $x_1 \doteqdot x_1^*$, $x_2 \doteqdot x_2^*$, and $\dfrac{E_1 E_2}{x_1 x_2} \doteqdot 0$, so that

$$E^R_{x_1 x_2} \doteqdot \frac{E_1}{x_1} + \frac{E_2}{x_2} = E^R_1 + E^R_2 \qquad (1.6)^2$$

In other words, the relative error of a product is approximately equal to the sum of the relative errors of the factors. The generalization to more than two multiplicands is obvious. Thus, maximum relative error for products grows linearly just as maximum absolute error does for sums.

Similarly, for division we have

$$E^R_{x_1 \div x_2} \doteqdot E^R_1 - E^R_2 \qquad (1.7)$$

that is, the relative error of a quotient is the difference between the relative errors of the numerator and the denominator. The derivation of formula (1.7) is left as Exercise 14.

The maximum roundoff error produced by formula (1.5) tends to be quite pessimistic. Roundoff errors usually do not accumulate nearly as fast as the formula indicates that they could. This is because the expected value of roundoff error in each individual term in the summation is only half the maximum error possible and also because positive and negative errors offset each other to a large extent.

The theory of statistics can be used to develop estimates and confidence intervals for roundoff error which tend to be much closer to the true error than the maximum error developed above. These results assume that roundoff errors are independently distributed from term to term and should not be applied if roundoff errors are correlated in some fashion. One set of assumptions[3] leads to the following useful point estimate, which is analogous to formula (1.5), for finding the estimated roundoff error in a summation of n terms

$$\text{estimated roundoff error} = .4607E\sqrt{n} \qquad (1.8)$$

The same assumptions can be used to produce a 95% confidence interval for roundoff error which is given by

$$0 \le \text{roundoff error} \le 1.1316E\sqrt{n} \qquad (1.9)$$

---

[2]The symbol $\doteqdot$ denotes approximate equality.

[3]The underlying assumptions and the statistical derivation of formulas (1.8) and (1.9) are contained in Appendix A.

Formulas (1.8) and (1.9) indicate that roundoff error is more nearly proportional to $\sqrt{n}$ than to n. To continue the example described after formula (1.5), in adding 100 numbers each of which has been rounded to three decimal places, the estimated roundoff error is equal to $(.4607)(.0005)(10) = .0023$ and the 95% probable maximum roundoff error is equal to $(1.1316)(.0005)(10) = .0057$. These results are certainly more optimistic than the maximum roundoff error of .05.

The student should be careful in using formula (1.8) to remember that the expression is a point estimate and not a bound; that is, the actual roundoff error can easily exceed that given by formula (1.8). Furthermore, if roundoff errors are not randomly distributed, then formula (1.8) and even formula (1.9) can seriously understate the roundoff error involved.

Formulas (1.8) and (1.9) can also be applied to products as well as sums, if relative error is used for products instead of actual error.

Example 1.6.   Compute the possible range of values of $\dfrac{ab-c}{d}$ using interval arithmetic, if $a = 7.31$, $b = 6.03$, $c = 4.21$, and $d = 1.13$, where the numbers are rounded to two decimal places.

Direct computation produces

$$\frac{(7.31)(6.03) - 4.21}{1.13} = 35.3$$

An upper bound is given by

$$\frac{(7.315)(6.035) - 4.205}{1.125} = 35.50$$

The student should carefully note that to maximize the function, maximum values of a and b are chosen, but minimum values of c and d are chosen. (Actually c can be viewed as the addition of the maximum negative number, that is, $-4.205 > -4.215$.) Similarly, a lower bound is given by

$$\frac{(7.305)(6.025) - 4.215}{1.135} = 35.06$$

Thus, the true answer A lies somewhere in the interval $35.06 \le A \le 35.50$.

Example 1.7.   Analyze the roundoff error in computing $\displaystyle\sum_{x=1}^{100} \log_e x$ and $\displaystyle\sum_{x=1}^{10} \log_e x$ using two-decimal arithmetic.

If each term is rounded to two decimal places and added, we find that

$$\sum_{x=1}^{100} \log_e x = 363.71$$

A more exact recomputation using five decimal places and rounding the result to three decimal places produces an answer of 363.739. This procedure guarantees three-decimal accuracy, since for $E < .000005$, $100E < .0005$, which will not affect the third decimal place.

Thus, the actual roundoff error is .029. Formula (1.5) produces a maximum error of $(.005)(100) = .500$. Formula (1.8) produces an estimated error of $(.4607)(.005)(10) = .023$, which is close to the actual roundoff error of .029. Although the actual roundoff error is slightly greater than estimated, it is well within the 95% probable maximum of $(1.1316)(.005)(10) = .056$ given by formula (1.9).

Similar computations for $\sum_{x=1}^{10} \log_e x$ produce a two-decimal answer of 15.11, an exact three-decimal answer of 15.104, for an actual roundoff error of .006. The maximum error is $(.005)(10) = .050$, and the estimated error is $(.4607)(.005)(\sqrt{10}) = .007$. The estimated error of .007 is remarkably close to the actual error of .006, while the upper bound of .050 is quite pessimistic.

The estimated error formulas indicate that roundoff errors tend to increase proportionally to $\sqrt{n}$ rather than to n. Since the first summation contains 10 times as many terms as the second summation, we would expect that roundoff error would be only $\sqrt{10}$ or about 3.2 times as great for the first summation than for the second summation. The actual roundoff errors of .029 and .006 are in a ratio of 4.8 to 1, which is somewhat greater than 3.2 to 1 but nowhere near a ratio of 10 to 1.

Example 1.8. ***Three measurements of some quantity*** A ***yield*** $A_1 = 2.9$, $A_2 = 3.0$, ***and*** $A_3 = 3.1$, ***where all results are rounded to two significant digits. Use interval arithmetic to find the maximum roundoff error to two significant digits in saying*** $A_1 A_2 A_3 = 27$. ***Compare your answer to that obtained using the approximate relative error formula (1.6).***

The absolute error with interval arithmetic would be $|(2.95)(3.05)(3.15) - 27| = 1.34$ using maximum values, and $|(2.85)(2.95)(3.05) - 27| = 1.36$ using minimum values.

The relative error formula (1.6) gives

$$\left| \frac{A_1 A_2 A_3 - A_1^* A_2^* A_3^*}{A_1 A_2 A_3} \right| \doteq \left| \frac{E_1}{A_1} \right| + \left| \frac{E_2}{A_2} \right| + \left| \frac{E_3}{A_3} \right|$$

Note that the given values of A should be written as A^* in the context of formula (1.6). Thus, we have

$$\left| A_1 A_2 A_3 - A_1^* A_2^* A_3^* \right| = \left| A_1 A_2 A_3 - 27 \right|$$

$$\doteq \left| A_1 A_2 A_3 \right| \left[\left| \frac{E_1}{A_1} \right| + \left| \frac{E_2}{A_2} \right| + \left| \frac{E_3}{A_3} \right| \right]$$

$$= \left| A_2 A_3 E_1 \right| + \left| A_1 A_3 E_2 \right| + \left| A_1 A_2 E_3 \right|$$

$$\doteq (3.0)(3.1)(.05) + (2.9)(3.1)(.05) + (2.9)(3.0)(.05)$$

$$= .465 + .4495 + .435$$

$$= 1.35$$

which is very close to the answers produced by the interval arithmetic method.

The student should note that the relative error of the product $1.35 \div 27 = .05$ is approximately the sum of the relative errors of the factors, that is,

$$\frac{.05}{2.9} + \frac{.05}{3.0} + \frac{.05}{3.1} = .05$$

1.6.   THE COLLOCATION POLYNOMIAL

In numerical analysis it is often necessary to approximate a function $f(x)$ to produce a numerical answer to a given problem. This will frequently happen, since direct analytical solutions for many functions are difficult or even impossible to implement.

There are obviously a great number of functions which could be used to approximate another function. One of the most popular and widely used approximating functions in numerical analysis is the *polynomial*. Polynomials are quite useful because of their ease of manipulation and their ability to approximate a wide variety of other functions successfully.

The standard form for a polynomial $p(x)$ of degree n is

$$p(x) = a_n x^n + a_{n-1} x^{n-1} + \cdots + a_1 x + a_0 \qquad (1.10)$$

where n is some positive integer and $a_n \neq 0$. It is assumed that the student is familiar with the basic properties of polynomials from previous algebra courses. Among these properties are the following:

1. $p(x)$ has at most n distinct, real *roots*, that is, values of x such that $p(x) = 0$. If complex roots are counted and if roots of multiplicity m are counted m times, then $p(x)$ has exactly n roots.

2.   It takes $n + 1$ independent pieces of information to determine $p(x)$, that is, to find the $n + 1$ coefficients $a_n, a_{n-1}, \ldots, a_1, a_0$. The polynomial thus determined is unique.

3. $p(x)$ can be written as $(x - r)q(x) + R$ where r is some number, $q(x)$ is a polynomial of degree $n - 1$, and R (for remainder) is some constant depending on r. This result is called the *division algorithm*.

   (a) A corollary of the above is that $p(r) = R$. This result is called the *remainder theorem*.

   (b) A second corollary of the above is that if $p(r) = 0$, that is, if r is a root of $p(x)$, then $x - r$ is a factor of $p(x)$ since $R = 0$. This leads to the result that if $r_1, r_2, \ldots, r_n$ are the n roots of $p(x)$, then

   $$p(x) = a_n(x - r_1)(x - r_2) \cdots (x - r_n)$$

   This result is called the *factor theorem*.

There are several approaches which can be used in fitting a polynomial to some functional data. One of the most common and useful methods is *collocation*. The *collocation polynomial*, $p(x)$, is the polynomial which takes on the same functional values at certain points that $f(x)$ does. These points are called *points of collocation*.

Property 2 above indicates that it will take $n + 1$ points to produce a collocation polynomial of degree n and that this collocation polynomial is unique. These $n + 1$ points of collocation are denoted by $x_0, x_1, \ldots, x_n$. For ease in handling we assume that these points are arranged in ascending order unless stated otherwise, that is, $x_0 < x_1 < x_2 < \cdots < x_n$.

In succeeding chapters numerous algorithms for determining the collocation polynomial $p(x)$ will be developed. For any of these methods it is important to be able to measure the truncation error involved in using $p(x)$ to approximate $f(x)$.

Let the kth derivative of $f(x)$ be denoted by $f^{(k)}(x)$ and define $\pi(x)$ as follows

$$\pi(x) = (x - x_0)(x - x_1) \cdots (x - x_n) \tag{1.11}$$

Then the truncation error is given by

$$f(x) - p(x) = \frac{f^{(n+1)}(\xi)\pi(x)}{(n + 1)!} \tag{1.12}$$

where $x_0 \le \xi \le x_n$, provided that $x_0 < x < x_n$. The number ξ is a function of x, that is, the required value of ξ to produce equality in formula (1.12) varies depending upon the value of x under consideration.

To derive formula (1.12) we note that $f(x) - p(x)$ is equal to zero at the points of collocation. Thus,

$$f(x) - p(x) = M(x - x_0)(x - x_1) \cdots (x - x_n) = M\pi(x)$$

If we define

$$g(x) = f(x) - p(x) - M\pi(x)$$

then $g(x) = 0$ for all the points of collocation. Choose a new distinct value x_{n+1} in the interval $x_0 < x_{n+1} < x_n$ such that $g(x_{n+1})$ is also equal to zero. This can be accomplished for any x_{n+1} by appropriately choosing M, that is, by setting

$$M = \frac{f(x_{n+1}) - p(x_{n+1})}{\pi(x_{n+1})}$$

Now $g(x) = 0$ has at least $n + 2$ roots; that is, $x_0, x_1, \ldots, x_n, x_{n+1}$; while $p(x)$ is a polynomial of degree n and $\pi(x)$ is a polynomial of degree $n + 1$ with a leading coefficient of 1.

Recall Rolle's theorem from calculus. This theorem guarantees at least $n + 1$ roots for $g^{(1)}(x) = 0$ between those of $g(x) = 0$. If it is applied a second time, it guarantees at least n roots for $g^{(2)}(x) = 0$ between those of $g^{(1)}(x) = 0$. If this process is continued $n + 1$ times, it guarantees at least one root for $g^{(n+1)}(x) = 0$. Let this root be at $x = \xi$. Clearly, $x_0 \le \xi \le x_n$.

Now

$$g(x) = f(x) - p(x) - M\pi(x)$$

so that if we differentiate $n + 1$ times and evaluate at $x = \xi$ we have

$$\begin{aligned} g^{(n+1)}(\xi) &= f^{(n+1)}(\xi) - p^{(n+1)}(\xi) - M\pi^{(n+1)}(\xi) \\ &= f^{(n+1)}(\xi) - 0 - M(n+1)! \\ &= 0 \end{aligned}$$

which gives

$$M = \frac{f^{(n+1)}(\xi)}{(n+1)!}$$

Thus,

$$f(x_{n+1}) - p(x_{n+1}) = M\pi(x_{n+1}) = \frac{f^{(n+1)}(\xi)\pi(x_{n+1})}{(n+1)!}$$

but since x_{n+1} is general, we have

$$f(x) - p(x) = \frac{f^{(n+1)}(\xi)\pi(x)}{(n+1)!}$$

It should be noted that formula (1.12) may be of only limited value in a practical problem, since there is usually no way of determining ξ. However, useful results can be obtained if it is possible to find an upper bound for $f^{(n+1)}(\xi)$ independent of ξ.

It should also be noted that formula (1.12) is valid for collocation polynomials determined by any of the methods to be developed in succeeding chapters.

Some methods of fitting polynomials other than by collocation will be discussed in Chapter 15.

Example 1.9. Find the maximum error involved in approximating $f(x) = x^4$ **over the interval** $-1 \le x \le 1$ **using** $x = -1, 0, 1$ **as points of collocation.**

Since we have three points given, the collocation polynomial is of second degree. It is apparent that the second-degree polynomial $p(x) = x^2$ collocates with $f(x)$ at $x = -1, 0, 1$.

In this example we do not even need to use formula (1.12). Let $E(x)$ be the error at x. Then we have

$$E(x) = f(x) - p(x) = x^4 - x^2$$

To maximize $|E(x)|$ we have

$$E'(x) = 4x^3 - 2x = 0$$

The root $x = 0$ is extraneous, and so we have extrema at

$$x = \pm\sqrt{\tfrac{1}{2}}$$

The absolute actual maximum error is

$$|E(\sqrt{\tfrac{1}{2}})| = |E(-\sqrt{\tfrac{1}{2}})| = |\tfrac{1}{4} - \tfrac{1}{2}| = \tfrac{1}{4}$$

Example 1.10. Find the maximum error involved in approximating $f(x) = \sin \dfrac{\pi}{2} x$ **over the interval** $0 \le x \le 1$ **using** $x = 0, 1$ **as points of collocation.**

The collocation polynomial is clearly $p(x) = x$, so that $n = 1$. We have the following:

$$\pi(x) = x(x - 1) = x^2 - x$$

$$f(x) = \sin \frac{\pi}{2} x$$

$$f'(x) = \frac{\pi}{2} \cos \frac{\pi}{2} x$$

$$f''(x) = -\frac{\pi^2}{4} \sin \frac{\pi}{2} x$$

Now applying formula (1.12)

$$|f(x) - p(x)| = \left| \frac{f''(\xi)(x^2 - x)}{2} \right|$$

$$\leq \frac{\pi^2}{8} |x^2 - x|$$

$$= \frac{\pi^2}{8}(x - x^2) \qquad \text{for } 0 \leq x \leq 1$$

since the sine function is bounded above by one for all x.

This answer is an upper bound for the error at each x value. It is often of interest to carry the problem one step further and try to obtain a bound independent of x for all x under consideration. To do this we must maximize

$$g(x) = x - x^2$$

Now

$$g'(x) = 1 - 2x = 0$$

so that the maximum value occurs at $x = \frac{1}{2}$, which gives $g(\frac{1}{2}) = \frac{1}{4}$.
Thus,

$$\frac{\pi^2}{8}(x - x^2) \leq \frac{\pi^2}{32}$$

a bound completely independent of x. This bound is likely to be quite pessimistic for most, if not all, x in the interval.

Example 1.11. *A linear interpolation between $f(x_0)$ and $f(x_1)$ is performed for some value of x between x_0 and x_1. It is known that the second derivative of $f(x)$ is bounded, that is, $|f''(x)| \leq K$ for all x in the interval $x_0 \leq x \leq x_1$. Find an expression for the maximum error in linear interpolation independent of x.*

Since the interpolation is linear, we have $n = 1$. Then formula (1.12) produces an error of

$$\left| \frac{f''(\xi)\pi(x)}{2} \right| \leq \left| \frac{K}{2}(x - x_0)(x - x_1) \right|$$

We seek to maximize

$$g(x) = (x - x_0)(x - x_1) = x^2 - (x_0 + x_1)x + x_0 x_1$$

Now

$$g'(x) = 2x - (x_0 + x_1) = 0$$

so that

$$x = \frac{x_0 + x_1}{2}$$

and

$$g\left(\frac{x_0 + x_1}{2}\right) = \left(\frac{x_0 + x_1}{2} - x_0\right)\left(\frac{x_0 + x_1}{2} - x_1\right)$$

$$= \left(\frac{x_1 = x_0}{2}\right)\left(\frac{x_0 - x_1}{2}\right)$$

Thus,

$$\left|\frac{K}{2}(x - x_0)(x - x_1)\right| \le \left|\frac{K}{2}\left(\frac{x_1 - x_0}{2}\right)\left(\frac{x_0 - x_1}{2}\right)\right|$$

$$= \frac{K}{8}(x_1 - x_0)^2$$

which is the maximum error in linear interpolation independent of x.

EXERCISES

1.1. Introduction; 1.2. Algorithms

1. In computing $(1.02)^{64}$, determine the number of multiplications which will be required using the following algorithms:
   (a) Multiplying successively by 1.02 until the answer is obtained.
   (b) Squaring 1.02, squaring the result, and so forth, until the answer is obtained.

2. You are asked to multiply 444 by 693 by conventional hand multiplication. Compute the answer with each number on top. Which algorithm is superior?

3. In computing $1 - \cos x$ for values of x close to zero, significant digits may be lost because of the subtraction of two nearly equal numbers. Develop an alternative algorithm which avoids this problem.

4. It is known that to six decimal places $\sqrt{7} = 2.645751$ and $\sqrt{8} = 2.828427$. It is required to find $x(\sqrt{x+1} - \sqrt{x})$ evaluated at $x = 7$.
   (a) Compute an answer accurate to three decimal places by direct substitution.
   (b) Rework (a) rounding off $\sqrt{7}$ and $\sqrt{8}$ to three decimal places before performing the computations.
   (c) Derive the alternative algorithm

$$\frac{x}{\sqrt{x+1} + \sqrt{x}}$$

(d) Compute the answer to (c) for $x = 7$ using the same values of $\sqrt{7}$ and $\sqrt{8}$ as in (b).

(e) How do you account for the greater accuracy in (d) than in (b)?

1.3. Types of error

5. (a) Find the absolute values of actual and relative error in Exercise 4(b).

(b) Find the absolute values of actual and relative error in Exercise 4(d).

6. The relative error in a computed answer A_1 is $+.2$. Find the relative error in another computed answer A_2 if—

(a) A_2 exceeds A_1 by 10% of A_1.

(b) A_2 exceeds A_1 by 10% of the true answer.

Assume A_1, A_2, and the true answer are all positive.

7. Let the error involved in performing a computational procedure on n numbers be denoted by $E(n)$. Error growth is said to be linear if

$$E(n) = kn$$

where $k > 0$. Error growth is said to be exponential if

$$E(n) = kc^n$$

where $k > 0$ and $c > 1$. It is known that the error in performing the computational procedure on two numbers is twice the error for one number.

(a) Find the ratio of the error in performing the computational procedure on six numbers to the error for three numbers, assuming linear error growth.

(b) Rework (a), assuming exponential error growth.

1.4. Truncation error

8. A more efficient algorithm for computing $\log_e 2$ than that given in Example 1.5 can be developed by rewriting the series as

$$1 - \frac{1}{2} + \frac{1}{3} - \cdots + (-1)^{n+1}\frac{1}{n} + \cdots$$

$$= \frac{1}{2} + \frac{1}{2}\left(1 - \frac{1}{2}\right) - \frac{1}{2}\left(\frac{1}{2} - \frac{1}{3}\right) + \frac{1}{2}\left(\frac{1}{3} - \frac{1}{4}\right) - \cdots$$

$$+ (-1)^n\frac{1}{2}\left(\frac{1}{n-1} - \frac{1}{n}\right) + \cdots$$

(a) Prove that the above series to n terms can be expressed as

$$\frac{1}{2} + \sum_{k=2}^{n} (-1)^k \frac{1}{2} \cdot \frac{1}{k(k-1)}$$

(b) Prove that only 32 terms of this series will be required to produce three-decimal accuracy as compared with 2000 terms in the original series.

9. The Taylor series expansion for $\sin x$ is given by

$$\sin x = x - \frac{x^3}{3!} + \frac{x^5}{5!} - \cdots$$

and is to be applied for $0 \le x \le \pi/4$. If three terms in the series are carried, how many decimal places of accuracy can be guaranteed for all x in the interval? It is known that $\pi/4 = .785398$.

10. Find the range of x for which the series

$$\frac{1}{1-x} = 1 + x + x^2 + x^3 + \cdots \qquad \text{for } -1 < x < 1$$

will produce three-decimal accuracy, if the expansion is carried to only two terms as in Example 1.1.

1.5. Roundoff error

11. Find the roundoff error in computing $\sum_{i=1}^{9} \frac{i}{9}$ in which each term in the summation is chopped to two decimal places.

12. Express $\prod_{i=1}^{5} i \cdot 10^i$ as a normalized floating-point number.

13. (a) Convert 165 in decimal notation to binary notation.
    (b) Convert 101101 in binary notation to decimal notation.

14. Derive formula (1.7).

15. Let E be the actual error and E^R the relative error in some quantity x. Let both the true value and computed value of x be multiplied by some constant k.
    (a) Express the actual error of the result in terms of E.
    (b) Express the relative error of the result in terms of E^R.

16. Using the data in Exercise 15, if each value of x is modified by the addition of some constant c, find the actual error of the result expressed in terms of E.

17. Using the data in Exercise 15, if each value of x is raised to the nth power, find the relative error of the result expressed in terms of E^R.

18. Find the maximum possible absolute actual error of the discriminant $b^2 - 4ac$ in using the quadratic formula to solve the equation $2.1x^2 - 5.7x + 3.0$, if each of the three coefficients is rounded off to the nearest .1.

19. In adding n numbers each rounded to five decimal places, what is the largest value of n such that if the maximum roundoff error occurs, the answer will still be accurate to three decimal places?

20. Rework Exercise 19 using estimated roundoff error instead of maximum roundoff error.

1.6. The collocation polynomial

21. Find the polynomial with leading coefficient equal to one whose roots are $-2, 1, 3$.

22. Find the quadratic which passes through the points $(0, 0)$, $(1, 1)$, $(2, 8)$. Determine the answer by solving three equations in three unknowns.[4]

23. The function x^3 also takes on the values specified in Exercise 22. Use formula (1.12) to find an expression for the error involved in approximating x^3 by the second-degree collocation polynomial determined in Exercise 22.

24. Pursuing Exercises 22 and 23 one step further, determine the maximum value of $|f(x) - p(x)|$ over the range of collocation.

25. Quadratic interpolation is used to interpolate values of e^x for $0 \leq x \leq 1$. The points of collocation are $x = 0, \frac{1}{2}, 1$. Use formula (1.12) to find an expression for the error involved in this quadratic interpolation.

26. Find an upper bound for the answer to Exercise 25 which depends only on x, that is, which is independent of ξ.

27. Find an upper bound for the answer to Exercise 26 which is independent of x (and of ξ).[5]

28. The collocation polynomial $p(x) = 7x^2 - 6x$ collocates with $f(x) = x^4$ for $x = 0, 1, 2$. The error term $f(x) - p(x) = x^4 - 7x^2 + 6x$ is equal to zero for $x = -3, 0, 1, 2$. Use formula (1.12) to determine the value of ξ corresponding to $x = -3$ which explains the vanishing of the error term at $x = -3$.

29. The collocation polynomial $p(x) = x$ collocates with $f(x) = x^4$ for $x = 0, 1$. Use formula (1.12) to find ξ corresponding to $x = \frac{1}{2}$.

Miscellaneous problems

30. (a) If $E = x - x^*$, show that

$$\sqrt{x} - \sqrt{x^*} \doteq \frac{E}{2\sqrt{x}}$$

    (b) Hence, show that the relative error of the square root of x is approximately equal to one half the relative error of x.

31. If $E = x - x^*$, show that

$$\log_e x - \log_e x^* \doteq \frac{E}{x}$$

32. (a) Show that

$$E^R_{x_1 + x_2} = \frac{x_1}{x_1 + x_2} E^R_{x_1} + \frac{x_2}{x_1 + x_2} E^R_{x_2}$$

    (b) Show that

$$E^R_{x_1 - x_2} = \frac{x_1}{x_1 - x_2} E^R_{x_1} - \frac{x_2}{x_1 - x_2} E^R_{x_2}$$

---

[4]More efficient algorithms for finding the collocation polynomial will be developed in succeeding chapters.

[5]Upper bounds found by the sequential process illustrated in Exercises 25–27 tend to be quite pessimistic. Tighter bounds can sometimes be found by finding $p(x)$ and working with $f(x) - p(x)$ directly, rather than using formula (1.12). This was the approach used in Example 1.9.

33. (a) Evaluate $f(x) = 3x^3 - 7x^2 + 5x - 2$ at $x = 2.3$ by direct substitution.
   (b) Evaluate $f(x)$ by the following successive substitution approach. Given an nth-degree polynomial

$$a_n x^n + a_{n-1}x^{n-1} + \cdots + a_1 x + a_0$$

which is to be evaluated at $x = k$, compute successively

$$
\begin{aligned}
b_n &= a_n k \\
b_{n-1} &= (b_n + a_{n-1})k \\
b_{n-2} &= (b_{n-1} + a_{n-2})k \\
&\vdots \\
b_2 &= (b_3 + a_2)k \\
b_1 &= (b_2 + a_1)k \\
b_0 &= b_1 + a_0
\end{aligned}
$$

The answer $f(k) = b_0$.[6]

34. (a) Find the maximum absolute actual error in evaluating $2x^2 - 41x + 211$ for $x = 10$, where 10 is rounded off to the nearest integer.
   (b) Find the maximum absolute actual error in evaluating $5x^3 - 3x + 2$ for $x = 0$, where 0 is rounded off to the nearest integer and it is known that no values of x can be negative.

35. The value of e^{-1} to three decimal places is .368.
   (a) Compute the series expansion to four terms

$$e^{-x} = 1 - x + \frac{x^2}{2!} - \frac{x^3}{3!}$$

   for $x = 1$.
   (b) Compute the series expansion to four terms

$$e^{-x} = \frac{1}{e^x} = \frac{1}{1 + x + \dfrac{x^2}{2!} + \dfrac{x^3}{3!}}$$

   for $x = 1$.
   (c) How do you account for the fact that the answer in (b) is superior to the answer in (a)?

36. (a) An approximation to 5! is computed by multiplying five numbers together, each of which is within .01 of the first five positive integers. Find the approximate maximum relative error of the product using formula (1.6).
   (b) What approximate maximum actual error corresponds to this maximum relative error?
   (c) What is the exact maximum actual error using interval arithmetic?

37. Find the maximum range of ξ for the truncation error in Exercise 29 for any x in the interval $0 \le x \le 1$. Does the answer to Exercise 29 lie in this interval?

---

[6]This method is sometimes called the *nested method* of evaluating polynomials. It is quite convenient for computation with either a desk calculator or a computer, since no intermediate results need to be stored.

2

Finite differences

2.1. INTRODUCTION

As EXPLAINED in Section 1.6, a frequent problem in numerical analysis is the determination of the collocation polynomial to fit various types of functions and data. Efficient algorithms for doing this are based upon a mathematical subject called *finite differences.*

The theory of finite differences forms a discrete analogue to the theory of infinitesimal calculus. The name "finite differences" indicates that the mathematical analysis proceeds by finite intervals rather than the infinitesimally small intervals which are used in calculus. A significant portion of the development of finite differences is similar to the corresponding development of calculus. These analogous results are of significance in helping the student develop a deeper understanding of the subject matter and a greater facility in applying it.

The theory of finite differences was developed long before the appearance of high-speed digital computers. However, many of the techniques used in finite differences are still quite important to the numerical analyst in the computer age. The student should not be misled by the title of this chapter, since the traditional theory of finite differences is not restricted to Chapter 2 but includes significant portions of later chapters as well.

In the development of finite differences various functional notations appear. The two most common notational patterns used to represent a function of x are $f(x)$ and u_x. The first has the advantage of familiarity, since it is widely used in other areas of mathematics. The second has the advantage of being simpler to write and will make many of the long expressions to be developed less cumbersome. Naturally different letters than f and u may be used, for example, $g(x)$ and v_x. The student should not let different notations which may be encountered in this book and in the literature confuse otherwise equivalent concepts and expressions.

2.2. DIFFERENCING AND DIFFERENCE TABLES

The first *difference* of a function $f(x)$ is found by subtracting two successive functional values and is defined by

$$\Delta f(x) = f(x + 1) - f(x) \tag{2.1}$$

Formula (2.1) assumes that the functional values are tabulated one unit apart.

In general the interval at which functional values are tabulated may be h, and we have the following more general formula

$$\Delta_h f(x) = f(x + h) - f(x) \tag{2.2}$$

The value h is called the *interval of differencing*. Unless stated otherwise it is assumed that the interval of differencing is equal to one. Also, it is assumed that $h > 0$, that is, negative intervals of differencing have no meaning. The notation Δ_h indicates the interval of differencing which is used. If $h = 1$, then the h is dropped as in formula (2.1).

Chapters 2, 3, and 4 assume that h is constant for all x, that is, that the values of x are equally spaced. A generalized approach for differencing when this is not the case is developed in Chapter 5.

Higher-order differences can be derived from lower-order differences. For example, the second difference of $f(x)$ is given by

$$\begin{aligned}
\Delta^2 f(x) &= \Delta[\Delta f(x)] \\
&= \Delta[f(x + 1) - f(x)] \\
&= [f(x + 2) - f(x + 1)] - [f(x + 1) - f(x)] \\
&= f(x + 2) - 2f(x + 1) + f(x) \tag{2.3}
\end{aligned}$$

It should be noted that Δ^2 does not mean the square of some quantity Δ, but rather means that the process of differencing is to be performed twice. Note particularly that

$$\Delta^2 f(x) \neq [\Delta f(x)]^2$$

Similarly, the third difference of $f(x)$ is given by

$$\Delta^3 f(x) = [\Delta\Delta^2 f(x)]$$
$$= \Delta[f(x + 2) - 2f(x + 1) + f(x)]$$
$$= f(x + 3) - 3f(x + 2) + 3f(x + 1) - f(x) \qquad (2.4)$$

This process can be continued indefinitely to find $\Delta^n f(x)$ for any positive integer n. It is important to note that any difference can be expressed in terms of functional values. The fact that the coefficients involve alternating signs and are equal to the binomial coefficients suggests that some simple procedure exists for evaluating higher-order differences. It will be shown later that this is indeed the case.

The differences of $f(x)$ are often arranged in a tabular format. Such an array is called a *difference table*. Table 2.1 is the standard format of a difference table.

TABLE 2.1

x	$f(x)$	$\Delta f(x)$	$\Delta^2 f(x)$	$\Delta^3 f(x)$	$\Delta^4 f(x)$
0	$f(0)$				
		$\Delta f(0)$			
1	$f(1)$		$\Delta^2 f(0)$		
		$\Delta f(1)$		$\Delta^3 f(0)$	
2	$f(2)$		$\Delta^2 f(1)$		$\Delta^4 f(0)$
		$\Delta f(2)$		$\Delta^3 f(1)$	
3	$f(3)$		$\Delta^2 f(2)$		
		$\Delta f(3)$			
4	$f(4)$				

It should be noted that each column after $f(x)$ is positioned halfway between the entries in the preceding column to clearly indicate which two values are subtracted to produce the difference. Also it should be noted that the values of $x = 0, 1, 2, 3, 4$ appear as labels on symbols along downward diagonals to the right.

A numerical example should help clarify the basic concepts and the construction of a difference table. Table 2.2 is a difference table for the function $f(x) = x^3$. It is no coincidence that the third differences of this third-degree polynomial are constant, as will be seen later.

The symbol Δ is not a number but is an *operator*, which when applied to a function changes it into some other function. The operator Δ is often called the *forward difference operator* or the *advancing difference operator*. The

TABLE 2.2

x	$f(x)$	$\Delta f(x)$	$\Delta^2 f(x)$	$\Delta^3 f(x)$
0	0			
		1		
1	1		6	
		7		6
2	8		12	
		19		6
3	27		18	
		37		6
4	64		24	
		61		
5	125			

usage of the terms "forward" and "advancing" will become clear when other types of difference operators are defined later.

The student should not let the term "operator" be a source of confusion. Although the term "operator" may not have been used, operators have appeared in previous mathematics courses. For example, the process of *differentiation* in calculus, which is the analogue of *differencing* in numerical analysis, can be viewed in terms of operators. The derivative operator will be denoted by D.

It is useful to define another operator E, such that

$$Ef(x) = f(x+1) \tag{2.5}$$

or in general

$$\underset{h}{E}f(x) = f(x+h) \tag{2.6}$$

Again h is assumed to equal one unless stated otherwise. This operator when applied to a function $f(x)$ increases the value of x by one interval of differencing and is called the *shifting operator*. This operator can also be applied more than once to produce results such as $E^6 f(-2) = f(4)$.[1] The student should note that $E^n f(x) \neq [Ef(x)]^n$.

Thus, formula (2.1) can be rewritten as

$$\Delta f(x) = f(x+1) - f(x) = Ef(x) - f(x) \tag{2.1}$$

---

[1] $E^6 f(-2)$ and $\Delta^6 f(-2)$ are compact ways of writing, respectively, $E^6 f(x)$ evaluated at $x = -2$ and $\Delta^6 f(x)$ evaluated at $x = -2$. In other words, the operator must be applied before the function is evaluated; otherwise, the operator would be applied to a constant.

Since the function $f(x)$ is perfectly general, the following operator equations are valid:

$$\Delta \equiv E - 1 \tag{2.7a}$$

$$E \equiv 1 + \Delta \tag{2.7b}^2$$

It should be carefully noted that the "1" which appears in formulas (2.7a) and (2.7b) is not a number but is itself an operator, the *identity operator*. When applied to a function the identity operator leaves the function unchanged.

An operator L is said to be a *linear operator* if for two functions u_x and v_x and for two constants a and b we have

$$L(au_x + bv_x) = aLu_x + bLv_x \tag{2.8}$$

In other words, formula (2.8) says the following:

1. An operator on a constant times a function produces the same result as the constant times the operator on the function.
2. An operator on the sum of two functions produces the same result as the sum of the operator on each function separately.

In Exercise 1 the student will show that both Δ and E are linear operators. In succeeding portions of this book several new operators will be introduced, all of which will be linear operators.

In algebraic terminology formula (2.8) indicates that a linear operator L commutes with respect to constants,[3] that is,

$$L(au_x) = (aL)u_x$$

and that the distributive law holds, that is,

$$L(u_x + v_x) = Lu_x + Lv_x$$

In addition linear operators possess the property of exponents if the operators are applied more than one time, that is,

$$L^m(L^n u_x) = L^{m+n} u_x$$

This latter result is clearly true for both the operators Δ and E.

Linear operators have the desirable property that algebraic manipulations may be performed on the operators themselves independently of any function. For example, if we take formula (2.7a) and "cube" (that is, operate three times) both sides algebraically, we obtain

$$\Delta^3 \equiv (E - 1)^3$$
$$\equiv E^3 - 3E^2 + 3E - 1$$

---

[2]Formal mathematical notation indicates that "$\equiv$" is preferable to "$=$" when operator equations are written.
[3]This property is sometimes called *homogeneity*.

and applying this result to $f(x)$ produces an alternative derivation of formula (2.4)

$$\Delta^3 f(x) = f(x+3) - 3f(x+2) + 3f(x+1) - f(x) \qquad (2.4)$$

The process of deriving identities by using operators is called *separation of symbols* and is examined more completely in Section 2.8.

The above discussion of linear operators considers the algebraic properties of one operator. It is also important to consider the properties of more than one operator in various combinations.

Let L_i denote a linear operator for $i = 1, 2, 3$. Two operators L_1 and L_2 are said to be equal if

$$L_1 u_x = L_2 u_x$$

for all u_x under consideration. The following algebraic properties are satisfied by linear operators:

1. Commutative law of addition: $L_1 + L_2 \equiv L_2 + L_1$.
2. Associative law of addition: $L_1 + (L_2 + L_3) \equiv (L_1 + L_2) + L_3$.
3. Associative law of multiplication: $L_1(L_2 L_3) \equiv (L_1 L_2)L_3$.
4. Distributive law: $L_1(L_2 + L_3) \equiv L_1 L_2 + L_1 L_3$.

One property noticeably missing from the above list is the commutative law of multiplication, which need not always be true, that is,

$$L_1 L_2 \neq L_2 L_1$$

It should be noted that this law is true for the operators already defined, that is,

$$E\Delta \equiv \Delta E \qquad (2.9)$$

and is true for all operators introduced through Chapter 5. However, an operator will be introduced in Chapter 6 which does not obey the commutative law of multiplication.

It is seen above in formula (2.8) that the difference of the sum of two functions is equal to the sum of the differences (that is, $\Delta(u_x + v_x) = \Delta u_x + \Delta v_x$), a result analogous to that for derivatives. It is also possible to derive a result for the difference of a product analogous to a well-known formula for derivatives

$$\begin{aligned}
\Delta(u_x v_x) &= u_{x+1} v_{x+1} - u_x v_x \\
&= u_{x+1} v_{x+1} - u_x v_{x+1} + u_x v_{x+1} - u_x v_x \\
&= u_x(v_{x+1} - v_x) + v_{x+1}(u_{x+1} - u_x) \\
&= u_x \Delta v_x + v_{x+1} \Delta u_x \qquad (2.10)
\end{aligned}$$

The label $x + 1$ in the second term should be carefully noted. Also it should be noted that u_x and v_x are interchangeable.

Similarly, the following formula for quotients is valid.

$$\Delta \frac{u_x}{v_x} = \frac{v_x \Delta u_x - u_x \Delta v_x}{v_x v_{x+1}} \tag{2.11}$$

Again the label $x + 1$ in the denominator should be noted. The derivation of formula (2.11) is left as Exercise 2(a).

Since polynomials are destined to play such a large role in numerical analysis, the differencing of them deserves special attention. Consider

$$\begin{aligned}
\Delta x^n &= (x + 1)^n - x^n \\
&= \left(x^n + nx^{n-1} + \cdots + 1\right) - x^n \\
&= nx^{n-1} + \cdots + 1
\end{aligned}$$

which is a polynomial of degree $n - 1$. If this process is continued for a total of n times, a polynomial of degree zero, that is, a constant, is obtained. Thus, the nth difference of a polynomial of degree n is constant and all higher differences are zero, a result illustrated in Table 2.2 for the simple cubic $f(x) = x^3$. The student should not find this result surprising in view of the analogous properties for the derivative of a polynomial.

Unfortunately, differencing polynomials by expansion as in the above expression is somewhat cumbersome. Sections 2.3, 2.4, and 2.5 discuss efficient algorithms for differencing polynomials.

Before proceeding to these next sections, however, let us examine one other function which is frequently encountered, the exponential function. We have

$$\begin{aligned}
\Delta a^x &= a^{x+1} - a^x \\
&= a^x(a - 1)
\end{aligned} \tag{2.12}$$

Thus, the difference of an exponential is the same exponential times a constant, again a result analogous to that for derivatives. Interestingly enough, if $a = 2$, we have $\Delta 2^x = 2^x$, a result not unlike $De^x = e^x$.

Functions other than polynomials and exponentials will occasionally be of interest. These can generally be handled by reverting to the definition of a difference. Some of these other functions and their differences appear in the exercises and succeeding portions of the book.

Example 2.1.   Find $f(x)$ given the following functional values: $f(0) = 3$, $f(1) = 0$, $f(2) = 5$, $f(3) = 34$, $f(4) = 135$, $f(5) = 452$.

If we construct a difference table, we have the following:

x	$f(x)$	$\Delta f(x)$	$\Delta^2 f(x)$
0	3		
		-3	
1	0		8
		5	
2	5		24
		29	
3	34		72
		101	
4	135		216
		317	
5	452		

By convention, we stop differencing at second differences, since a geometric progression appears in the difference table. Formula (2.12) indicates that once a geometric progression appears it will continue to perpetuate geometric progressions and further differencing will accomplish nothing.

Since second differences are the first order of differences to become a geometric progression, we assume $f(x)$ is the sum of a first-degree polynomial and an exponential, that is, that

$$f(x) = a + bx + c \cdot 3^x$$

since the common ratio in the second difference column is three.

Finding the differences, we have the following:

$$f(x) = a + bx + \quad c \cdot 3^x$$
$$\Delta f(x) = \qquad b + 2c \cdot 3^x$$
$$\Delta^2 f(x) = \qquad\qquad 4c \cdot 3^x$$

Substituting $x = 0$ leads to three equations in the three unknowns, a, b, c, as follows:

$$3 = a \quad + c$$
$$-3 = \quad b + 2c$$
$$8 = \qquad 4c$$

The solution is $a = 1$, $b = -7$, $c = 2$. Thus,

$$f(x) = 1 - 7x + 2 \cdot 3^x$$

The student should verify that this equation does produce the numerical values of $f(x)$ for the given values of x.

2.3. FACTORIAL NOTATION

Polynomials written in standard notation, that is, in terms of decreasing powers of x, are not convenient when working with finite difference operators. This section develops an alternative notation for polynomials, called *factorial notation*, which is quite convenient when working with finite difference operators.

Define $x^{(m)}$ as

$$x^{(m)} = x(x - 1)(x - 2) \cdots (x - m + 1) \tag{2.13}$$

in which there appears the product of m successive decreasing factors starting with x and each differing by one. It is assumed that m is a positive integer. For the general case with interval of differencing h, we have

$$x_h^{(m)} = x(x - h)(x - 2h) \cdots [x - (m - 1)h] \tag{2.14}$$

The expression $x^{(m)}$ has a simple first difference

$$
\begin{aligned}
\Delta x^{(m)} &= (x + 1)^{(m)} - x^{(m)} \\
&= [(x + 1)x(x - 1) \cdots (x - m + 2)] \\
&\quad - [x(x - 1)(x - 2) \cdots (x - m + 1)] \\
&= [(x + 1) - (x - m + 1)]x(x - 1) \cdots (x - m + 2) \\
&= m[x(x - 1) \cdots (x - m + 2)] \\
&= m x^{(m-1)}
\end{aligned}
\tag{2.15}
$$

The analogy with $Dx^m = mx^{m-1}$ is apparent. The general case with interval of differencing h gives

$$\Delta x_h^{(m)} = hmx_h^{(m-1)} \tag{2.16}$$

that is, a constant h appears each time the function is differenced.

The student should be careful to note that formula (2.16) is valid only if the interval of differencing in the Δ operator is the same as the interval of differencing in the factorial notation. If these two intervals of differencing are not equal, then no "nice" results like formula (2.16) appear (see, however, Exercise 51).

Thus, if a polynomial can be expressed in factorial notation, its differences can be obtained quite readily. From the linearity property of the Δ operator, each term in the polynomial can be differenced in turn and the results combined. Moreover, formulas (2.15) and (2.16) can be applied more than once to produce higher-order differences. Section 2.4 discusses an efficient algorithm to convert any polynomial from standard notation to factorial notation.

The above definition of $x^{(m)}$ assumes that m is a positive integer. It is possible to produce a consistent and useful definition for $x^{(m)}$ when m is zero or a negative integer.

From formula (2.13), we have

$$x^{(m)} = x^{(m-1)}(x - m + 1)$$

which can be translated to

$$x^{(m-1)} = \frac{x^{(m)}}{x - m + 1}$$

If we let $m = 1$, we immediately have

$$x^{(0)} = 1 \qquad (2.17)$$

If we let $m = 0$, we have

$$x^{(-1)} = \frac{1}{x + 1}$$

Continuing this process, we have for $m = -1$

$$x^{(-2)} = \frac{1}{(x + 1)(x + 2)}$$

and so forth until, in general,

$$x^{(-m)} = \frac{1}{(x + 1)(x + 2) \cdots (x + m)} = \frac{1}{(x + m)^{(m)}} \qquad (2.18)$$

It should be noted that $x^{(-m)}$ is undefined if any of the factors in the denominator is equal to zero.

It is possible to show that the first difference of $x^{(-m)}$ is the hoped-for result

$$\Delta x^{(-m)} = -m x^{(-m-1)} \qquad (2.19)$$

The derivation of formula (2.19) is left as Exercise 14.

In the preceding section it is seen that the nth differences of a polynomial of degree n are constant and higher-order differences are equal to zero. It is now possible to determine the magnitude of these constant differences.

From the definition of $x^{(n)}$ we see that

$$x^n = x^{(n)} + \text{terms of degree } n - 1 \text{ and lower}$$

The terms of degree $n - 1$ and lower will all have nth differences equal to zero. If we combine this result with n applications of formula (2.15), we have

$$\Delta^n x^n = \Delta^n x^{(n)} = n(n - 1)(n - 2) \cdots (1) x^{(0)} = n! \qquad (2.20)$$

This result is illustrated in Table 2.2, in which the constant third differences for the cubic x^3 are seen to equal $3! = 6$.

For the general case in which h is the interval of differencing, n applications of formula (2.16) gives

$$\underset{h}{\Delta^n} x^n = \underset{h}{\Delta^n} x^{(n)} = h^n n! \qquad (2.21)$$

The symbol $x^{(m)}$ is sometimes said to be expressed in *upper factorial notation* to distinguish it from another symbol $\begin{pmatrix} x \\ m \end{pmatrix}$, which is expressed in *lower factorial notation*. These symbols are called "x upper m" and "x lower m," respectively. The expression $\begin{pmatrix} x \\ m \end{pmatrix}$ is defined by

$$\begin{pmatrix} x \\ m \end{pmatrix} = \frac{x^{(m)}}{m!} \qquad (2.22)$$

It is possible to show that $\begin{pmatrix} x \\ m \end{pmatrix}$ has the first difference

$$\Delta \begin{pmatrix} x \\ m \end{pmatrix} = \begin{pmatrix} x \\ m-1 \end{pmatrix} \qquad (2.23)$$

The derivation of formula (2.23) is left as Exercise 15.

The student has previously encountered lower factorials in connection with the binomial theorem, for example,

$$(1 + r)^x = 1 + \begin{pmatrix} x \\ 1 \end{pmatrix} r + \begin{pmatrix} x \\ 2 \end{pmatrix} r^2 + \begin{pmatrix} x \\ 3 \end{pmatrix} r^3 + \cdots$$

in which the lower factorials are equivalent to the binomial coefficients.

In fact, it is interesting to note that formula (2.23) can be written as

$$\begin{pmatrix} x+1 \\ m \end{pmatrix} - \begin{pmatrix} x \\ m \end{pmatrix} = \begin{pmatrix} x \\ m-1 \end{pmatrix}$$

which leads to the well-known formula[4]

$$\begin{pmatrix} x+1 \\ m \end{pmatrix} = \begin{pmatrix} x \\ m-1 \end{pmatrix} + \begin{pmatrix} x \\ m \end{pmatrix}$$

The above formulas have assumed that m is a positive integer, but they are valid for all values of x.[5]

---

[4] This formula also appears in elementary combinatorial theory. If x is a positive integer greater than or equal to m, then $x^{(m)}$ is the number of permutations of x objects taken m at a time, while $\begin{pmatrix} x \\ m \end{pmatrix}$ is the corresponding number of combinations.

[5] The student should note that although $x^{(m)}$ is defined for $h \neq 1$ (see formula (2.14)), our definitions require that $\begin{pmatrix} x \\ m \end{pmatrix}$ always be based on $h = 1$.

Although the student should be able to work with both upper and lower factorials, the somewhat ambiguous term "factorial notation" will refer to upper factorials unless stated otherwise.

Example 2.2.   Find $\Delta x_3^{(2)}$.

In this case the interval of differencing in the Δ operator is two, whereas the interval of differencing in the factorial is three, so that no result is immediate. Resorting to the definition of $x_3^{(2)}$

$$\Delta_2 x_3^{(2)} = \Delta_2 x(x-3)$$

$$= \Delta_2 [x(x-2) - x]$$

$$= \Delta_2 [x_2^{(2)} - x_2^{(1)}]$$

$$= 4x - 2$$

using formula (2.16) with $h = 2$.

Example 2.3.   Find $\Delta^2 \dfrac{x+3}{x^2 + 9x + 20}$.

It is always possible in a problem of this type to resort to the definition of Δ^2. However, in this example the definitional approach involves an extensive amount of algebra, and a better approach exists.

We can manipulate the expression to give

$$\frac{x+3}{x^2 + 9x + 20} = \frac{x+3}{(x+4)(x+5)}$$

$$= \frac{(x+4) - 1}{(x+4)(x+5)}$$

$$= \frac{1}{x+5} - \frac{1}{(x+4)(x+5)}$$

$$= (x+4)^{(-1)} - (x+3)^{(-2)}$$

and now using formula (2.19)

$$\Delta^2 \left[\frac{x+3}{x^2 + 9x + 20} \right] = \Delta^2 [(x+4)^{(-1)} - (x+3)^{(-2)}]$$

$$= 2(x+4)^{(-3)} - 6(x+3)^{(-4)}$$

The student should note the use of $x^{(-m)}$ in this example. Although any polynomial is capable of being expressed in factorial notation, not all reciprocals of polynomials (or, as in this example, ratios of polynomials) are capable of such expression. The use of $x^{(-m)}$ requires the presence of a cooperative denominator.

2.4. CONVERTING STANDARD POLYNOMIALS TO FACTORIAL POLYNOMIALS

The importance of expressing polynomials in factorial notation is discussed in Section 2.3. This section presents an efficient algorithm, called *synthetic division*, by which any polynomial in standard notation can readily be converted to factorial notation.

The method of solution relies upon successive applications of the division algorithm given in Section 1.6. The method can best be illustrated by example.

Suppose that we wish to express $5x^4 - 7x^3 + 3x^2 + x - 4$ in factorial notation. We first set

$$5x^4 - 7x^3 + 3x^2 + x - 4 = Ax^{(4)} + Bx^{(3)} + Cx^{(2)} + Dx^{(1)} + E$$
$$= Ax(x - 1)(x - 2)(x - 3) + Bx(x - 1)(x - 2)$$
$$+ Cx(x - 1) + Dx + E$$

and the problem is to find the five coefficients A, B, C, D, E.

If we divide both sides of the equation by x, we have on the left-hand side

$$5x^3 - 7x^2 + 3x + 1 \text{ with a remainder of } -4$$

and on the right-hand side

$$A(x - 1)(x - 2)(x - 3) + B(x - 1)(x - 2) + C(x - 1) + D$$

with a remainder of E

Since the remainders must be equal, we have $E = -4$.

If we repeat the procedure and divide by $x - 1$, we have on the left-hand side

$$
\begin{array}{r}
5x^2 - 2x\ + 1 \\
x - 1\ \overline{)\ 5x^3 - 7x^2 + 3x + 1} \\
5x^3 - 5x^2 \\
\hline
-2x^2 + 3x \\
-2x^2 + 2x \\
\hline
x + 1 \\
x - 1 \\
\hline
2
\end{array}
$$

and on the right-hand side

$$A(x - 2)(x - 3) + B(x - 2) + C \text{ with a remainder of } D$$

Since the remainders must be equal, we have $D = 2$.

If we repeat the procedure and divide by $x - 2$, we have on the left-hand side

$$
\begin{array}{r}
5x \;+\; 8 \\[2pt]
x - 2\,\overline{\big)\,5x^2 \;-\; 2x \;+\; 1\,} \\
5x^2 \;-\; 10x \\ \hline
8x \;+\; 1 \\
8x \;-\; 16 \\ \hline
17
\end{array}
$$

and on the right-hand side

$$A(x - 3) + B \text{ with a remainder of } C$$

Since the remainders must be equal, we have $C = 17$.

If we repeat the procedure one last time and divide by $x - 3$, we have on the left-hand side

$$
\begin{array}{r}
5 \\[2pt]
x - 3\,\overline{\big)\,5x \;+\; 8\,} \\
5x \;-\; 15 \\ \hline
23
\end{array}
$$

and on the right-hand side

$$A \text{ with a remainder of } B$$

Since the remainders and quotients must be equal, we have $B = 23$ and $A = 5$.

Thus, our answer is $5x^4 - 7x^3 + 3x^2 + x - 4 = 5x^{(4)} + 23x^{(3)} + 17x^{(2)} + 2x^{(1)} - 4$. It is interesting to note that the coefficients of the highest and lowest degree terms are equal in both expressions, that is, 5 and -4, respectively. It is seen from the above procedure that this will always be the case.

Synthetic division is a more efficient method of performing the above divisions and obtaining the required remainders by working on just the coefficients. The procedure is illustrated in Table 2.3. The following is a description of how Table 2.3 is constructed:

1. List the coefficients of the polynomial in standard notation across the top in descending powers of x. Be sure to insert any zeros for missing powers of x, so that one number appears for each power.
2. List numbers down the side of the page $1, 2, \ldots$, stopping when the degree of the polynomial is reached, in this case 4. These numbers will be referred to as *index numbers*.

3. Each pair of rows in the table is constructed by starting with a zero in the first position, adding, taking the result and multiplying by the index number corresponding to that pair of rows, entering the result, and repeating the process. For example, in the second pair of rows we have in succession: 0 entered, $5 + 0 = 5$, $5 \times 2 = 10$, $-2 + 10 = 8$, $8 \times 2 = 16$, and $1 + 16 = 17$.

TABLE 2.3

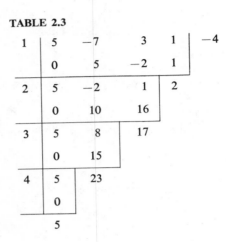

The coefficients of the polynomial after each division appear directly in the table. After the first division we have by the long division process

$$5x^3 - 7x^2 + 3x + 1 \text{ with a remainder of } -4$$

after the second

$$5x^2 - 2x + 1 \text{ with a remainder of } 2$$

after the third

$$5x + 8 \text{ with a remainder of } 17$$

and after the fourth

$$5 \text{ with a remainder of } 23$$

The answers are the remainders appearing at the ends of an upward diagonal to the right. The method uses $+1$, $+2$, $+3$, $+4$ down the side instead of -1, -2, -3, -4, so that additions can be performed instead of subtractions. With some practice the student should be able to use synthetic division to convert polynomials to factorial notation quite rapidly.

If it is required to convert a polynomial to lower factorials rather than to upper factorials, then the following procedure should be used:

1. Convert the polynomial to upper factorials as above.
2. Convert the upper factorials to lower factorials using the relation

$$x^{(m)} = \binom{x}{m} m!$$

In the above example,

$$5x^{(4)} + 23x^{(3)} + 17x^{(2)} + 2x^{(1)} - 4 = 120\binom{x}{4} + 138\binom{x}{3} + 34\binom{x}{2} + 2\binom{x}{1} - 4$$

It is possible to generalize the technique of synthetic division for intervals of differencing other than one, that is, to convert polynomials from standard notation to factorial notation with an interval of differencing h. The only change necessary is to multiply the index numbers by h. For example, in Table 2.3 if it is required to convert to factorials with $h = 3$, then the index numbers are 3, 6, 9, 12 instead of 1, 2, 3, 4. The rest of the table is constructed as before.

2.5. CONVERTING FACTORIAL POLYNOMIALS TO STANDARD POLYNOMIALS

In many cases it is important to be able to convert polynomials which are in factorial notation back to standard notation. For example, a polynomial in standard notation may be converted to factorial notation for differencing and then converted back again to standard notation.

Techniques for doing this are best illustrated by example. Let us use the same example considered in Section 2.4, only this time we are going in the other direction. We have

$$5x^{(4)} + 23x^{(3)} + 17x^{(2)} + 2x^{(1)} - 4 = Ax^4 + Bx^3 + Cx^2 + Dx + E$$

and the problem is to determine A, B, C, D, E.

There are two approaches to this type of problem. The first is a straightforward algebraic approach. The left-hand side of the above equation is written out using the definition of $x^{(m)}$, then the terms are expanded and grouped by powers of x, and finally the coefficients are equated to those on the right-hand side. This method is excellent for low-order polynomials but becomes unwieldy for high-order polynomials.

A second more sophisticated approach, which works well for a polynomial of any degree, is the method of *inverse synthetic division*. This method can be derived from the first approach by an appropriate expansion of the factorials.

We have

$$5x^{(4)} + 23x^{(3)} + 17x^{(2)} + 2x^{(1)} - 4 = 5x(x-1)(x-2)(x-3)$$
$$+ 23x(x-1)(x-2)$$
$$+ 17x(x-1) + 2x - 4$$

Now, we break the first term into two terms by splitting $x - 3$, leave the x portion behind, and combine the -3 portion with the second term. Then we break the new second term into two terms by splitting $x - 2$, leave the x portion behind, and combine the -2 portion with the third term. Then we break the new third term into two terms by splitting $x - 1$, leave the x portion behind, and combine the -1 portion with the fourth term. This completes the first cycle. Now we perform as many more cycles as are necessary.

In the above example we have

$$5x(x-1)(x-2)(x-3) + 23x(x-1)(x-2) + 17x(x-1) + 2x - 4$$
$$= [5x(x-1)(x-2)x] + [(-15x + 23x)(x-1)x]$$
$$+ [(-16x + 17x)x] + (-1 + 2)x - 4$$
$$= 5x^2(x-1)(x-2) + 8x^2(x-1) + x^2 + x - 4 \qquad \text{after one cycle}$$
$$= 5x^3(x-1) - 2x^3 + 3x^2 + x - 4 \qquad \text{after two cycles}$$
$$= 5x^4 - 7x^3 + 3x^2 + x - 4 \qquad \text{after three cycles}$$

which we know is the correct answer.

The arithmetical work may again be performed on just the coefficients and arranged in a tabular format. Table 2.4 is a table for inverse synthetic division analogous to Table 2.3. The following is a description of how Table 2.4 is constructed:

TABLE 2.4

3	5	23	17	2	−4
	0	15	16	1	
2	5	8	1	1	
	0	10	−2		
1	5	−2	3		
	0	5			
0	5	−7			
	0				
	5				

1. List the coefficients of the polynomial in factorial notation across the top.
2. List index numbers down the side in descending order, starting with 1 less than the degree of the polynomial and stopping at 0.
3. The construction of the table is similar to Table 2.3 with two exceptions:
   (a) Subtractions are performed instead of additions.
   (b) The first term is multiplied by the index number on the immediate left, and then each successive term is multiplied by the next index number down until the cycle is completed. No number is ever multiplied by the zero index number. For example, in the second pair of rows we have in succession: 0 entered, $5 - 0 = 5$, $5 \times 2 = 10$, $8 - 10 = -2$, $-2 \times 1 = -2$, and $1 - (-2) = 3$.

The coefficients of the terms after each cycle in the longer method appear directly in the table. The answers appear at the ends on an upward diagonal to the right.

If it is required to convert a polynomial from lower factorials to standard form, it can either be done directly by expansion; or alternatively, the lower factorials can be converted to upper factorials, and then inverse synthetic division can be used.

It is also possible to generalize inverse synthetic division to other intervals of differencing by multiplying the index numbers by h and constructing the rest of the table as before.

A completely general type of synthetic division for the *transformation of polynomial forms* is possible for which the methods described in Sections 2.4 and 2.5 are special cases. Since this method is not required for any further development and is rather complex, it is described in Appendix B.

2.6. BACKWARD DIFFERENCES

It is sometimes more convenient to find differences in reverse order. For this purpose we introduce a new difference operator ∇, called the *backward difference operator*. It is defined by

$$\nabla f(x) = f(x) - f(x - 1) \tag{2.24a}$$

or in terms of operators

$$\nabla \equiv 1 - E^{-1} \tag{2.24b}$$

The backward difference operator ∇ is closely related to the forward difference operator Δ by

$$\nabla \equiv 1 - E^{-1} \equiv (E - 1)E^{-1} \equiv \Delta E^{-1} \tag{2.25}$$

Higher-order backward differences are built up similarly to forward differences, for example,

$$\nabla^2 f(x) = (1 - E^{-1})^2 f(x) = (1 - 2E^{-1} + E^{-2}) f(x)$$
$$= f(x) - 2f(x - 1) + f(x - 2)$$

and

$$\nabla^3 f(x) = f(x) - 3f(x - 1) + 3f(x - 2) - f(x - 3)$$

It is possible to construct a difference table analogous to Table 2.1 using backward differences. Table 2.5 is such a table.

TABLE 2.5

x	$f(x)$	$\nabla f(x)$	$\nabla^2 f(x)$	$\nabla^3 f(x)$	$\nabla^4 f(x)$
0	$f(0)$				
		$\nabla f(1)$			
1	$f(1)$		$\nabla^2 f(2)$		
		$\nabla f(2)$		$\nabla^3 f(3)$	
2	$f(2)$		$\nabla^2 f(3)$		$\nabla^4 f(4)$
		$\nabla f(3)$		$\nabla^3 f(4)$	
3	$f(3)$		$\nabla^2 f(4)$		
		$\nabla f(4)$			
4	$f(4)$				

It should be noted that the values of $x = 4, 3, 2, 1, 0$ appear as labels on symbols along upward diagonals to the right. This is indicated by the term "backward differences" in constrast to the "forward differences" in Table 2.1.

In any given problem the numerical values of the numbers in Tables 2.1 and 2.5 are equal. Forward differences and backward differences are two alternative notations for the same differences. Thus comparing Tables 2.1 and 2.5, we have, for example, $\Delta f(0) = \nabla f(1)$, $\Delta^2 f(1) = \nabla^2 f(3)$, etc.

2.7. ERROR DETECTION

It is useful to determine the effect which errors in functional values produce in the difference table. In many cases the construction of a difference table proves to be a convenient method to detect and correct errors.

First, consider data which contain mistakes, as the term was defined in Section 1.3. Often mistakes occur only in isolated functional values. Thus, it is useful to consider the effect on the difference table produced by an error in one functional value. From the linearity property the computed difference table is composed of the sum of the correct difference table and the difference table arising from the error. Table 2.6 is the difference table produced by an error of one unit in one functional value.

TABLE 2.6

u_x	Δu_x	$\Delta^2 u_x$	$\Delta^3 u_x$	$\Delta^4 u_x$
0				
	0			
0				
	0	0		
0				
	0	0	0	
0				
	0	0	0	0
0				
	1	1	1	1
1				
	-1	-2	-3	-4
0				
	0	1	3	6
0				
	0	0	-1	-4
0				
	0	0	0	1
0				
	0	0	0	0
0				

The effect of the error is to spread outward through the difference table and to magnify in size. The pattern of binomial coefficients with alternating signs should be carefully noted. This pattern is quite helpful in spotting mistakes, if the differences in the correct difference table are equal to zero after a certain order of differences. Even if the function being approximated is nonpolynomial and correct differences never become exactly zero, the technique may still be quite useful as long as the correct differences are smooth.

Other patterns of error in the difference table are produced if more than one mistake is involved, for example, two successive values may be in error. Difference tables for other patterns of error can be constructed similarly to Table 2.6. Examples of this are given in Exercises 33 and 34.

The second type of error to be considered is roundoff error. The effect of roundoff error is illustrated by comparing Tables 2.7a and 2.7b. Both tables are based on the function $u_x = x^3$. However, Table 2.7a is exact to three decimal places, whereas Table 2.7b is rounded off to two decimal places.

TABLE 2.7a

x	u_x	Δu_x	$\Delta^2 u_x$	$\Delta^3 u_x$
1.0	1.000			
		.331		
1.1	1.331		.066	
		.397		.006
1.2	1.728		.072	
		.469		.006
1.3	2.197		.078	
		.547		.006
1.4	2.744		.084	
		.631		.006
1.5	3.375		.090	
		.721		.006
1.6	4.096		.096	
		.817		.006
1.7	4.913		.102	
		.919		
1.8	5.832			

TABLE 2.7b*

x	u_x	Δu_x	$\Delta^2 u_x$	$\Delta^3 u_x$	$\Delta^4 u_x$	$\Delta^5 u_x$	$\Delta^6 u_x$
1.0	1.00						
		.33					
1.1	1.33		.07				
		.40		.00			
1.2	1.73		.07		.00		
		.47		.00		.03	
1.3	2.20		.07		.03		−.11
		.54		.03		−.08	
1.4	2.74		.10		−.05		.16
		.64		−.02		.08	
1.5	3.38		.08		.03		−.10
		.72		.01		−.02	
1.6	4.10		.09		.01		
		.81		.02			
1.7	4.91		.11				
		.92					
1.8	5.83						

* To be precise Δ should be written as $\Delta_{.1}$ in both Tables 2.7a and 2.7b. However to avoid cluttering the symbols, the h is often dropped from a symbol in a given problem, if there is no possibility of ambiguity.

In Table 2.7a third differences are constant as they must be. However, in Table 2.7b third differences are not constant because of the effect of roundoff error. Thus, fourth and higher differences are spurious, since they arise entirely from roundoff error. As might have been expected after seeing Table 2.6, the errors become progressively larger for the higher-order differences, that is, the difference table magnifies roundoff error.

The worst possible accumulation of roundoff error occurs if each term has a maximum roundoff error of one half a unit and if the signs alternate. Table 2.8 is the error table for this situation. In this case alternating signs and

TABLE 2.8

u_x	Δu_x	$\Delta^2 u_x$	$\Delta^3 u_x$	$\Delta^4 u_x$
.5				
	−1			
−.5		2		
	1		−4	
.5		−2		8
	−1		4	
−.5		2		−8
	1		−4	
.5		−2		8
	−1		4	
−.5		2		−8
	1		−4	
.5		−2		
	−1			
−.5				

powers of two are encountered. Hopefully, roundoff errors will seldom occur in this unfortunate pattern.

In approximating data with the collocation polynomial, it is often maintained that increasing the degree of the polynomial will improve the accuracy of the approximation. For most types of functions and data to be approximated, increasing the degree of the collocation polynomial will decrease truncation error. However, as demonstrated above, increasing the degree of the collocation polynomial will increase roundoff error, if roundoff error is present in the problem. In certain cases the accumulation of roundoff error may even become severe enough to be counterproductive and more than offset the lessening of truncation error, in which case a lower-degree polynomial will produce more accuracy than a higher-degree polynomial! Fortunately these cases are rare, if enough significant digits are carried in the computations.

The theory of statistics offers some insight into how rapidly roundoff errors magnify through the difference table. A statistical argument shows that the variance in the kth differences is approximately $\dfrac{(2k)!}{(k!)^2}$ times as great as the variance in the original functional values. Results for the first few values of k are given in Table 2.9.

TABLE 2.9

Order of difference (k)	Ratio of variances	Ratio of standard deviations
0	1	1.00
1	2	1.41
2	6	2.45
3	20	4.47
4	70	8.37
5	252	15.87
6	924	30.40

For example, in Table 2.7b each functional value is rounded off to two decimal places. If we multiply $E = .005$ by the ratio of standard deviations given above, we have for fourth, fifth, and sixth differences, respectively

$$.005 \times\ \ 8.37 = .04$$
$$.005 \times 15.87 = .08$$
$$.005 \times 30.40 = .15$$

These numbers appear quite reasonable in comparison with the spurious fourth, fifth, and sixth differences appearing in Table 2.7b.

It is also interesting to consider the correlation of the differences arising strictly from roundoff error. A statistical argument shows that each pair of successive kth differences has a correlation coefficient of $-\dfrac{k}{k+1}$ in the presence of independently distributed random errors. The high negative correlation between successive differences is apparent in the fourth, fifth, and sixth difference columns of Table 2.7b.[6]

Example 2.4.   You are given the following set of data but are told that one of the values may have been recorded in error. Assuming that the data follow a

---

[6]The derivations of these two statistical results are given in Appendix A.

polynomial, determine which one, if any, of the functional values is in error and what it should be.

7   10   17   33   63   121   185   287   423   598   817

Constructing a difference table, we have the following:

u_x	Δu_x	$\Delta^2 u_x$	$\Delta^3 u_x$	$\Delta^4 u_x$
7				
	3			
10		4		
	7		5	
17		9		0
	16		5	
33		14		9
	30		14	
63		28		−36
	58		−22	
121		6		54
	64		32	
185		38		−36
	102		−4	
287		34		9
	136		5	
423		39		0
	175		5	
598		44		
	219			
817				

The familiar pattern shown in Table 2.6 appears in the fourth difference column, in which the binomial coefficients with alternating signs 1, −4, 6, −4, 1 appear multiplied by 9. Working backward to see which functional value affected these five differences, we see that 121 must be in error by 9. The student should verify that if 121 is replaced with 112, the third differences become constant at 5. This type of mistake, in which two digits are reversed, is very common when numbers are written by hand, and is often called a *transposition*.

2.8. SEPARATION OF SYMBOLS

Section 2.2 indicates that *separation of symbols* is often a useful method of proving identities involving finite difference operators. In this method the operators are manipulated as algebraic quantities independently of the function on which they are operating.

As examples of this approach, two important identities can be derived from formulas (2.7a) and (2.7b) using the binomial theorem. The first of these for any positive integer n is

$$\Delta^n \equiv (E - 1)^n$$

$$\equiv E^n - \binom{n}{1}E^{n-1} + \binom{n}{2}E^{n-2} - \cdots + (-1)^n$$

which when applied to a function yields

$$\Delta^n f(x) = f(x + n) - \binom{n}{1}f(x + n - 1) + \binom{n}{2}f(x + n - 2) - \cdots + (-1)^n f(x)$$

$$(2.26)$$

This formula indicates how any forward difference can be evaluated in terms of the functional values which compose it. Formulas (2.3) and (2.4) are special cases of this more general equation.

The second of these for any positive integer n is

$$E^n \equiv (1 + \Delta)^n$$

$$\equiv 1 + \binom{n}{1}\Delta + \binom{n}{2}\Delta^2 + \cdots + \binom{n}{n}\Delta^n$$

which when applied to a function yields

$$f(x + n) = f(x) + \binom{n}{1}\Delta f(x) + \binom{n}{2}\Delta^2 f(x) + \cdots + \binom{n}{n}\Delta^n f(x) \quad (2.27)$$

This formula indicates how the functional value at one point can be evaluated in terms of the functional value and successive differences at some other point. This formula is quite important in the study of interpolation and will be discussed further in Chapter 3.

The student can prove formulas (2.26) and (2.27) for positive integers n by the more rigorous process of mathematical induction, if the separation of symbols approach seems mathematically too lax.

Some identities proved by separation of symbols involve infinite series, so that convergence must be considered. A finite difference identity involving such a series is valid only if it converges for the function in question. If the infinite series involves Δ's of increasing degree and if the operator series is being applied to a polynomial, then there is no problem, since differences of degree $n + 1$ and greater are zero for a polynomial of degree n.

Example 2.5.   *Derive a formula analogous to formula (2.27) using backward differences.*

We have

$$\nabla \equiv 1 - E^{-1}$$

or

$$E \equiv (1 - \nabla)^{-1}$$

Thus,

$$E^n \equiv (1 - \nabla)^{-n}$$

$$\equiv 1 + \binom{n}{1}\nabla + \binom{n+1}{2}\nabla^2 + \binom{n+2}{3}\nabla^3 + \cdots + \binom{2n-1}{n}\nabla^n$$

which when applied to a function yields

$$f(x+n) = f(x) + \binom{n}{1}\nabla f(x) + \binom{n+1}{2}\nabla^2 f(x)$$

$$+ \binom{n+2}{3}\nabla^3 f(x) + \cdots + \binom{2n-1}{n}\nabla^n f(x) \quad (2.28)$$

Example 2.6. **Show that if** $-1 < x < 1$, **then** $xu_1 + x^2 u_2 + x^3 u_3 + \cdots =$

$$\frac{x}{1-x}u_1 + \frac{x^2}{(1-x)^2}\Delta u_1 + \frac{x^3}{(1-x)^3}\Delta^2 u_1 + \cdots.$$

It is necessary to establish the equality of the operators on both sides of the equation.

The left-hand operator on u_1 is[7]

$$x + x^2 E + x^3 E^2 + \cdots$$

which can be summed as an infinite geometric progression to give[8]

$$\frac{x}{1 - xE} \equiv \frac{x}{1 - x - x\Delta}$$

The right-hand operator on u_1 is

$$\frac{x}{1-x} + \left(\frac{x}{1-x}\right)^2 \Delta + \left(\frac{x}{1-x}\right)^3 \Delta^2 + \cdots$$

which can also be summed as an infinite geometric progression to give

$$\frac{\dfrac{x}{1-x}}{1 - \dfrac{x}{1-x}\Delta} \equiv \frac{x}{1 - x - x\Delta}$$

---

[7]The student should be very careful to write $x^2 E$ and not Ex^2, that is, $E(x^2)$. The quantity $x^2 E$ is an operator, whereas the quantity $E(x^2)$ is the number $(x+1)^2$.

[8]This step is disturbing, since an infinite geometric progression with common ratio r converges if and only if $-1 < r < 1$. In this example the common ratio is an operator and is not even a number! The question of convergence depends upon the nature of the function to which the operator is applied.

Since the two operators are equal, the equality holds for any u_x for which both sides of the equation converge. If $-1 < x < 1$ and if u_x is a polynomial, then convergence will occur. This is clear on the right-hand side, since Δ's are zero after a certain point for any polynomial. It is also true on the left-hand side, since the exponential factor given by the increasing powers of x, for $-1 < x < 1$, will ultimately dominate any polynomial values for which the degree of the polynomial is given in advance, that is, the limit of successive terms on the left-hand side goes to zero.

EXERCISES

2.1. Introduction; 2.2. Differencing and difference tables

1. Show that formula (2.8) holds for both the operator Δ and the operator E, that is, that Δ and E are both linear operators.
2. (a) Derive formula (2.11).
    (b) Also show that the following formula is valid for quotients.

$$\Delta \frac{u_x}{v_x} = \frac{v_{x+1}\Delta u_x - u_{x+1}\Delta v_x}{v_x v_{x+1}}$$

3. Find the sixth term of the series $2, 0, 1, 5, 12, \ldots$.
4. Show that

$$\underset{h}{\Delta} \log_e x = \log_e \left(1 + \frac{h}{x}\right)$$

5. Show that $u_3 = u_2 + \Delta u_1 + \Delta^2 u_0 + \Delta^3 u_0$.
6. Find $\Delta^3 x^x$ evaluated at $x = 1$.
7. Find $\Delta^2 u_1$ given that $u_0 = 1$, $u_2 = 2u_1 + 1$, and $\Delta^3 u_0 = 1$.
8. If u_x is an exponential function for which $\Delta^n u_2 = 2^{n+1}$ for $n = 2, 3, 4, 5, \ldots$, find $\Delta^4 u_{-2}$.
9. Show that $\Delta x! = x \cdot x!$.
10. Find $\left(\dfrac{\Delta^2}{E}\right) 3^x x!$ evaluated at $x = 2$.
11. Find $\sum\limits_{i=1}^{20} \Delta^i a^{bx}$.
12. (a) Show that $\underset{h}{\Delta} \sin x = 2 \sin \left(\tfrac{1}{2}h\right) \cos \left(x + \tfrac{1}{2}h\right)$.
    (b) Show that $\underset{h}{\Delta} \cos x = -2 \sin \left(\tfrac{1}{2}h\right) \sin \left(x + \tfrac{1}{2}h\right)$.
13. Express the interval of differencing h as a function of a, if $Da^x = \underset{h}{\Delta} a^x$.

2.3. Factorial notation

14. Derive formula (2.19).
15. Derive formula (2.23).

16. Find $\Delta^n[ax + b][a(x + 1) + b] \cdots [a(x + m) + b]$ where $m > n$.
17. Find $\Delta^2(2x + 3)^{(3)}$ evaluated at $x = -1$.
18. Show that

$$\Delta^2[(2x - 3)(2x - 1)]^{-1} = 24[(2x - 3)(2x - 1)(2x + 1)(2x + 3)]^{-1}$$

19. Find $\Delta^3(x + 2)(x + 1)x(x - 2)$.
20. Simplify

$$\binom{y + 1}{r + 1}\Delta^r u_{x+1} - \binom{y}{r + 1}\Delta^r u_{x+1} - \binom{y}{r}\Delta^r u_x$$

21. Find $\Delta^3 \binom{6}{x}$ evaluated at $x = 0$.
22. Find $\Delta^3 x(x + 4)(x + 8)$.

2.4. Converting standard polynomials to factorial polynomials

23. Convert $8x^4 - 13x^3 + 9x - 3$ to factorial notation.
24. Convert the polynomial given in Exercise 23 to factorial notation in which the interval of differencing is equal to 2.
25. Find $\Delta^{10} \prod_{i=1}^{4} (1 - ix^i)$.
26. Express $\Delta(x^4 - 2x^2 + 5x + 2)$ in lower factorial notation.
27. Find $\underset{3}{\Delta^2} x^4$.

2.5. Converting factorial polynomials to standard polynomials

28. Use inverse synthetic division to convert the answer to Exercise 23 back to standard polynomial notation.
29. Use inverse synthetic division to convert the answer to Exercise 24 back to standard polynomial notation.

2.6. Backward differences

30. (a) Express ∇ as a series expansion in Δ.
    (b) Express Δ as a series expansion in ∇.
31. Express

$$\nabla^2 \frac{1}{x^3 + 3x^2 + 2x}$$

in factorial notation.

32. Find $\nabla^2 x^4$.

2.7. Error detection

33. Set up a difference table analogous to Table 2.6 for the situation in which two successive functional values are in error by $+1$ each.

34. Set up a difference table analogous to Table 2.6 for the situation in which two successive functional values are in error by $+1$, -1, respectively.[9]

35. You are given the following set of data but are told that one of the values may have been recorded in error. Assuming that the data follow a polynomial, determine which one, if any, of the functional values is in error and what it should be.

$$10 \quad 12 \quad 15 \quad 21 \quad 32 \quad 50 \quad 79 \quad 115 \quad 166$$

2.8. Separation of symbols

36. Derive a formula analogous to formula (2.26) using backward differences.

37. Show that

$$u_0 + \binom{x}{1}\Delta u_1 + \binom{x}{2}\Delta^2 u_2 + \binom{x}{3}\Delta^3 u_3 + \cdots$$

$$= u_x + \binom{x}{1}\Delta^2 u_{x-1} + \binom{x}{2}\Delta^4 u_{x-2} + \cdots$$

38. Show that

$$u_0 + \Delta u_0 + \Delta^2 u_0 + \Delta^3 u_0 + \cdots = \tfrac{1}{2}[u_0 + \tfrac{1}{2}u_1 + \tfrac{1}{4}u_2 + \tfrac{1}{8}u_3 + \cdots]$$

39. Show that

$$\left[\frac{\Delta^{-1}}{2} - \frac{1}{4} + \frac{\Delta}{8} - \frac{\Delta^2}{16} + \frac{\Delta^3}{32} - \cdots\right]^{-1} f(x) = f(x+2) - f(x)$$

Miscellaneous problems

40. If $f(x) = x^3$, find the following:

(a) $\left(\dfrac{\Delta^2}{E^2}\right) f(x)$.

(b) $\dfrac{\Delta^2 f(x)}{E^2 f(x)}$.

41. Find

$$\Delta^5 (1-x)\left(1 - \frac{x}{2}\right)\left(1 - \frac{x}{3}\right)\cdots\left(1 - \frac{x}{10}\right)$$

evaluated at $x = -1$.

42. Find

$$\Delta^9 \left[\frac{1}{(x+1)(x+3)}\right]$$

evaluated at $x = 0$.

---

[9]This type of error can occur if data are tabulated into categories and one value is tabulated incorrectly into a category adjacent to the correct one.

43. Find $x\Delta(x\Delta - 1)x^2$.

44. Find

$$\Delta^2 \log_{\sqrt[h]{x}} x^{x^2}$$

45. Find

$$\left(1 + \frac{\Delta^2}{E}\right)^{1/2} x^{(5)}$$

evaluated at $x = 4$.

46. Express $\Delta^4 u_4$ in terms of u_4, u_8, Δu_4, $\Delta^2 u_5$, and $\Delta^3 u_4$.

47. Simplify $u_1 - \Delta u_1 + \Delta^2 u_1 - \cdots + (-1)^n \Delta^n u_1$, if u_x is a polynomial of degree n.

48. Show that

$$\Delta(u_x v_x w_x) = u_{x+1} v_{x+1} \Delta w_x + u_{x+1} w_x \Delta v_x + v_x w_x \Delta u_x$$

49. If n is a positive integer, show that $[E^{-n} x^{(-n)}]^{-1}$ can be written as $k^{(n)}$, where k is a function of x. Find k.

50. Complete the missing entries in the following difference table:

x	u_x	Δu_x	$\Delta^2 u_x$	$\Delta^3 u_x$	$\Delta^4 u_x$
0					
1	1		12		
2				24	
3			108		
4	241				

51. Show that $\Delta(ax + b)_a^{(n)} = na(ax + b)_a^{(n-1)}$.[10]

52. Some older works in finite differences discussed *differences of zero*. These are defined as follows.

$$\Delta^n 0^m = \Delta^n x^m \text{ evaluated at } x = 0$$

(a) (1) Find $\Delta^5 0^2$.
    (2) Find $\Delta^5 0^5$.
    (3) Find $\Delta^2 0^5$.
(b) Show that $\Delta^n 0^m = n(\Delta^{n-1} 0^{m-1} + \Delta^n 0^{m-1})$.

53. *Stirling's numbers of the first kind* are the coefficients when $x^{(n)}$ is expressed in standard polynomial notation. They are denoted by $S_i^{(n)}$, so that

$$x^{(n)} = S_n^{(n)} x^n + S_{n-1}^{(n)} x^{n-1} + \cdots + S_1^{(n)} x$$

(a) Find $S_2^{(4)}$.
(b) Show that $S_i^{(n+1)} = S_{i-1}^{(n)} - n S_i^{(n)}$ by using induction.

---

[10]This result is analogous to the application of the chain rule in calculus which gives $D(ax + b)^n = na(ax + b)^{n-1}$.

54. *Stirling's numbers of the second kind* are the coefficients when x^n is expressed in factorial notation. They are denoted by $s_i^{(n)}$, so that

$$x^n = s_n^{(n)}x^{(n)} + s_{n-1}^{(n)}x^{(n-1)} + \cdots + s_1^{(n)}x^{(1)} .$$

(a) Find $s_2^{(4)}$.

(b) Show that $s_i^{(n+1)} = s_{i-1}^{(n)} + is_i^{(n)}$ by using induction.

3

Interpolation with equal intervals

3.1. INTRODUCTION

AN IMPORTANT PROBLEM in numerical analysis is to develop techniques of *interpolation*. It is often required to find an unknown functional value corresponding to a value of x which lies between other values of x for which functional values are known. Interpolation is often used for this type of problem, since an exact computation of the required functional value may be difficult or even impossible.

The interpolation techniques to be developed in this chapter rely on the use of the collocation polynomial. If points of collocation $x_0, x_1, \ldots, x_n$ are chosen such that functional values are known for each chosen point, then interpolation is the process of estimating $f(x)$ based on these known functional values for $x_0 < x < x_n$, where $x \neq x_i$ for $i = 1, 2, \ldots, n - 1$. The estimated value of $f(x)$ is $p(x)$, that is, the collocation polynomial evaluated at x. The error involved in this approximation is given by formula (1.12).

There are also numerous methods of interpolation not based on the collocation polynomial. Some of these methods are examined in Chapter 15.

Interpolation should be distinguished from *extrapolation*, which is the process of estimating functional values for points lying outside the *range of collocation*, that is, for $x < x_0$ or $x > x_n$. Extrapolation is a safe procedure to use on polynomials. For example, we can find $f(100)$ based on $f(0)$, $f(1)$,

61

$f(2)$, $f(3)$ with complete accuracy, if $f(x)$ is a third-degree polynomial and if no roundoff error is present.

However, if $f(x)$ is not exactly a third-degree polynomial or if roundoff error is present, then extrapolation in this example is much more subject to error than interpolation. In general, the further away from the range of collocation that the extrapolation is being performed, the poorer the results are likely to be. The student should analyze this mathematically based on formula (1.12)

$$f(x) - p(x) = \frac{f^{(n+1)}(\xi)\pi(x)}{(n+1)!} \tag{1.12}$$

As values of x move away from the range of collocation in either direction, $\pi(x)$ becomes large in absolute value without bound. Unless $f^{(n+1)}(\xi)$ goes to zero, which rarely happens, the error term will become large in absolute value without bound.

For instance, in Example 1.10, $\pi(x)$ for the points of collocation $x = 0, 1$ is seen to be $x^2 - x$. As x moves to the left of 0 or to the right of 1, $x^2 - x$ becomes large without bound. Since there is nothing to force $f''(\xi)$ to go to zero for all x, the entire error term $f(x) - p(x)$ becomes large without bound.

There are numerous methods of extrapolation not based on numerical analysis techniques. For example, there are statistical approaches involving regression or time series analysis. However, these and other methods of extrapolation are beyond the scope of this book.

The simplest collocation polynomial of practical value is the one through just two points. This leads to the familiar procedure of *linear interpolation*. The truncation error involved in linear interpolation is derived in Example 1.11.

This chapter will develop methods of interpolation based on polynomials of any degree, which will often lead to improved results over linear interpolation. The basic theory of finite differences discussed in Chapter 2 will be quite useful in developing efficient algorithms for fitting collocation polynomials.

It is assumed in this chapter that equally spaced values of x are used in the interpolation. The more general case in which values of x are not equally spaced will be considered in Chapter 5.

3.2. NEWTON'S ADVANCING DIFFERENCE FORMULA

The most basic formula for interpolation with equal intervals is *Newton's advancing difference formula*. This formula is also called *Newton's forward difference formula* or the *Newton-Gregory formula*[1] and is given by

---

[1] There is an interesting historical debate as to whether Isaac Newton or James Gregory was the first to originate this formula in the 17th century.

$$f(x + n) = f(x) + \binom{n}{1} \Delta f(x) + \binom{n}{2} \Delta^2 f(x) + \cdots \qquad (3.1a)$$

The name of the formula indicates that values down the top diagonal of the difference table are used.

This formula is the same as formula (2.27) developed in the preceding chapter. However, formula (2.27) is developed under the assumption that n is a positive integer. If this is the case, then the formula has a finite number of terms and is valid for any type of function.

If the formula is to be used for interpolation, then n will not be a positive integer and the formula becomes an infinite series. However, if the collocation polynomial is being used as an approximation, formula (3.1a) will again have only a finite number of terms. This is because all higher differences past the degree of the polynomial are equal to zero.

It is interesting to note that for $0 < n < 1$, if formula (3.1a) is carried to only two terms, we have

$$f(x + n) = f(x) + \binom{n}{1} \Delta f(x)$$

$$= f(x) + n[f(x + 1) - f(x)]$$

$$= (1 - n)f(x) + nf(x + 1) \qquad (3.2)$$

which is the standard formula for *linear interpolation*. Each new term introduced in formula (3.1a) brings in one additional degree of accuracy in the collocation polynomial.

Example 3.1. **Estimate** $f(2.3)$ **given that** $f(1) = 1, f(2) = 8, f(3) = 27,$ **and** $f(4) = 64$ **using Newton's advancing difference formula.**

The four given values of $f(x)$ agree with the third-degree polynomial x^3. Thus, we hope our procedures will produce the answer $(2.3)^3 = 12.167$.

Setting up the difference table, we have the following:

x	$f(x)$	$\Delta f(x)$	$\Delta^2 f(x)$	$\Delta^3 f(x)$
1	1			
		7		
2	8		12	
		19		6
3	27		18	
		37		
4	64			

Now applying Newton's advancing difference formula using the numbers down the top diagonal in the difference table, we have[2]

$$f(2.3) = E^{1.3}f(1)$$

$$= (1 + \Delta)^{1.3}f(1)$$

$$= \left[1 + 1.3\Delta + \frac{(1.3)(.3)}{2}\Delta^2 + \frac{(1.3)(.3)(-.7)}{6}\Delta^3\right]f(1)$$

$$= [1 + 1.3\Delta + .195\Delta^2 - .0455\Delta^3]f(1)$$

$$= 1 + (1.3)(7) + (.195)(12) - (.0455)(6)$$

$$= 12.167$$

the hoped-for result.

Although this example contains a very simple cubic, for polynomials in general the student should note how much more efficient this algorithm is than the straightforward algebraic approach of fitting a cubic to the four functional values illustrated in Example 22 in Chapter 1. This latter approach necessitates setting up four equations in four unknowns, solving them, and substituting back the value of $x = 2.3$. In using the difference formula approach it is not necessary to determine the underlying collocation polynomial to estimate $f(2.3)$.

The student should be careful not to think of the answer 12.167 as of necessity being the correct answer. *It is an estimate of the correct answer based on the assumption that $f(x)$ is a third-degree polynomial.* However, there are any number of other functions which could be chosen to pass through the four points in question, and if any of these other functions is the correct one, then there will be a truncation error involved.

3.3. NEWTON'S BACKWARD DIFFERENCE FORMULA

Another formula for interpolation with equal intervals is *Newton's backward difference formula* and is given by

$$f(x + n) = f(x) + \binom{n}{1}\nabla f(x) + \binom{n+1}{2}\nabla^2 f(x) + \cdots \qquad (3.3a)$$

The name of the formula indicates that values up the bottom diagonal of the difference table are used.

---

[2]Complete precision in the notation would substitute $p(x)$ for $f(x)$ at this point in the problem. However, for simplicity this is not usually done. It should be remembered that $f(0), f(1), f(2),$ and $f(3)$ are exact, whereas $f(2.3)$ is an estimate

This formula is the same as formula (2.28) developed in Example 2.5. The same discussion presented in Section 3.2 for Newton's forward difference formula applies to Newton's backward difference formula as well.

Example 3.2. Rework Example 3.1 using Newton's backward difference formula.

The difference table given in Example 3.1 can be used. Applying Newton's backward difference formula using the numbers up the bottom diagonal in the difference table, we have

$$f(2.3) = E^{-1.7}f(4)$$

$$= (1 - \nabla)^{1.7}f(4)$$

$$= \left[1 - 1.7\nabla + \frac{(1.7)(.7)}{2}\nabla^2 - \frac{(1.7)(.7)(-.3)}{6}\nabla^3 \right]f(4)$$

$$= [1 - 1.7\nabla + .595\nabla^2 + .0595\nabla^3]f(4)$$

$$= 64 - (1.7)(37) + (.595)(18) + (.0595)(6)$$

$$= 12.167$$

which agrees with the answer obtained in Example 3.1. The student should verify that the third line above can be directly written down using formula (3.3a) with $n = -1.7$.

In working numerical problems similar to either Example 3.1 or 3.2, it is advisable to set up a fundamental expression involving the operators E, Δ, ∇, and then to directly expand with the binomial theorem. This is preferable to applying formulas (3.1a) and (3.3a) by rote memorization, since it not only eliminates the need to memorize these rather long involved expressions but it also is less subject to error.

3.4. CHANGE IN ORIGIN AND SCALE

In interpolation problems it is possible to make a change in origin and scale before performing the interpolation. Any change in origin and scale does not affect the answer obtained, but it may result in a simpler procedure less subject to error.

A *change in origin* involves a linear shift along the x-axis so that the interpolation formula starts at $x = 0$. A *change in scale* involves a stretching or shrinking of the x-axis so that the interval of differencing $h = 1$. In theory, other changes in origin and scale than these would be possible, but they would have little practical value in interpolation.

Interpolation formulas are often written in a standardized form in which it is assumed that any changes in origin and scale have been made. Interpolation formulas to be introduced in later portions of this book will generally be in this standardized form.

The standardized form for Newton's advancing difference formula can be obtained from formula (3.1a) by using the u-notation for functions, letting $x = 0$, and letting $n = x$ to produce

$$u_x = u_0 + \binom{x}{1}\Delta u_0 + \binom{x}{2}\Delta^2 u_0 + \cdots \tag{3.1b}$$

Similarly, Newton's backward difference formula can be obtained from formula (3.3a) to produce

$$u_x = u_0 + \binom{x}{1}\nabla u_0 + \binom{x+1}{2}\nabla^2 u_0 + \cdots \tag{3.3b}$$

The student should note that Examples 3.1 and 3.2 were worked without making any change in origin (no change in scale would even be considered, since the interval of differencing is already equal to one). However, the same answer is obtained whether or not the change in origin is made.

Let x apply before the change in origin and scale, and let y apply after the change. Then in Example 3.1 the transformation $y = x - 1$ would be made

x		y
1	$\longrightarrow$	0
2	$\longrightarrow$	1
3	$\longrightarrow$	2
4	$\longrightarrow$	3

and the interpolated value $2.3 \rightarrow 1.3$. It is evident that the same difference table and the same numbers in the interpolation formula appear.

Similarly, in Example 3.2 the transformation $y = x - 4$ would be made

x		y
1	$\longrightarrow$	-3
2	$\longrightarrow$	-2
3	$\longrightarrow$	-1
4	$\longrightarrow$	0

and the interpolated value $2.3 \rightarrow -1.7$. Again the same difference table and the same numbers in the interpolation formula appear.

Although changes in scale are not necessary to produce correct answers, it is usually desirable to make them. This avoids the complexity of working with intervals of differencing other than one and is probably a more error-free procedure for most students.

It should be carefully noted that although a change in origin and scale does not affect the numerical answer in interpolating for any specific value of *x*, it does not give a formula for general *x*, since the formula after the change is based on *y*. If it is required to find the general formula for *x*, another linear transformation from *y* back to *x* must be made. This procedure is illustrated in Example 3.4.

Example 3.3. **Estimate $f(4)$ given that $f(3) = 1$, $f(5) = 11$, and $f(7) = 29$.**

Making the transformation $y = \dfrac{x - 3}{2}$, setting up the difference table, and letting f apply to x and g to y, we have the following:

$x \longrightarrow$	y	$g(y)$	$\Delta g(y)$	$\Delta^2 g(y)$
3 $\longrightarrow$	0	1		
			10	
5 $\longrightarrow$	1	11		8
			18	
7 $\longrightarrow$	2	29		

Now $4 \to .5$, and using Newton's advancing difference formula, we have

$$f(4) = g(.5) = 1 + (.5)(10) + \frac{(.5)(-.5)}{2} (8)$$
$$= 5$$

Example 3.4. **Rework Example 3.3 to find the general formula for $f(x)$ assuming it to be a second-degree polynomial.**

In this example we seek to find the form of $f(x)$ for general x. There are two approaches which can be used.

The first approach is to make the change in origin and scale, find the form of the answer, and then convert the answer back to the origin and scale in the original problem.

Using Newton's advancing difference formula and the difference table in Example 3.3

$$g(y) = 1 + 10y + \tfrac{8}{2} y(y - 1)$$
$$= 4y^2 + 6y + 1$$

We then have

$$f(x) = g(y) = g\left(\frac{x-3}{2}\right) = 4\left(\frac{x-3}{2}\right)^2 + 6\left(\frac{x-3}{2}\right) + 1 = x^2 - 3x + 1$$

As a check, the given values of $f(3)$, $f(5)$, and $f(7)$ are reproduced by this formula. Also the interpolated value for $f(4)$ found in Example 3.3 is reproduced by this formula.

The second approach is to work the problem directly without making the change in origin and scale. In this case Newton's advancing difference formula starts at $x = 3$ and has $h = 2$.

We then have

$$f(x) = E^{\frac{x-3}{2}}_2 f(3)$$

$$= \left(1 + \underset{2}{\Delta}\right)^{\frac{x-3}{2}} f(3)$$

$$= \left[1 + \frac{x-3}{2}\underset{2}{\Delta} + \frac{\left(\dfrac{x-3}{2}\right)\left(\dfrac{x-5}{2}\right)}{2}\underset{2}{\Delta^2}\right] f(3)$$

$$= 1 + \left(\frac{x-3}{2}\right)10 + \left(\frac{x^2 - 8x + 15}{8}\right)8$$

$$= x^2 - 3x + 1$$

The student should note that the third line above can be written down directly if it is noticed that in moving down the top diagonal in the difference table, the first value of x brought in is 3, the second is 5, and if it is remembered that the Δ must be divided by its interval of differencing. Thus, we have

$$f(x) = \left[1 + (x-3)\left(\frac{\Delta}{2}\right) + \frac{(x-3)(x-5)}{2}\left(\frac{\Delta}{2}\right)^2\right] f(3)$$

This last approach illustrates that interpolation formulas can directly be written down based on the order of the values of x brought into the formula. *Sheppard's rules* provide a systematic method of directly expressing formulas in this fashion and will be discussed in Section 3.6.

3.5. SUBDIVISION OF INTERVALS

It has been assumed thus far that the interval of differencing h is given in a problem and that it has a fixed value. However, in practice there may be some latitude in the choice of h.

If the value of h is chosen too large, then interpolation formulas may not be sufficiently accurate. On the other hand, if the value of h is chosen too small, then excessive amounts of computation may needlessly be performed. In addition, the effect of roundoff error should be considered in choosing h.

A common problem encountered is to have functional values available at a certain interval of differencing and to need functional values at some smaller interval of differencing. For example, a maker of mathematical tables is preparing a table of values of $\sqrt{x}$. Values of $\sqrt{x}$ computed in steps of .1, for example, 1.0, 1.1, 1.2, . . . are available, but values in steps of .01, for example, 1.00, 1.01, 1.02, . . . are required.

One straightforward approach to obtain functional values for a smaller interval of differencing is to compute all the intermediate values directly. The disadvantage of this approach is that the volume of computations may become excessive. Furthermore, in certain cases it may not even be possible to compute the values directly. However, the advent of high-speed digital computers has increased the usefulness of this approach substantially.

A second approach which was developed in the days of desk calculators is *subdivision of intervals* or *subtabulation*. This approach requires much less computation than the first approach above. Even in the computer age, this approach is still useful on occasion.

Subdivision of intervals requires that the relationship between Δ_m and Δ_n be known, where m and n are two unequal intervals of differencing. A general formula relating Δ_m and Δ_n can be derived by noting that

$$E_n^m \equiv E_m^n$$

since each side of the equation is an operator which will increment the value of x in a function by nm units. Thus, we have

$$\left(1 + \Delta_n\right)^m \equiv \left(1 + \Delta_m\right)^n \tag{3.4a}$$

which gives

$$\Delta_n \equiv \left(1 + \Delta_m\right)^{\frac{n}{m}} - 1 \tag{3.4b}$$

The above expressions are valid for any positive numbers m and n; thus m and n are interchangeable in formula (3.4b). The formulas can then be manipulated with the binomial theorem in working problems.

The idea behind subdivision of intervals is to construct a difference table based on the available functional values at the larger interval of differencing and then to fill in the required functional values at the smaller interval of differencing using formula (3.4b) in which $n < m$. The technique requires that the function be closely enough approximated by a polynomial so that higher-order differences in the binomial expansion of formula (3.4b) can safely be ignored. Usually more decimal places are carried in the functional values on which the subtabulation is based than are required in the answers in order to minimize or eliminate roundoff error. The technique is illustrated in Example 3.5.

The choice of h may have a significant impact on whether the difference table for interpolation of nonpolynomials converges or diverges. A difference table is said to *converge* if successive orders of differences diminish in absolute value and approach zero, and is said to *diverge* otherwise. The convergence or divergence involved is independent of any roundoff error which may be present. As illustrated in Table 2.7b, if roundoff error were to be considered, then divergence would always result ultimately.

Interpolations for nonpolynomials must be based on convergent difference tables in order to produce reasonable answers. One of the reasons that inter-polation techniques developed in numerical analysis have proven so success-ful in practice is that many nonpolynomials do have convergent difference tables. However, the choice of h may influence whether convergence or diver-gence occurs. This will be illustrated in Example 3.6.

Example 3.5. *Illustrate the technique of subdivision of intervals to produce square roots of x at intervals of .01 in the interval $1.00 \leq x \leq 1.05$ assuming that square roots at intervals of .05 are available. Five-decimal accuracy is required in the answers.*

Assume that the starting values are computed to six decimal places to eliminate the effect of roundoff error. Proceeding as far as third differences we have the following:

x	$\sqrt{x}$	$\Delta\sqrt{x}$ .05	$\Delta^2\sqrt{x}$ .05	$\Delta^3\sqrt{x}$ .05
1.00	1.000000			
		.024695		
1.05	1.024695		−.000581	
		.024114		.000039
1.10	1.048809		−.000542	
		.023572		
1.15	1.072381			

From formula (3.4b), we have

$$\underset{.01}{\Delta} \equiv \left(1 + \underset{.05}{\Delta}\right)^{.2} - 1$$

$$\equiv \left(1 + .2\underset{.05}{\Delta} + \frac{(.2)(-.8)}{2}\underset{.05}{\Delta^2} + \frac{(.2)(-.8)(-1.8)}{6}\underset{.05}{\Delta^3}\right) - 1$$

$$\equiv .2\underset{.05}{\Delta} - .08\underset{.05}{\Delta^2} + .048\underset{.05}{\Delta^3}$$

We can find $\underset{.01}{\Delta^2}$ by squaring the above expression and eliminating any terms higher than third differences, which gives

$$\underset{.01}{\Delta^2} \equiv .04\underset{.05}{\Delta^2} - .032\underset{.05}{\Delta^3}$$

Finally, $\underset{.01}{\Delta^3}$ can be similarly found by taking cross products in $\underset{.01}{\Delta^2} \times \underset{.01}{\Delta}$ and eliminating any terms higher than third differences, which gives

$$\underset{.01}{\Delta^3} \equiv .008\underset{.05}{\Delta^3}$$

Now computing the advancing differences based on $x = 1.00$ and $h = .01$, we have the following:

$$\underset{.01}{\Delta}\sqrt{1.00} = (.2)(.024695) - (.08)(-.000581) + (.048)(.000039) = .004987$$

$$\underset{.01}{\Delta^2}\sqrt{1.00} = (.04)(-.000581) - (.032)(.000039) = -.000024$$

$$\underset{.01}{\Delta^3}\sqrt{1.00} = (.008)(.000039) = 0 \text{ to six decimal places}$$

The difference table based on $h = .01$ is now completed using these advancing differences:

x	$\sqrt{x}$	$\underset{.01}{\Delta\sqrt{x}}$	$\underset{.01}{\Delta^2\sqrt{x}}$
1.00	1.000000		
		.004987	
1.01	1.004987		−.000024
		.004963	
1.02	1.009950		−.000024
		.004939	
1.03	1.014889		−.000024
		.004915	
1.04	1.019804		−.000024
		.004891	
1.05	1.024695		

The difference table agrees with $\sqrt{1.05}$ with no roundoff error even in the sixth decimal place. The values of $\sqrt{x}$ for $x = 1.01, 1.02, 1.03,$ and 1.04 are all correct to five decimal places. In fact, the only error in the sixth decimal place is $\sqrt{1.01}$, which is actually 1.004988.

In applying formula (3.4b) to this example, the student should note that in subdividing the interval by a factor of five, the following approximate equalities hold:

$$\underset{.01}{\Delta} \equiv .200 \underset{.05}{\Delta} \equiv \frac{\underset{.05}{\Delta}}{5}$$

$$\underset{.01}{\Delta^2} \equiv .040 \underset{.05}{\Delta^2} \equiv \frac{\underset{.05}{\Delta^2}}{5^2}$$

$$\underset{.01}{\Delta^3} \equiv .008 \underset{.05}{\Delta^3} \equiv \frac{\underset{.05}{\Delta^3}}{5^3}$$

As a general rule, in subdividing intervals by a factor of n, it is seen that kth differences are approximately reduced by a factor of $\frac{1}{n^k}$. That is,

$$\underset{h/n}{\Delta^k} \equiv \frac{1}{n^k} \underset{h}{\Delta^k} \tag{3.5}$$

Thus, subdivision of intervals results in a much more rapidly converging difference table.

Example 3.6.   Find the largest interval of differencing for which the difference table of e^x will converge.

We have the following differences of e^x:

$$\underset{h}{\Delta e^x} = e^x(e^h - 1)$$

$$\underset{h}{\Delta^2 e^x} = e^x(e^h - 1)^2$$

$$\vdots$$

$$\underset{h}{\Delta^n e^x} = e^x(e^h - 1)^n$$

Thus, the difference table will converge if

$$e^h - 1 < 1$$

or

$$e^h < 2$$

or

$$h < \log_e 2 = .693$$

to three decimal places, and it will diverge if $h \geq \log_e 2$.

Example 3.7. *Find the largest interval of differencing that can be used in a table of $\sqrt{x}$, $x \geq 1$, which will guarantee five-decimal accuracy, if interpolations are to be performed with Newton's advancing difference formula accurate for second-degree polynomials.*

This type of problem is of considerable interest to makers of mathematical tables.

Assume points of collocation $x_0, x_1, x_2, x_3, \ldots$ where $h = x_1 - x_0 = x_2 - x_1 = x_3 - x_2 = \cdots$. In interpolating for $\sqrt{x}$, if $x_0 < x < x_1$, use points of collocation x_0, x_1, x_2; if $x_1 < x < x_2$, use points of collocation x_1, x_2, x_3; and so forth.

From formula (1.12) the truncation error is

$$f(x) - p(x) = \frac{f^{(3)}(\xi)\pi(x)}{3!}$$

where

$$f(x) = \sqrt{x}$$

Now

$$f(x) = x^{1/2}$$

and

$$f^{(3)}(x) = \left(\tfrac{1}{2}\right)\left(-\tfrac{1}{2}\right)\left(-\tfrac{3}{2}\right)x^{-5/2}$$

For $x \geq 1$

$$\left|f^{(3)}(\xi)\right| = \left|\tfrac{3}{8}\xi^{-5/2}\right| \leq \tfrac{3}{8}$$

Also,

$$\pi(x) = (x - x_0)(x - x_1)(x - x_2)$$

Let $x_1 - x = hk$ where $0 < k < 1$. The term hk can be viewed as a proportion k of one interval of differencing h. Also using the fact that $h = x_1 - x_0 = x_2 - x_1$, we have

$$\pi(x) = (-hk + h)(-hk)(-hk - h) = -h^3(k + 1)k(k - 1) = -h^3(k^3 - k)$$

We now seek to maximize $|g(k)| = |k^3 - k|$ for $0 < k < 1$. We have

$$g(k) = k^3 - k$$

and

$$g'(k) = 3k^2 - 1 = 0$$

Thus, an extremum occurs at $k = \sqrt{\tfrac{1}{3}}$, and

$$|g(k)| \leq \left|g(\sqrt{\tfrac{1}{3}})\right| = \tfrac{2}{3}\sqrt{\tfrac{1}{3}}$$

The student should verify that if $x - x_0 = hk$, the same upper bound is obtained, only the details are slightly more complex.

Now putting the pieces together, we have

$$|f(x) - p(x)| = \left| \frac{f^{(3)}(\xi)\pi(x)}{3!} \right|$$

$$\leq \left| (\tfrac{1}{6})(\tfrac{3}{8})(\tfrac{2}{3})(\sqrt{\tfrac{1}{3}})h^3 \right|$$

$$= .024056h^3 \leq .000005$$

or

$$h^3 \leq .000208$$

giving

$$h \leq .059$$

Thus far only truncation error has been considered. Since it pays to be on the safe side because of roundoff error in the quadratic interpolation formula, a convenient choice of h would appear to be .05. Note that more than five decimal places will still have to be carried in the values tabulated at intervals of .05, if all roundoff errors in the fifth decimal place for interpolated values are to be avoided.

3.6. SHEPPARD'S RULES

Newton's advancing difference formula is characterized by a path through the difference table down the top diagonal, while Newton's backward difference formula goes up the bottom diagonal. These are illustrated in Figure 3.1.

FIGURE 3.1

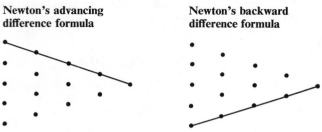

Newton's advancing difference formula

Newton's backward difference formula

We now show that it is possible to take any path through the difference table as long as each step to the next higher difference column takes place to one of the two adjacent differences in that column. Figure 3.2 is one of several possible paths.

FIGURE 3.2

The student should verify that for five starting values, as in Figure 3.2, there are 16 paths through the difference table. Since each step in the difference table can be made in two ways and since there are four steps to get to fourth differences, it is seen that the total number of possible paths is $2^4 = 16$. In general, if there are $n + 1$ starting values and the difference table proceeds to nth differences, then the number of possible paths is 2^n.

Sheppard's rules are a set of procedures for directly writing down the collocation polynomial based on any of these paths through the difference table. The rules are sometimes given the picturesque name *zigzag rules*. The following is a summary of the procedure:

1. The formula starts at any selected functional value.
2. The formula then proceeds through the difference table with each step to the next higher difference column branching to either of the two adjacent differences in that column.
3. The kth difference in the formula has a coefficient of k factors of the form $(x - x_i)(x - x_j) \ldots$, where x_i, x_j, $\ldots$ are the first k points of collocation brought into the formula.
4. Rule 3 can be implemented by noting that each value of x brought into the formula determines the new factor introduced in the next following term. When the formula proceeds along an upper diagonal in the difference table, the next value of x in the difference table appearing above those already used is introduced in the following term; and similarly, when the formula proceeds along a lower diagonal, the next value of x below those already used is introduced.
5. The kth difference in the formula is divided by $k! \, h^k$.

The standardized form of Newton's advancing difference formula given by formula (3.1b) can be written down directly by applying Sheppard's rules

$$u_x = u_0 + x\Delta u_0 + \frac{x(x - 1)}{2!} \Delta^2 u_0 + \frac{x(x - 1)(x - 2)}{3!} \Delta^3 u_0 + \cdots \quad (3.1b)$$

The student should verify that Newton's backward difference formula can similarly be written down directly.

The use of Sheppard's rules will be quite important in developing the central difference formulas to be discussed in Chapter 4.

Example 3.8.   Use Sheppard's rules to find the formula for the collocation polynomial given by the path displayed in Figure 3.2. Assume the points of collocation are u_0, u_1, u_2, u_3, u_4. Verify that the answer given is the collocation polynomial.

The difference table is given in Table 3.1. Sheppard's rules give

$$u_x = u_2 + (x - 2)\Delta u_1 + \frac{(x - 1)(x - 2)}{2!} \Delta^2 u_1 + \frac{(x - 1)(x - 2)(x - 3)}{3!} \Delta^3 u_1$$

$$+ \frac{(x - 1)(x - 2)(x - 3)(x - 4)}{4!} \Delta^4 u_0$$

TABLE 3.1

x	u_x	Δu_x	$\Delta^2 u_x$	$\Delta^3 u_x$	$\Delta^4 u_x$
0	u_0				
		Δu_0			
1	u_1		$\Delta^2 u_0$		
		Δu_1		$\Delta^3 u_0$	
2	u_2		$\Delta^2 u_1$		$\Delta^4 u_0$
		Δu_2		$\Delta^3 u_1$	
3	u_3		$\Delta^2 u_2$		
		Δu_3			
4	u_4				

To verify that the fourth-degree polynomial just stated is the collocation polynomial, we must show that it passes through u_0, u_1, u_2, u_3, u_4. In demonstrating this result, it is easiest to take the values of x in the order in which they are introduced.

If $x = 2$, $u_x = u_2$ immediately

If $x = 1$, $u_x = u_2 - \Delta u_1 = u_2 - (u_2 - u_1) = u_1$

If $x = 3$, $u_x = u_2 + \Delta u_1 + \Delta^2 u_1 = u_2 + (u_2 - u_1) + (u_3 - 2u_2 + u_1) = u_3$

If $x = 4$, $u_x = u_2 + 2\Delta u_1 + 3\Delta^2 u_1 + \Delta^3 u_1$

$\qquad\qquad = u_2 + 2(u_2 - u_1) + 3(u_3 - 2u_2 + u_1) + (u_4 - 3u_3 + 3u_2 - u_1)$

$\qquad\qquad = u_4$

If $x = 0$, $u_x = u_2 - 2\Delta u_1 + \Delta^2 u_1 - \Delta^3 u_1 + \Delta^4 u_0$

$\qquad\qquad = u_2 - 2(u_2 - u_1) + (u_3 - 2u_2 + u_1) - (u_4 - 3u_3 + 3u_2 - u_1)$

$\qquad\qquad + (u_4 - 4u_3 + 6u_2 - 4u_1 + u_0)$

$\qquad\qquad = u_0$

Formulas for any other path may be similarly verified. A general proof of Sheppard's rules depends upon the properties of divided differences to be developed in Chapter 5 and will be given at that time.

3.7. FUNCTIONS OF MORE THAN ONE VARIABLE

The interpolation formulas developed in this chapter can be generalized to functions of more than one variable. In this section we develop interpolation techniques for functions of two variables. The generalization to n variables is similar, but the details are more complex.

Consider a function of two variables $u_{x:y}$. Let Δ_x and E_x be operators applicable to the variable x, and let Δ_y and E_y be the corresponding operators for the variable y.[3] The operators for the variable x are defined by

$$\Delta_x u_{x:y} = u_{x+1:y} - u_{x:y} \tag{3.6a}$$

$$E_x u_{x:y} = u_{x+1:y} \tag{3.7a}$$

while the operators for the variable y similarly are

$$\Delta_y u_{x:y} = u_{x:y+1} - u_{x:y} \tag{3.6b}$$

$$E_y u_{x:y} = u_{x:y+1} \tag{3.7b}$$

It is assumed that the given functional values in an interpolation problem have been standardized, that is, changes in origin and scale for both variables x and y have been made.

A generalized version of Newton's advancing difference formula is given by

$$
\begin{aligned}
u_{x:y} &= E_x^x E_y^y u_{0:0} \\
&= (1 + \Delta_x)^x (1 + \Delta_y)^y u_{0:0} \\
&= \left[1 + x\Delta_x + \frac{x(x-1)}{2!} \Delta_x^2 + \cdots \right]\left[1 + y\Delta_y + \frac{y(y-1)}{2!} \Delta_y^2 + \cdots \right] u_{0:0}
\end{aligned}
\tag{3.8}
$$

In applying formula (3.8) cross products which can be evaluated are computed. When intuitively interpreting various orders of differences, the student should add the exponents on Δ_x and Δ_y, for example, $\Delta_x^2 \Delta_y u_{0:0}$ is considered to be a third difference. In expanding formula (3.8) it is not necessary to carry the expansions for x and y to the same order of differences. For example, the x-expansion could be carried to third differences and the y-expansion to second differences, if this is appropriate for the data involved.

---

[3]Care should be taken in positioning labels so as not to confuse symbols such as Δ and Δ_x.

The above approach assumes that x and y are independent variables. If y is a function of x, then the function of two variables can be reduced to a function of one variable. This type of transformation should always be made, if appropriate, before performing the interpolation to simplify the problem.

Example 3.9. **Estimate** $f(2, 3)$, **given that** $f(1, 2) = 2$, $f(3, 2) = 18$, $f(5, 2) = 50$, $f(1, 5) = 5$, $f(3, 5) = 45$, **and** $f(5, 5) = 125$.

A tabular array of the functional values together with a change of origin and scale is as follows:

x	1 $\downarrow$ 0	3 $\downarrow$ 1	5 $\downarrow$ 2
y			
$2 \to 0$	2	18	50
$5 \to 1$	5	45	125

where $x \to \dfrac{x-1}{2}$

and $y \to \dfrac{y-2}{3}$

The functional values given agree with $f(x, y) = x^2 y$. Thus, we hope our procedures will produce the answer $2^2 \cdot 3 = 12$.

Let the standardized functional values after the change in origin and scale be denoted by $g(x, y)$, so that $f(2, 3) = g(\frac{1}{2}, \frac{1}{3})$. The required differences are as follows:

$$\Delta_x g(0, 0) = (E_x - 1)g(0, 0) = 18 - 2 = 16$$
$$\Delta_x^2 g(0, 0) = (E_x^2 - 2E_x + 1)g(0, 0) = 50 - 36 + 2 = 16$$
$$\Delta_y g(0, 0) = (E_y - 1)g(0, 0) = 5 - 2 = 3$$
$$\Delta_x \Delta_y g(0, 0) = (E_x - 1)(E_y - 1)g(0, 0) = (E_x E_y - E_x - E_y + 1)g(0, 0)$$
$$= 45 - 18 - 5 + 2 = 24$$
$$\Delta_x^2 \Delta_y g(0, 0) = (E_x^2 - 2E_x + 1)(E_y - 1)g(0, 0)$$
$$= (E_x^2 E_y - 2E_x E_y + E_y - E_x^2 + 2E_x - 1)g(0, 0)$$
$$= 125 - 90 + 5 - 50 + 36 - 2 = 24$$

Now,
$$f(2, 3) = g(\tfrac{1}{2}, \tfrac{1}{3})$$
$$= (1 + \Delta_x)^{1/2}(1 + \Delta_y)^{1/3}g(0, 0)$$
$$= (1 + \tfrac{1}{2}\Delta_x - \tfrac{1}{8}\Delta_x^2)(1 + \tfrac{1}{3}\Delta_y)g(0, 0)$$
$$= (1 + \tfrac{1}{2}\Delta_x - \tfrac{1}{8}\Delta_x^2 + \tfrac{1}{3}\Delta_y + \tfrac{1}{6}\Delta_x \Delta_y - \tfrac{1}{24}\Delta_x^2 \Delta_y)g(0, 0)$$
$$= 2 + \tfrac{1}{2}(16) - \tfrac{1}{8}(16) + \tfrac{1}{3}(3) + \tfrac{1}{6}(24) - \tfrac{1}{24}(24)$$
$$= 12$$

the anticipated result.

Example 3.10. Estimate $u_{44:51}$, **given that** $u_{40:55} = 10.135$, $u_{45:50} = 10.763$, **and** $u_{50:45} = 10.763$.

Although $u_{x:y}$ appears to be a function of two variables, for all cases $y = 95 - x$. Thus, we can work the problem as a function of one variable. The difference table is as follows:

(x, y)		$u_{x:y}$	$\Delta u_{x:y}$	$\Delta^2 u_{x:y}$
$(40, 55)$	$\longrightarrow$ 0	10.135		
			.628	
$(45, 50)$	$\longrightarrow$ 1	10.763		$-.628$
			.000	
$(50, 45)$	$\longrightarrow$ 2	10.763		

Now $(44, 51) \to .8$, and applying Newton's advancing difference formula

$$u_{44:51} = 10.135 + (.8)(.628) + \frac{(.8)(-.2)}{2}(-.628)$$

$$= 10.688$$

EXERCISES

3.1. Introduction; 3.2. Newton's advancing difference formula

1. Estimate $f(.5)$ given that $f(-1) = 3$, $f(0) = 1$, $f(1) = 7$, and $f(2) = 21$ using Newton's advancing difference formula.
2. Rework Exercise 1 using linear interpolation based upon—
   (a) $f(0)$ and $f(1)$.
   (b) $f(-1)$ and $f(2)$.
   (c) $f(0)$ and $f(2)$.
   (d) $f(-1)$ and $f(1)$.
3. If second differences of the sequence that begins with the terms $f(0) = 5$, $f(1) = 19$, $f(2) = 43$ are assumed to be constant, what is $f(19)$?
4. Find $f(x)$, given that it is a third-degree polynomial and that $f(2) = -5$, $f(3) = 5$, $f(4) = 29$, $f(5) = 73$.
5. Show that $u_5 = u_0 + 5\Delta u_0 + 10\Delta^2 u_{-1} + 20\Delta^3 u_{-1}$ as far as third differences.
6. Find the second-degree polynomial which collocates with $f(x) = x^4$ for $x = 0, 1, 2$. (This collocation polynomial was given in Exercise 28 in Chapter 1.)
7. Show that if fifth differences are constant, then

$$u_3 = \tfrac{1}{20}\left[(u_0 + u_6) - 6(u_1 + u_5) + 15(u_2 + u_4)\right]$$

8. If $u_4 = pu_1 + qu_0 + ru_{-1} + su_{-2}$, find $p - q + r - s$.
9. Find u_{51} given that $u_{49} = 13$, $u_{50} = 18$, $u_{52} = 25$, and $u_{53} = 33$ assuming that third differences are constant.

3.3. Newton's backward difference formula

10. Rework Exercise 1 using Newton's backward difference formula.
11. Using the data given in Exercise 3, find $\nabla f(-2)$.
12. If $\bar{x} = 1 - x$, derive the following interpolation formula.

$$f(x) = \left[1 - \bar{x}\nabla + \frac{\bar{x}(\bar{x} - 1)}{2!} \nabla^2 - \cdots \right] f(1)$$

3.4. Change in origin and scale

13. Use Newton's advancing difference formula to estimate q_{63} to three decimal places from the following data:

x	q_x	$\underset{5}{\Delta q_x}$	$\underset{5}{\Delta^2 q_x}$	$\underset{5}{\Delta^3 q_x}$
65	.0380			
		.0120		
70	.0500		.0116	
		.0236		.0013
75	.0736		.0129	
		.0365		.0016
80	.1101		.0145	
		.0510		.0015
85	.1611		.0160	
		.0670		
90	.2281			

14. Use Newton's backward difference formula to estimate q_{92} to three decimal places from the data given in Exercise 13.
15. Find the second-degree polynomial which collocates with $f(x) = x^4$ for $x = 0$, 2, 4.
16. If $Ku_0 + Lu_1 + Mu_2 + Nu_3 = u_{4/3}$, where u_x is any third-degree polynomial and K, L, M, and N are constants, find c if

$$Ku_3 + Lu_1 + Mu_{-1} + Nu_{-3} = u_c$$

3.5. Subdivision of intervals

17. (a) Express $\underset{5}{\Delta}$ in terms of $\underset{2}{\Delta}$ accurate to third differences.

     (b) Express $\underset{2}{\Delta}$ in terms of $\underset{5}{\Delta}$ accurate to third differences.

18. If intervals are bisected three times, fourth differences are approximately reduced by what factor?

19.   It can be shown by extending Exercise 12 in Chapter 2 that

$$\Delta^n_h \sin x = \pm 2^n \sin^n\left(\frac{h}{2}\right) \sin\left(x + \frac{hn}{2}\right)$$

or

$$\pm 2^n \sin^n\left(\frac{h}{2}\right) \cos\left(x + \frac{hn}{2}\right)$$

depending upon whether n is even or odd, respectively. Find the largest interval of differencing, h, for which the difference table of $\sin x$ will converge.

20.   Values of u_x, a third-degree polynomial, are given below with an interval of differencing $h = 5$. If the interval is subdivided to an interval of differencing $h = 1$, what is the second advancing difference at $x = 5$ for the new interval?

x	u_x
0	1
5	7
10	8
15	9

21.   The voting population of a certain city at five-year intervals from 1955 to 1970 was as follows:

Year	Population (in thousands)
1955	200
1960	230
1965	245
1970	260

Find the increase (in thousands) in the voting population from 1960 to 1961 assuming third differences are constant.

22.   Find the largest interval of differencing that can be used in a table of $\sin x$ which will guarantee four-decimal accuracy using linear interpolation. Ignore roundoff errors.

23.   A maker of mathematical tables is preparing a table of values of $\log_e x$. If $h = .01$, how many decimal places of accuracy can be guaranteed using linear interpolation? Assume $x \geq 1$. Ignore roundoff errors.

24.   Quadratic interpolation for the function $f(x) = x^{10/3}$ is being performed using the points of collocation $-h, 0, h$. Find how small h must be to guarantee four-decimal accuracy. Assume that the values x_0 being interpolated all satisfy $-h \leq x_0 \leq h$ for whatever h is found to be the answer. Ignore roundoff errors.

3.6. Sheppard's rules

25. Express u_x as a function of u_1, ∇u_2, $\nabla^2 u_2$, $\nabla^3 u_2$, $\nabla^4 u_3$.

26. If $u_x = C_1 u_3 + C_2 \Delta u_3 + C_3 \Delta^2 u_3 + C_4 \Delta^3 u_2 + C_5 \Delta^4 u_1$, find the coefficients C_1, C_2, C_3, C_4, C_5.

3.7. Functions of more than one variable

27. If $f(x, y) = x^3 y - 3y^2$, find $\Delta_x \Delta_y f(0, 1)$.

28. Estimate $u_{13:12}$ from the following table of values of $u_{x:y}$:

x \ y	10	15
10	3	0
15	8	7

29. A polynomial in two variables is used to interpolate values for $u_{x:y}$ from the values $u_{0:0} = 1$, $u_{1:0} = 4$, $u_{0:1} = 2$, $u_{1:1} = 1$. Find the interpolated value at the point (x, x) on the line segment joining $(0, 0)$ and $(1, 1)$.

30. Find $\Delta_x \Delta_r^2 \begin{pmatrix} x \\ r \end{pmatrix}$ evaluated at $x = 5$ and $r = 2$.

31. Estimate $f(3, 3)$, given that $f(1, 1) = 1$, $f(1, 5) = 5$, $f(1, 9) = 17$, $f(4, 1) = 7$, $f(4, 5) = 20$, and $f(7, 1) = 22$.

Miscellaneous problems

32. If $f(x)$ is a second-degree polynomial, such that $f(0) = (c + 1)^{-1}$, $f(1) = c^{-1}$, and $f(2) = (c - 1)^{-1}$, find $f(c)$.

33. If $f(x)$ is a fourth-degree polynomial which equals 0 for $x = 1, 2, 3, 4$ and equals 6 for $x = 5$, what is its value at $x = 10$?

34. If $u_0 = -8$, $u_1 = 12$, $u_2 = 36$, $u_3 = 72$, find the value of

$$\left(\frac{1 - E^{-1/2}}{3\Delta - 2 - 3E} \right)^2 u_2$$

35. It is known that u_x is a polynomial of degree n, that $u_x = x$ for $x = 1, 2, \ldots, n$, and that $u_{n+1} = n + 2$. Find u_0.

36. Find B in terms of x if $u_x = u_0 + x \Delta u_1 + B \Delta^2 u_2$.

4

Central difference
formulas

4.1. INTRODUCTION

THE TWO INTERPOLATION formulas developed in Chapter 3, that is, Newton's advancing difference formula and Newton's backward difference formula, are characterized by paths in the difference table down the top diagonal and up the bottom diagonal, respectively. In Section 3.6 Sheppard's rules provide a convenient method of writing down formulas for other paths as well.

In practical work in interpolation it is often desirable to use paths more horizontal through the central portion of the difference table rather than the two extreme diagonals. There are five such formulas in common use which are called *central difference formulas*. These are:

1. The Gauss forward formula.
2. The Gauss backward formula.
3. Stirling's formula.
4. Bessel's formula.
5. Everett's formula.

A rationale for the use of central difference formulas will not be given at this time. However, Section 4.8 contains a general comparison of these five formulas together with Newton's two formulas, at which point the rationale will be discussed.

The student is advised to learn each central difference formula as it is introduced in terms of its path through the difference table. This approach not only clarifies the characteristics of the formula but it also eliminates much needless rote memorization. As each formula is introduced, its path is clearly illustrated as in Figures 3.1 and 3.2.

4.2. CENTRAL DIFFERENCE NOTATION

Before developing the five central difference formulas it is useful to develop two new operators which are commonly used in expressing these formulas. Formulas expressed in terms of these two operators are said to be expressed in *central difference notation*.

The first operator is a new difference operator, δ, called the *central difference operator*. It is defined by

$$\delta f(x) = f(x + \tfrac{1}{2}) - f(x - \tfrac{1}{2}) \qquad (4.1a)$$

or in terms of operators

$$\delta \equiv E^{1/2} - E^{-1/2} \qquad (4.1b)$$

The central difference operator δ is closely related to the forward difference operator Δ by

$$\delta \equiv \Delta E^{-1/2} \qquad (4.2)$$

Higher-order central differences can be immediately built up from formula (4.2), for example,

$$\delta^n \equiv \Delta^n E^{-\frac{1}{2}n} \qquad (4.3)$$

It is possible to construct a difference table analogous to Tables 2.1 and 2.5 using central differences. Table 4.1 is such a table.

It should be noted that the values of $x = 0, 1, 2, 3, 4$ appear as labels on symbols along horizontal lines. This contrasts with labels for the forward and

TABLE 4.1

x	$f(x)$	$\delta f(x)$	$\delta^2 f(x)$	$\delta^3 f(x)$	$\delta^4 f(x)$
0	$f(0)$				
		$\delta f(\tfrac{1}{2})$			
1	$f(1)$		$\delta^2 f(1)$		
		$\delta f(\tfrac{3}{2})$		$\delta^3 f(\tfrac{3}{2})$	
2	$f(2)$		$\delta^2 f(2)$		$\delta^4 f(2)$
		$\delta f(\tfrac{5}{2})$		$\delta^3 f(\tfrac{5}{2})$	
3	$f(3)$		$\delta^2 f(3)$		
		$\delta f(\tfrac{7}{2})$			
4	$f(4)$				

backward difference operators, Δ and ∇, in which labels lie along diagonal lines. The presence of labels halfway between integers on the odd differences should also be noted.

In any given problem the numerical values of the numbers in Tables 2.1, 2.5, and 4.1 are equal. The three difference operators Δ, ∇, and δ are merely three alternative notations for the same differences.

The second operator is called the *averaging operator* and is defined by

$$\mu f(x) = \tfrac{1}{2}[f(x + \tfrac{1}{2}) + f(x - \tfrac{1}{2})] \tag{4.4a}$$

or in terms of operators

$$\mu \equiv \tfrac{1}{2}(E^{1/2} + E^{-1/2}) \tag{4.4b}$$

This operator applied to a function at a certain point averages two functional values half a unit on each side of that point.

The averaging operator can be applied to differences of a function as well as the function itself. For example,

$$\mu\delta \equiv \tfrac{1}{2}(E^{1/2} + E^{-1/2})(E^{1/2} - E^{-1/2})$$
$$\equiv \tfrac{1}{2}(E - E^{-1}) \tag{4.5}$$

which, interestingly enough, appears to be one half of a first difference in which the interval of differencing is equal to two.

It is also possible to apply the averaging operator to higher-order differences as well. This leads to simplified expressions for certain of the central difference formulas to be introduced.

Example 4.1. *Compute* $\mu\delta^3 f(1)$, *if* $f(x) = x^4$.

Setting up a difference table, we have the following:

x	$f(x)$	$\delta f(x)$	$\delta^2 f(x)$	$\delta^3 f(x)$	$\delta^4 f(x)$
-1	1				
		-1			
0	0		2		
		1		12	
1	1		14		24
		15		36	
2	16		50		
		65			
3	81				

Now,

$$\mu\delta^3 f(1) = \tfrac{1}{2}[\delta^3 f(\tfrac{3}{2}) + \delta^3 f(\tfrac{1}{2})] = \tfrac{1}{2}(36 + 12) = 24$$

The fact that $\mu\delta^3 f(1) = \delta^4 f(1)$ in this example is coincidence.

Example 4.2.   Show that $\mu^2 \equiv 1 + \frac{1}{4}\delta^2$.

In operators the left-hand side is

$$\mu^2 \equiv \left[\frac{1}{2}(E^{1/2} + E^{-1/2})\right]^2 \equiv \frac{1}{4}(E + 2 + E^{-1})$$

Similarly, the right-hand side is

$$1 + \frac{1}{4}\delta^2 \equiv 1 + \frac{1}{4}(E^{1/2} - E^{-1/2})^2 \equiv 1 + \frac{1}{4}(E - 2 + E^{-1})$$
$$\equiv \frac{1}{4}(E + 2 + E^{-1})$$

4.3.   THE GAUSS FORWARD FORMULA

The *Gauss forward formula* follows the path through the difference table illustrated in Figure 4.1.

FIGURE 4.1

The Gauss forward formula

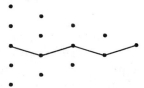

In finding an expression for the Gauss forward formula it is usual to have the formula start at u_0 and have the interval of differencing equal to one. If this is not the case, then a change in origin and scale should be made, so that the standardized formula can be used.

Applying Sheppard's rules directly to the above path gives

$$u_x = u_0 + \binom{x}{1}\Delta u_0 + \binom{x}{2}\Delta^2 u_{-1} + \binom{x+1}{3}\Delta^3 u_{-1} + \binom{x+1}{4}\Delta^4 u_{-2} + \cdots$$

$$(4.6a)$$

It should be noted that as the formula moves from term to term, we alternately decrease the label on u_x by one and increase the largest factor in the coefficient by one, starting with the third term.

The Gauss forward formula can alternatively be expressed in central difference notation as

$$u_x = u_0 + \binom{x}{1}\delta u_{1/2} + \binom{x}{2}\delta^2 u_0 + \binom{x+1}{3}\delta^3 u_{1/2} + \binom{x+1}{4}\delta^4 u_0 + \cdots$$

$$(4.6b)$$

Example 4.3. Estimate $u_{2.5}$ **given that** $u_0 = 0$, $u_1 = 1$, $u_2 = 16$, $u_3 = 81$, **and** $u_4 = 256$ **using the Gauss forward formula.**

The five given values of u_x agree with the fourth-degree polynomial x^4. Thus, we hope the formula will produce the answer $(2.5)^4 = 39.0625$.

Setting up the difference table, we have the following:

$x \longrightarrow$	y	v_y	δv_y	$\delta^2 v_y$	$\delta^3 v_y$	$\delta^4 v_y$
$0 \longrightarrow$	-2	0				
			1			
$1 \longrightarrow$	-1	1		14		
			15		36	
$2 \longrightarrow$	0	16		50		24
			65		60	
$3 \longrightarrow$	1	81		110		
			175			
$4 \longrightarrow$	2	256				

Now applying the Gauss forward formula with $2.5 \rightarrow .5$ gives

$$u_{2.5} = v_{.5} = 16 + (.5)(65) + \frac{(.5)(-.5)}{2}(50) + \frac{(1.5)(.5)(-.5)}{6}(60)$$

$$+ \frac{(1.5)(.5)(-.5)(-1.5)}{24}(24)$$

$$= 39.0625$$

the anticipated answer.

4.4. THE GAUSS BACKWARD FORMULA

The *Gauss backward formula* follows the path through the difference table illustrated in Figure 4.2. Again it is assumed that the formula starts at u_0 and has the interval of differencing equal to one.

FIGURE 4.2

The Gauss backward formula

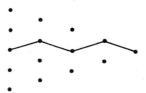

Applying Sheppard's rules directly to the above path gives

$$u_x = u_0 + \binom{x}{1}\Delta u_{-1} + \binom{x+1}{2}\Delta^2 u_{-1} + \binom{x+1}{3}\Delta^3 u_{-2} + \binom{x+2}{4}\Delta^4 u_{-2} + \cdots$$

$$(4.7a)$$

It should be noted that just as with the Gauss forward formula, we alternately decrease the label on u_x by one and increase the largest factor in the coefficient by one, only in this case starting with the second term.

The Gauss backward formula can alternatively be expressed in central difference notation as

$$u_x = u_0 + \binom{x}{1}\delta u_{-1/2} + \binom{x+1}{2}\delta^2 u_0 + \binom{x+1}{3}\delta^3 u_{-1/2} + \binom{x+2}{4}\delta^4 u_0 + \cdots$$

$$(4.7b)$$

Example 4.4.   Rework Example 4.3 using the Gauss backward formula.

We can use the same difference table as in Example 4.3. Applying the Gauss backward formula gives

$$u_{2.5} = v_{.5} = 16 + (.5)(15) + \frac{(1.5)(.5)}{2}(50) + \frac{(1.5)(.5)(-.5)}{6}(36)$$

$$+ \frac{(2.5)(1.5)(.5)(-.5)}{24}(24)$$

$$= 39.0625$$

the anticipated answer.

4.5. STIRLING'S FORMULA

Stirling's formula follows the path through the difference table illustrated in Figure 4.3. This formula is derived by taking the arithmetic mean of the two

FIGURE 4.3

Stirling's formula

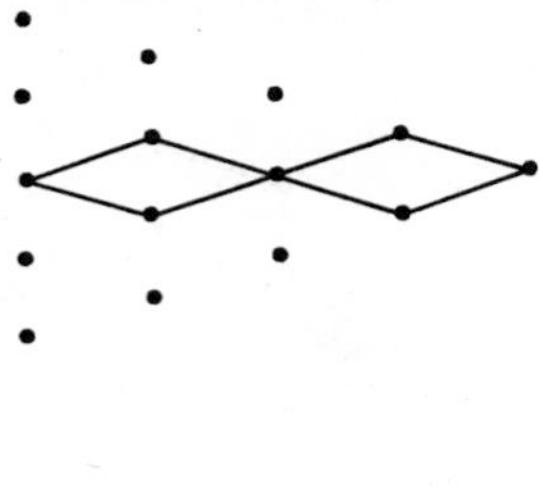

Gauss formulas. Thus, we have

$$u_x = u_0 + x \cdot \tfrac{1}{2}(\Delta u_0 + \Delta u_{-1}) + \frac{x^2}{2!}\Delta^2 u_{-1}$$

$$+ \frac{x(x^2 - 1)}{3!} \cdot \tfrac{1}{2}(\Delta^3 u_{-1} + \Delta^3 u_{-2}) + \frac{x^2(x^2 - 1)}{4!}\Delta^4 u_{-2} + \cdots \quad (4.8a)$$

It should be noted that each odd difference term is the mean of two consecutive differences of that order. The averaging operator μ provides a convenient method of denoting these terms. Thus, Stirling's formula expressed in central difference notation is

$$u_x = u_0 + x\mu\delta u_0 + \frac{x^2}{2!}\delta^2 u_0 + \frac{x(x^2 - 1)}{3!}\mu\delta^3 u_0 + \frac{x^2(x^2 - 1)}{4!}\delta^4 u_0 + \cdots$$

$$(4.8b)$$

Example 4.5.   Rework Example 4.3 using Stirling's formula.

We can use the same difference table as in Example 4.3. Applying Stirling's formula gives

$$u_{2.5} = v_{.5} = 16 + (.5)\frac{65 + 15}{2} + \frac{(.5)^2}{2}(50) + \frac{(.5)(.5^2 - 1)}{6} \cdot \frac{60 + 36}{2}$$

$$+ \frac{(.5^2)(.5^2 - 1)}{24}(24)$$

$$= 39.0625$$

the anticipated answer.

4.6.   BESSEL'S FORMULA

Bessel's formula follows the path through the difference table illustrated in Figure 4.4. This is the first central difference formula discussed which is not

FIGURE 4.4

Bessel's formula

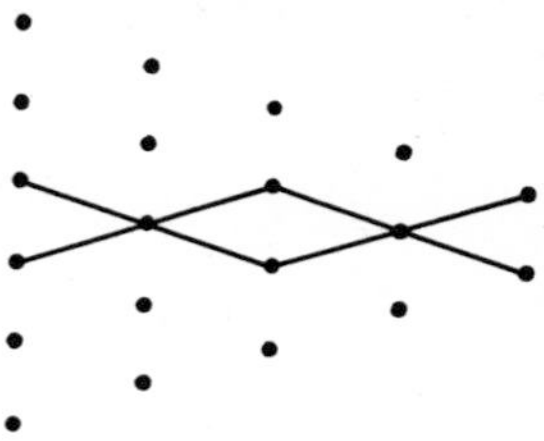

centered at $x = 0$ but is instead centered at $x = \frac{1}{2}$. Although $u_{1/2}$ does not itself appear in the difference table or the formula, the path taken moves horizontally halfway between $x = 0$ and $x = 1$.

Bessel's formula is derived by taking the mean of the Gauss forward formula starting at u_0 and the Gauss backward formula starting at u_1. The former is given by formula (4.6a), while the latter is given by

$$u_x = u_1 + \binom{x-1}{1}\Delta u_0 + \binom{x}{2}\Delta^2 u_0 + \binom{x}{3}\Delta^3 u_{-1} + \binom{x+1}{4}\Delta^4 u_{-1} + \cdots$$

The mean of formula (4.6a) and the above formula is

$$u_x = \tfrac{1}{2}(u_1 + u_0) + (x - \tfrac{1}{2})\Delta u_0 + \frac{x(x-1)}{2!} \cdot \tfrac{1}{2}(\Delta^2 u_0 + \Delta^2 u_{-1})$$

$$+ \frac{(x - \tfrac{1}{2})x(x - 1)}{3!}\Delta^3 u_{-1}$$

$$+ \frac{(x + 1)x(x - 1)(x - 2)}{4!} \cdot \tfrac{1}{2}(\Delta^4 u_{-1} + \Delta^4 u_{-2}) + \cdots \qquad (4.9a)$$

It should be noted that each even difference term is the mean of two consecutive differences of that order. Again the averaging operator μ can be used so that Bessel's formula in central difference notation is

$$u_x = \mu u_{1/2} + (x - \tfrac{1}{2})\delta u_{1/2} + \frac{x(x-1)}{2!}\mu\delta^2 u_{1/2} + \frac{(x - \tfrac{1}{2})x(x - 1)}{3!}\delta^3 u_{1/2}$$

$$+ \frac{(x + 1)x(x - 1)(x - 2)}{4!}\mu\delta^4 u_{1/2} + \cdots \qquad (4.9b)$$

Example 4.6.   *Rework Example 4.3 using Bessel's formula.*

We can use the same difference table as in Example 4.3. In using Bessel's formula the odd difference terms vanish because of the factor $(x - \tfrac{1}{2})$. Also, $\delta^4 u_3$ is missing from the difference table, since u_5 was not given. We shall assume that fourth differences are constant, so that $\delta^4 u_3 = 24$.

Thus, applying Bessel's formula gives

$$u_{2.5} = v_{.5} = \frac{81 + 16}{2} + \frac{(.5)(-.5)}{2} \cdot \frac{110 + 50}{2} + \frac{(1.5)(.5)(-.5)(-1.5)}{24} \cdot \frac{24 + 24}{2}$$

$$= 39.0625$$

the anticipated answer.

4.7. EVERETT'S FORMULA

Everett's formula follows the path through the difference table illustrated in Figure 4.5. Everett's formula, like Bessel's formula, is centered at $x = \frac{1}{2}$ instead of at $x = 0$. This formula is unlike the other central difference formulas in that only even differences appear.

FIGURE 4.5

Everett's formula

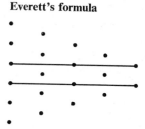

Everett's formula to fourth differences can be derived from Bessel's formula to fifth differences by expressing the odd differences in terms of even differences of order one less.[1] We have from Bessel's formula

$$u_x = \tfrac{1}{2}(u_1 + u_0) + (x - \tfrac{1}{2})\delta u_{1/2} + \frac{x(x-1)}{2} \cdot \tfrac{1}{2}(\delta^2 u_1 + \delta^2 u_0)$$

$$+ \frac{(x - \tfrac{1}{2})x(x-1)}{6}\,\delta^3 u_{1/2} + \frac{(x+1)x(x-1)(x-2)}{24} \cdot \tfrac{1}{2}(\delta^4 u_1 + \delta^4 u_0)$$

$$+ \frac{(x - \tfrac{1}{2})(x+1)x(x-1)(x-2)}{120}\,\delta^5 u_{1/2} + \cdots$$

Now make the following substitutions:

$$\delta u_{1/2} = u_1 - u_0$$
$$\delta^3 u_{1/2} = \delta^2 u_1 - \delta^2 u_0$$
$$\delta^5 u_{1/2} = \delta^4 u_1 - \delta^4 u_0$$

and simplify to obtain Everett's formula

$$u_x = \binom{x}{1}u_1 + \binom{x+1}{3}\delta^2 u_1 + \binom{x+2}{5}\delta^4 u_1 + \cdots$$

$$+ \binom{\bar{x}}{1}u_0 + \binom{\bar{x}+1}{3}\delta^2 u_0 + \binom{\bar{x}+2}{5}\delta^4 u_0 + \cdots \qquad (4.10)$$

where $\bar{x} = 1 - x$.

---

[1] Note that Everett's formula to fourth differences will be accurate for fifth-degree polynomials, since odd difference terms vanish.

The details of transforming Bessel's formula into Everett's formula (4.10) are left as Exercise 22. The student should note carefully the two symmetrical portions of Everett's formula, one with x centered on u_1 and the other with $\bar{x}$ centered on u_0.

It is possible to derive an analogous formula in which, except for an initial starting value, only odd difference terms appear. The path is indicated in Figure 4.6. The formula for this path is called *Everett's second formula* or

FIGURE 4.6

Everett's second formula

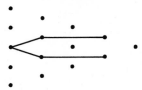

Steffensen's formula. This formula has appeared in the literature, but will not be further discussed here. A statement of the formula and its derivation are contained in Appendix C.

Example 4.7.   Rework Example 4.3 using Everett's formula.

We can use the difference table contained in Example 4.3. As in Example 4.6, assume that fourth differences are constant so that $\delta^4 u_3 = 24$. Applying Everett's formula gives

$$u_{2.5} = v_{.5} = (.5)(81) + \frac{(1.5)(.5)(-.5)}{6}(110) + \frac{(2.5)(1.5)(.5)(-.5)(-1.5)}{120}(24)$$

$$+ (.5)(16) + \frac{(1.5)(.5)(-.5)}{6}(50) + \frac{(2.5)(1.5)(.5)(-.5)(-1.5)}{120}(24)$$

$$= 39.0625$$

the anticipated answer.

4.8.   COMPARISON OF FORMULAS

In performing interpolations it is advisable to use the most relevant data possible. For most situations the value to be interpolated should be centrally located if possible, so that functional values on either side are brought into the formula with approximately equal weight. For example, in interpolating for $f(2.5)$ it is usually preferable to use values of $x = 0, 1, 2, 3, 4, 5$ rather than $x = 2, 3, 4, 5, 6, 7$.

The above argument is the primary rationale for using central difference formulas, since these formulas automatically bring in functional values on each side. However, it should be carefully noted that once the points on which the interpolation is to be based are determined, it makes no difference which interpolation formula is used (including even the two Newton formulas). This is because there is one unique collocation polynomial through a given set of points, and all of the interpolation formulas are merely different versions of it.

Another advantage of using central difference formulas is that the coefficients in the formulas are smaller and convergence to the desired value is more rapid than with the two Newton formulas. This is because the Newton formulas bring in functional values on only one side, which is a somewhat lopsided arrangement. The central difference formulas bring in functional values on both sides, which produces smaller coefficients and thus more rapid convergence.

In many interpolation problems the data are arranged in tabular form. There is a problem in using central difference formulas for interpolations near the two ends of the table, since required values are not directly available. Furthermore, it may be impractical or even impossible to extend the table further. In these situations Newton's advancing difference formula is often used at the top of the table, Newton's backward difference formula at the bottom of the table, and a central difference formula in between.

In certain cases it may be necessary to perform extrapolations outside the range of tabulated values. As indicated in Section 3.1, this is usually a risky procedure. However, if it is to be done, Newton's advancing difference formula should be used for values of x before the start of the table, while Newton's backward difference formula should be used for values of x after the end of the table.

Once it has been decided to use a central difference formula, there is still a decision concerning which of the formulas developed in this chapter is most appropriate. Fortunately, the choice is usually not all that critical, since any of the formulas will generally produce comparable answers. However, there are some guidelines that are generally useful to follow:

1. The two Gauss formulas are not as widely used as Stirling's, Bessel's, or Everett's formulas in the actual interpolations. The primary role of the Gauss formulas is in deriving other central difference formulas.
2. Stirling's and Bessel's formulas are often considered to be a pair of formulas, in which Stirling's formula should be used for $-\frac{1}{4} < x < \frac{1}{4}$ and Bessel's formula should be used for $\frac{1}{4} < x < \frac{3}{4}$. Together the two formulas cover an entire unit interval and can be used to interpolate for any x.

3. Some investigations have indicated that Bessel's formula to even mean differences is more accurate than Stirling's formula to the same even differences, whereas Stirling's formula to odd mean differences is more accurate than Bessel's formula to the same odd differences.[2]

4. As seen in Example 4.6, Bessel's formula is particularly useful in bisecting intervals, since all the odd difference terms vanish.

5. Everett's formula is useful in the following situations:

   (a) In some mathematical tables a secondary column of δ^2 is tabulated along with the functional values. Since all odd difference terms vanish, Everett's formula can be directly applied to only these two columns and still produce third-degree accuracy in any interpolations.

   (b) The coefficients of Everett's formula are often tabulated. This is particularly easy to do for this formula because of the symmetry and the missing odd differences.

   (c) In using Everett's formula advantage can be taken of the symmetry in the terms in the two portions of the formula. In doing a large number of interpolations through a complete table, it develops that each computation can be used in two different interpolations. This property was particularly useful when interpolations were being performed on desk calculators, since only half as many computations were required. Since this property is of substantially less importance in the age of high-speed digital computers, it will not be discussed further. However, a numerical example illustrating the process is contained in Appendix C.

   (d) Everett's formula is the foundation upon which much of the theory of osculatory interpolation is developed. The type of interpolation formulas developed so far are called *piecewise-polynomial interpolation formulas* when they are applied to successive intervals. The term "piecewise" refers to the fact that the interpolating polynomials differ from interval to interval. For example, if a cubic is used to interpolate for values of x, $1 < x < 2$, based upon $x = 0, 1, 2, 3$ and also to interpolate for values of x, $2 < x < 3$, based upon $x = 1, 2, 3, 4$, then the cubics will generally differ for the two intervals. The result will be a continuous curve which is "rough" at the points of juncture. *Osculatory interpolation* is a method of modifying the two interpolating curves to produce greater smoothness at the points of juncture and will be discussed in Chapter 15.

---

[2]See Harry Freeman, *Finite Differences for Actuarial Students* (Cambridge, England: Cambridge University Press, 1960), pp. 71–72.

6. Bessel's formula and Everett's formula are both useful in applying *throwback*, which is an approximate technique of obtaining a higher degree of accuracy than the degree of the polynomial being used. Throwback is primarily of historical interest, since with computers it is simpler to obtain the higher degree of accuracy directly using a higher-degree collocation polynomial. A discussion of throwback is contained in Appendix C.

EXERCISES

4.1. Introduction;   4.2. Central difference notation

1. Find $(\mu\delta)^2 x^3$.
2. Find $\delta^4(x+2)^{(6)}$ evaluated at $x = 3$.
3. It is known that $\mu f(-\frac{1}{2}) = 5$, $\mu f(\frac{1}{2}) = 11$, and $\delta^2 f(0) = 4$. If $f(x)$ is known to be a second-degree polynomial, find $\Delta f(1)$.
4. Find $(1 - \mu + \mu^2 - \mu^3 + \cdots)^{-2} x^2$.
5. If $(E\mu\delta)x! = f(x)x!$, find $f(x)$.
6. Show that the following two operators are equivalent to $\mu\delta$:

   (a) $\dfrac{\Delta(1 + \frac{1}{2}\Delta)}{1 + \Delta}$.

   (b) $\dfrac{\nabla(1 - \frac{1}{2}\nabla)}{1 - \nabla}$.

7. (a) Show that $E^{1/2} \equiv \mu + \frac{1}{2}\delta$.
   (b) Show that $E^{-1/2} \equiv \mu - \frac{1}{2}\delta$.
8. Express $\delta^2 - E^{1/2}\delta$ in terms of Δ.
9. Simplify $\Delta^3 u_{x-\frac{1}{2}} + 2\delta^2 u_{x+\frac{1}{2}} + \delta u_x - \delta u_{x+2}$.

4.3. The Gauss forward formula

10. Given the following functional values, use the Gauss forward formula centered at $x = 3$ to third differences to approximate $\log_e 3.25$ to four decimal places:

x	$\log_e x$
1	.0000
2	.6931
3	1.0986
4	1.3863
5	1.6094

11. The general form of the Gauss forward formula to third differences can be written as

$$Au_0 + B\Delta u_0 + C\Delta^2 u_{-1} + D\Delta^3 u_{-1}$$

The formula can then be expressed in terms of functional values to give

$$au_{-1} + bu_0 + cu_1 + du_2$$

Solve for A, B, C, D in terms of a, b, c, d.

12. Express the Gauss forward formula in central difference notation accurate to third differences for an interpolation for $f(x)$ in which the starting value is $f(4)$ and in which the interval of differencing $h = 3$.

4.4.   The Gauss backward formula

13. Rework Exercise 10 using the Gauss backward formula centered at $x = 3$.
14. The general form of the Gauss backward formula to third differences can be written as

$$Au_0 + B\Delta u_{-1} + C\Delta^2 u_{-1} + D\Delta^3 u_{-2}$$

The formula can then be expressed in terms of functional values to give

$$au_{-2} + bu_{-1} + cu_0 + du_1$$

Solve for A, B, C, D in terms of a, b, c, d.

15. Express the Gauss backward formula in central difference notation accurate to third differences for an interpolation for $f(x)$ in which the starting value is $f(3)$ and in which the interval of differencing $h = 2$.

4.5.   Stirling's formula

16. Rework Exercise 10 using Stirling's formula centered at $x = 3$.
17. Stirling's formula to fourth differences can be expressed as

$$f(x) = f(0) + Ax + Bx^2 + Cx^3 + Dx^4$$

Find the coefficients A, B, C, D given the following difference table:

x	$f(x)$	$\Delta f(x)$	$\Delta^2 f(x)$	$\Delta^3 f(x)$	$\Delta^4 f(x)$
-2	29				
		-22			
-1	7		17		
		-5		-5	
0	2		12		-2
		7		-7	
1	9		5		
		12			
2	21				

4.6. Bessel's formula

18. Rework Exercise 10 using Bessel's formula centered at $x = 3\frac{1}{2}$.
19. Show that if $y = x - \frac{1}{2}$, Bessel's formula can be written in the more symmetrical form

$$u_x = u_{y+\frac{1}{2}} = \mu u_{1/2} + y\delta u_{1/2} + \frac{y^2 - \frac{1}{4}}{2!}\mu\delta^2 u_{1/2}$$

$$+ \frac{y(y^2 - \frac{1}{4})}{3!}\delta^3 u_{1/2} + \frac{(y^2 - \frac{1}{4})(y^2 - \frac{9}{4})}{4!}\mu\delta^4 u_{1/2} + \cdots$$

20. Use Bessel's formula to third differences to show that

$$f(\tfrac{1}{2}) = \tfrac{1}{16}[-f(-1) + 9f(0) + 9f(1) - f(2)]$$

4.7. Everett's formula

21. Rework Exercise 10 using Everett's formula centered at $x = 3\frac{1}{2}$.
22. Complete the derivation of formula (4.10) given in the text.

4.8. Comparison of formulas

23. The answers to Exercises 10, 13, 16, 18, and 21 are compared in Table 4.2.

TABLE 4.2

	Computed answer	True answer	Error
Gauss forward	1.1795	1.1787	.0008
Gauss backward	1.1749	1.1787	.0038
Stirling	1.1772	1.1787	.0015
Bessel	1.1795	1.1787	.0008
Everett	1.1795	1.1787	.0008

Why do the Gauss forward, Bessel's, and Everett's formulas produce the same answer?

24. Using the results summarized in Exercise 23, explain why the Gauss forward formula produces a superior result to the Gauss backward formula.

Miscellaneous problems

25. Find u_4, if $u_0 = 3$, $u_1 = 5$, $\delta^2 u_1 = 6$, $\delta^2 u_3 = 28$, and $\delta^4 u_2 = 10$.
26. Show that for fifth-degree polynomials

$$u_{2\frac{1}{2}} = \frac{3a - 25b + 150c}{256}$$

where $a = u_0 + u_5$, $b = u_1 + u_4$, and $c = u_2 + u_3$.

27. Show that

$$\Delta \equiv \tfrac{1}{2}\delta^2 + \delta\sqrt{1 + \tfrac{1}{4}\delta^2}$$

28. Show that

$$\mu^{-1} \equiv 1 - \tfrac{1}{8}\delta^2 + \tfrac{3}{128}\delta^4 - \cdots$$

29. Show that—

(a) $\delta(u_x v_x) = (\mu u_x)(\delta v_x) + (\mu v_x)(\delta u_x)$.

(b) $\delta\left(\dfrac{u_x}{v_x}\right) = \dfrac{(\mu v_x)(\delta u_x) - (\mu u_x)(\delta v_x)}{v_{x-\frac{1}{2}}v_{x+\frac{1}{2}}}$.

30. Show that—

(a) $\mu(u_x v_x) = (\mu u_x)(\mu v_x) + \tfrac{1}{4}(\delta u_x)(\delta v_x)$.

(b) $\mu\left(\dfrac{u_x}{v_x}\right) = \dfrac{(\mu u_x)(\mu v_x) - \tfrac{1}{4}(\delta u_x)(\delta v_x)}{v_{x-\frac{1}{2}}v_{x+\frac{1}{2}}}$.

31. (a) Find the truncation error in using Stirling's formula to fourth differences centered at $x = 0$.

(b) Find an upper bound for the truncation error assuming $\left|f^{(5)}(x)\right| \le K$ for all x such that $-2 \le x \le 2$.

(c) Find four values of x, $-2 < x < 2$, such that the upper bound developed in (b) attains an extremum. Show that one extremum occurs between -2 and -1, one between -1 and 0, one between 0 and 1, and one between 1 and 2.

(d) Show that the maximum absolute error for the extrema between -2 and -1 and between 1 and 2 exceeds the maximum absolute error for the extrema between -1 and 0 and between 0 and 1. The implication of this result is that Stirling's formula tends to produce poorer results for interpolations for $x < -1$ or $x > 1$ than for $-1 < x < 1$.

32. (a) Find the truncation error in using Bessel's formula to third differences centered at $x = \frac{1}{2}$.

(b) Find an upper bound for the truncation error assuming $\left|f^{(4)}(x)\right| \le K$ for all x such that $-1 \le x \le 2$.

(c) Find three values of x, $-1 < x < 2$, such that the upper bound developed in (b) attains an extremum. Show that one extremum occurs between -1 and 0, one between 0 and 1, and one between 1 and 2.

(d) Show that the maximum absolute error for the extrema between -1 and 0 and between 1 and 2 exceeds the maximum absolute error for the extremum between 0 and 1. The implication of this result is that Bessel's formula tends to produce poorer results for interpolations for $x < 0$ or $x > 1$ than for $0 < x < 1$.

5

Interpolation with unequal intervals

5.1. INTRODUCTION

THE INTERPOLATION formulas developed in the preceding chapters assume that the points of collocation x_0, x_1, x_2, ..., x_n are spaced at equal intervals, that is, h exists such that $h = x_1 - x_0 = x_2 - x_1 = \cdots = x_n - x_{n-1}$. In this chapter more general techniques of interpolation are developed for cases in which the points of collocation are not equally spaced.

These more general techniques can also be applied to interpolation with equal intervals. In fact, many of the results developed in preceding chapters are special cases of the more general results developed in this chapter.

Four distinct methods of interpolation are considered:

1. Lagrange's formula.
2. Divided differences.
3. A determinant method.
4. The Aitken-Neville technique.

A general comparison of these methods is contained in Section 5.10.

In practical work the first two methods are more widely used than the latter two, and most of the emphasis will be on them. However, the latter two methods are useful algorithms in certain cases, as discussed in Section 5.10, and are included for completeness.

5.2. LAGRANGE'S FORMULA

Lagrange's formula for the collocation polynomial of degree n based on the $n + 1$ points of collocation $x_0, x_1, x_2, \ldots, x_n$ is derived by writing

$$
\begin{aligned}
f(x) = {} & A_0(x - x_1)(x - x_2) \cdots (x - x_n) \\
& + A_1(x - x_0)(x - x_2) \cdots (x - x_n) \\
& \vdots \\
& + A_i(x - x_0) \cdots (x - x_{i-1})(x - x_{i+1}) \cdots (x - x_n) \\
& \vdots \\
& + A_n(x - x_0)(x - x_1) \cdots (x - x_{n-1})
\end{aligned}
$$

Since $f(x)$ is a polynomial of degree n, all that remains is to appropriately determine $A_0, A_1, A_2, \ldots, A_n$.

If we substitute $x = x_0$, we have

$$
f(x_0) = A_0(x_0 - x_1)(x_0 - x_2) \cdots (x_0 - x_n)
$$

so that

$$
A_0 = \frac{f(x_0)}{(x_0 - x_1)(x_0 - x_2) \cdots (x_0 - x_n)}
$$

Similar results hold for $A_1, A_2, \ldots, A_n$.

Lagrange's formula is then obtained by substituting $A_0, A_1, A_2, \ldots, A_n$ back into the expression for $f(x)$, which gives

$$
\begin{aligned}
f(x) = {} & f(x_0) \frac{(x - x_1)(x - x_2) \cdots (x - x_n)}{(x_0 - x_1)(x_0 - x_2) \cdots (x_0 - x_n)} \\
& + f(x_1) \frac{(x - x_0)(x - x_2) \cdots (x - x_n)}{(x_1 - x_0)(x_1 - x_2) \cdots (x_1 - x_n)} \\
& \vdots \\
& + f(x_i) \frac{(x - x_0) \cdots (x - x_{i-1})(x - x_{i+1}) \cdots (x - x_n)}{(x_i - x_0) \cdots (x_i - x_{i-1})(x_i - x_{i+1}) \cdots (x_i - x_n)} \\
& \vdots \\
& + f(x_n) \frac{(x - x_0)(x - x_1) \cdots (x - x_{n-1})}{(x_n - x_0)(x_n - x_1) \cdots (x_n - x_{n-1})}
\end{aligned}
\tag{5.1}
$$

The fact that formula (5.1) is the collocation polynomial is seen by noting that it is a polynomial of degree n, and that if x_i is substituted for x on the right-hand side, where $i = 0, 1, 2, \ldots, n$, then $f(x) = f(x_i)$. This follows because the coefficient of $f(x_i)$ is equal to one, while the coefficients of all the other $f(x_j)$, $j \neq i$, are equal to zero.

It is important to note that Lagrange's formula expresses a functional value at one point as a linear combination of functional values at other points. Lagrange's formula is the most direct method of doing this.

From a somewhat different point of view, Lagrange's formula is equivalent to splitting a fractional expression into partial fractions. This can be seen by dividing both sides of formula (5.1) by $\pi(x)$, where

$$\pi(x) = (x - x_0)(x - x_1) \cdots (x - x_n) \tag{1.11}$$

The details are left to the student in Exercise 1(a).

It is convenient to express formula (5.1) in a more condensed form. Denote the coefficient of $f(x_i)$ by $L_i(x)$, that is,

$$L_i(x) = \frac{(x - x_0) \cdots (x - x_{i-1})(x - x_{i+1}) \cdots (x - x_n)}{(x_i - x_0) \cdots (x_i - x_{i-1})(x_i - x_{i+1}) \cdots (x_i - x_n)} \tag{5.2}$$

Then Lagrange's formula becomes

$$f(x) = \sum_{i=0}^{n} L_i(x) f(x_i) \tag{5.3}$$

The numerator of formula (5.2) is the same as $\pi(x)$, except that the factor $(x - x_i)$ is missing. This expression is often written as $F_i(x)$, that is,

$$F_i(x) = (x - x_0) \cdots (x - x_{i-1})(x - x_{i+1}) \cdots (x - x_n) \tag{5.4}$$

Then we have

$$L_i(x) = \frac{F_i(x)}{F_i(x_i)} \tag{5.5}$$

from which Lagrange's formula follows using formula (5.3).

Another version of the Lagrange coefficients is given by

$$L_i(x) = \frac{\pi(x)}{(x - x_i)\pi'(x_i)} \tag{5.6}$$

This can be derived by recalling

$$\pi(x) = (x - x_0)(x - x_1) \cdots (x - x_n)$$

and differentiating the product of the $n + 1$ factors gives

$$\begin{aligned}
\pi'(x) = \ & (x - x_1)(x - x_2) \cdots (x - x_n) \\
& + (x - x_0)(x - x_2) \cdots (x - x_n) \\
& \ \vdots \\
& + (x - x_0) \cdots (x - x_{i-1})(x - x_{i+1}) \cdots (x - x_n) \\
& \ \vdots \\
& + (x - x_0)(x - x_1) \cdots (x - x_{n-1}) \\
= \ & F_0(x) + F_1(x) + \cdots + F_i(x) + \cdots + F_n(x)
\end{aligned}$$

Now

$$\pi'(x_i) = F_i(x_i)$$

since $F_j(x_i) = 0$ for $j \neq i$. Furthermore

$$\frac{\pi(x)}{x - x_i} = F_i(x)$$

by definition, so that

$$\frac{\pi(x)}{(x - x_i)\pi'(x_i)} = \frac{F_i(x)}{F_i(x_i)} = L_i(x)$$

from formula (5.5).

Lagrange's formula reproduces any polynomial of degree n or less. Thus, it must be able to reproduce the polynomial of degree zero $f(x) = 1$. From formula (5.3) this leads to the interesting property of the Lagrange coefficients

$$\sum_{i=0}^{n} L_i(x) = 1 \qquad (5.7)$$

This result is useful for checking purposes.

Example 5.1. ***Estimate*** $f(3)$ ***given that*** $f(1) = 1, f(2) = 8, f(5) = 125,$ ***and*** $f(7) = 343$ ***using Lagrange's formula.***

The four given values of $f(x)$ agree with the third-degree polynomial x^3. Thus, we hope our procedures will produce the answer $3^3 = 27$.

Lagrange's formula for this example is expressed as follows:

$$f(3) = 1 \cdot \frac{(3-2)(3-5)(3-7)}{(1-2)(1-5)(1-7)} + 8 \cdot \frac{(3-1)(3-5)(3-7)}{(2-1)(2-5)(2-7)}$$

$$+ 125 \cdot \frac{(3-1)(3-2)(3-7)}{(5-1)(5-2)(5-7)} + 343 \cdot \frac{(3-1)(3-2)(3-5)}{(7-1)(7-2)(7-5)}$$

$$= 1\left(-\tfrac{1}{3}\right) + 8\left(\tfrac{16}{15}\right) + 125\left(\tfrac{1}{3}\right) + 343\left(-\tfrac{1}{15}\right)$$

$$= 27$$

the anticipated result. The student should verify that the sum of the coefficients does equal one, as indicated by formula (5.7).

5.3.   DIVIDED DIFFERENCES

A *divided difference* is a more general type of difference than the *ordinary difference* developed in the preceding chapters. The first divided difference of $f(x)$ based upon the points x_0 and x_1 is defined by

$$\underset{x_1}{\triangle} f(x_0) = \frac{f(x_1) - f(x_0)}{x_1 - x_0} \qquad (5.8)[1]$$

---

[1]There is no standard notation for divided differences. The $\triangle$-notation which we shall use was suggested by Aitken. It has the advantage of being most similar to the Δ-notation for ordinary differences. It is convenient, since the order of the difference and the functional values on which the given difference is based are themselves part of the notation.

Higher-order differences can be derived from lower-order differences. For example,

$$\underset{x_1, x_2}{\triangle^2} f(x_0) = \frac{\underset{x_2}{\triangle} f(x_1) - \underset{x_1}{\triangle} f(x_0)}{x_2 - x_0} \tag{5.9}$$

and, in general,

$$\underset{x_1, x_2, \dots, x_n}{\triangle^n} f(x_0) = \frac{\underset{x_2, x_3, \dots, x_n}{\triangle^{n-1}} f(x_1) - \underset{x_1, x_2, \dots, x_{n-1}}{\triangle^{n-1}} f(x_0)}{x_n - x_0} \tag{5.10}$$

It should be noted that it takes $n + 1$ functional values to determine an nth divided difference and that the $n + 1$ values of x are listed as part of the symbol.

Divided differences are often arranged in a difference table similarly to ordinary differences. Table 5.1 is the standard format of a divided difference table.

TABLE 5.1

x	$f(x)$	$\triangle f(x)$	$\triangle^2 f(x)$	$\triangle^3 f(x)$	$\triangle^4 f(x)$
x_0	$f(x_0)$				
		$\underset{x_1}{\triangle} f(x_0)$			
x_1	$f(x_1)$		$\underset{x_1,x_2}{\triangle^2} f(x_0)$		
		$\underset{x_2}{\triangle} f(x_1)$		$\underset{x_1,x_2,x_3}{\triangle^3} f(x_0)$	
x_2	$f(x_2)$		$\underset{x_2,x_3}{\triangle^2} f(x_1)$		$\underset{x_1,x_2,x_3,x_4}{\triangle^4} f(x_0)$
		$\underset{x_3}{\triangle} f(x_2)$		$\underset{x_2,x_3,x_4}{\triangle^3} f(x_1)$	
x_3	$f(x_3)$		$\underset{x_3,x_4}{\triangle^2} f(x_2)$		
		$\underset{x_4}{\triangle} f(x_3)$			
x_4	$f(x_4)$				

A numerical example should help clarify the basic concepts and the construction of a divided difference table. Table 5.2 is a divided difference table for the data given in Example 5.1. In each case the numerator is the difference of two adjacent differences of one degree less, just as with ordinary differences, while the denominator is the difference of the two values of x at the end points of the range of values of x upon which the difference is based.

TABLE 5.2

x	$f(x)$	$\triangle f(x)$	$\triangle^2 f(x)$	$\triangle^3 f(x)$
1	1			
		$\dfrac{8-1}{2-1} = 7$		
2	8		$\dfrac{39-7}{5-1} = 8$	
		$\dfrac{125-8}{5-2} = 39$		$\dfrac{14-8}{7-1} = 1$
5	125		$\dfrac{109-39}{7-2} = 14$	
		$\dfrac{343-125}{7-5} = 109$		
7	343			

It is instructive to express divided differences directly in terms of the functional values which determine them. For first differences, we have

$$\underset{x_1}{\triangle} f(x_0) = \frac{f(x_1) - f(x_0)}{x_1 - x_0}$$

$$= \frac{f(x_0)}{x_0 - x_1} + \frac{f(x_1)}{x_1 - x_0}$$

For second differences, we have

$$\underset{x_1, x_2}{\triangle^2} f(x_0) = \frac{\underset{x_2}{\triangle} f(x_1) - \underset{x_1}{\triangle} f(x_0)}{x_2 - x_0}$$

$$= \frac{1}{x_0 - x_2} \left[\frac{f(x_0)}{x_0 - x_1} + \frac{f(x_1)}{x_1 - x_0} \right] + \frac{1}{x_2 - x_0} \left[\frac{f(x_1)}{x_1 - x_2} + \frac{f(x_2)}{x_2 - x_1} \right]$$

$$= \frac{f(x_0)}{(x_0 - x_1)(x_0 - x_2)} + \frac{f(x_1)}{(x_1 - x_0)(x_1 - x_2)} + \frac{f(x_2)}{(x_2 - x_0)(x_2 - x_1)}$$

If this process is continued n times, we have in general

$$\underset{x_1, x_2, \ldots, x_n}{\triangle^n} f(x_0) = \frac{f(x_0)}{(x_0 - x_1)(x_0 - x_2) \cdots (x_0 - x_n)}$$

$$+ \frac{f(x_1)}{(x_1 - x_0)(x_1 - x_2) \cdots (x_1 - x_n)}$$

$$+ \cdots$$

$$+ \frac{f(x_n)}{(x_n - x_0)(x_n - x_1) \cdots (x_n - x_{n-1})} \tag{5.11}$$

A more compact version of formula (5.11) can be written using the definition of $F_i(x)$ given in formula (5.4). This yields

$$\underset{x_1, x_2, \ldots, x_n}{\triangle^n} f(x_0) = \sum_{i=0}^{n} \frac{f(x_i)}{F_i(x_i)} \tag{5.12}$$

It is possible to give formulas (5.11) and (5.12) a more rigorous proof using mathematical induction.

One very important property of divided differences is that such differences are equal under all permutations of the values of x. For example,

$$\underset{x_1, x_2}{\triangle^2} f(x_0) = \underset{x_2, x_1}{\triangle^2} f(x_0) = \underset{x_0, x_2}{\triangle^2} f(x_1) = \underset{x_2, x_0}{\triangle^2} f(x_1) = \underset{x_0, x_1}{\triangle^2} f(x_2) = \underset{x_1, x_0}{\triangle^2} f(x_2)$$

This result is apparent from formula (5.11). Any permutation of the values of x merely involves interchanges of these values in the symbol on the left-hand side. On the right-hand side the corresponding terms involving the functional values are also interchanged, but the value of the sum is unchanged.

The above property is sometimes called the *symmetrical property* of divided differences. The implication is that the points of collocation $x_0, x_1, \ldots, x_n$ need not be arranged in ascending order but can be arranged in any order that we please.

The property that nth ordinary differences of a polynomial of degree n are constant and all higher differences are zero also holds for divided differences. A formula analogous to formula (2.21) is

$$\triangle^n x^n = 1 \tag{5.13}$$

for any $n + 1$ values of x under consideration. Thus, the one which appears in the third difference column of Table 5.2 is no coincidence.

This result can be derived by noting that the first divided difference of x^n for the points x and x_0 is

$$\underset{x_0}{\triangle} x^n = \frac{x^n - x_0^n}{x - x_0} = x^{n-1} + x^{n-2}x_0 + \cdots + xx_0^{n-2} + x_0^{n-1}$$

which is a polynomial of degree $n - 1$ with a leading coefficient of one. If this result is differenced $n - 1$ more times, all that will be left is the leading coefficient of one.

A nother property which should be explored is the relationship between ordinary differences and divided differences, if the points of collocation happen to be equally spaced, that is, if

$$h = x_1 - x_0 = x_2 - x_1 = \cdots = x_n - x_{n-1}$$

In forming first divided differences the denominators are all equal to h, so that

$$\underset{x_{i+1}}{\triangle} f(x_i) = \frac{\Delta f(x_i)}{h} \quad \text{for } i = 0, 1, \ldots, n-1.$$ We shall abbreviate this expression

as $\triangle \equiv \dfrac{\Delta}{h}$ although the student must be careful to apply both operators to the

same functional values. In forming second divided differences the denominators are all equal to $2h$, so that

$$\triangle^2 \equiv \frac{\Delta^2_h}{2!\,h^2}$$

This process can be continued to give in general

$$\triangle^k \equiv \frac{\Delta^k_h}{k!\,h^k} \qquad \text{for } k = 1, 2, \ldots, n \tag{5.14}$$

It must be reemphasized that formula (5.14) is valid only if the operators Δ and $\triangle$ are applied to a function which has $k+1$ values of x tabulated at an equal interval h. No relationship exists between ordinary and divided differences if the intervals are not equal, since Δ^k_h is defined only for equally spaced intervals. However, formula (5.14) is true in general for any polynomial of degree k or less, since kth differences are constant for all values of x.

Example 5.2.   Show that $\underset{b,\,c}{\triangle^2}\dfrac{1}{a} = \dfrac{1}{abc}.$

Setting up a difference table gives the answer in the second difference column:

x	$\dfrac{1}{x}$	$\triangle\dfrac{1}{x}$	$\triangle^2\dfrac{1}{x}$
a	$\dfrac{1}{a}$		
		$\dfrac{\frac{1}{b}-\frac{1}{a}}{b-a} = -\dfrac{1}{ab}$	
b	$\dfrac{1}{b}$		$\dfrac{-\frac{1}{bc}+\frac{1}{ab}}{c-a} = \dfrac{1}{abc}$
		$\dfrac{\frac{1}{c}-\frac{1}{b}}{c-b} = -\dfrac{1}{bc}$	
c	$\dfrac{1}{c}$		

Example 5.3.   Find $\underset{3,\,7}{\triangle^2} f(2)$ **for the data given in Table 5.2.**

Because of the symmetrical property we can rearrange the values of x as follows:

x	$f(x)$	$\triangle f(x)$	$\triangle^2 f(x)$	$\triangle^3 f(x)$
1	1			
		31		
5	125		8	
		39		1
2	8		14	
		67		
7	343			

Note that the third divided difference is equal to one regardless of the order in which the values of x are used.

The required value of $\underset{3,\,7}{\triangle^2 f(2)}$ can be obtained by adding the value $x = 3$ at the bottom of the difference table and assuming that we can extend third differences as a constant. Then from the method of construction of a divided difference table

$$\frac{\underset{3,\,7}{\triangle^2 f(2)} - 14}{3 - 5} = 1$$

or

$$\underset{3,\,7}{\triangle^2 f(2)} = 12$$

An interesting alternative approach is to observe that $\triangle^n x^{n+1}$ is equal to the sum of the $n + 1$ values of x. This is illustrated in the above difference table, since $1 + 5 + 2 = 8$ and $5 + 2 + 7 = 14$. Thus, the answer is $2 + 3 + 7 = 12$. The general proof of this result is left as Exercise 7.

Although this second approach is substantially shorter for this particular example, it should not be considered superior to the first approach. The first approach is general for any order of differences and any type of function. Thus, it illustrates the principles in the construction of the table and the symmetrical property. The second approach is valid only for the nth-order differences for a polynomial of degree $n + 1$.

5.4. INTERPOLATION WITH DIVIDED DIFFERENCES

The problem is to express $f(x)$ in terms of divided differences based on the $n + 1$ points of collocation $x_0, x_1, \ldots, x_n$. By definition we have

$$f(x) = f(x_0) + (x - x_0)\underset{x_0}{\triangle f(x)}$$

$$\underset{x_0}{\triangle f(x)} = \underset{x_1}{\triangle f(x_0)} + (x - x_1)\underset{x_0, x_1}{\triangle^2 f(x)}$$

$$\underset{x_0, x_1}{\triangle^2 f(x)} = \underset{x_1, x_2}{\triangle^2 f(x_0)} + (x - x_2)\underset{x_0, x_1, x_2}{\triangle^3 f(x)}$$

$$\vdots$$

$$\underset{x_0, x_1, \ldots, x_{n-1}}{\triangle^n f(x)} = \underset{x_1, x_2, \ldots, x_n}{\triangle^n f(x_0)} + (x - x_n)\underset{x_0, x_1, \ldots, x_n}{\triangle^{n+1} f(x)}$$

Successively substituting each identity into the one preceding it, we obtain *Newton's divided difference formula*

$$f(x) = f(x_0) + (x - x_0)\underset{x_1}{\triangle} f(x_0) + (x - x_0)(x - x_1)\underset{x_1,\,x_2}{\triangle^2} f(x_0) + \cdots$$

$$+ (x - x_0)(x - x_1)\cdots(x - x_{n-1})\underset{x_1,\,x_2,\,\ldots,\,x_n}{\triangle^n} f(x_0) + R_{n+1}(x) \qquad (5.15)$$

where the remainder term $R_{n+1}(x)$ is given by

$$R_{n+1}(x) = (x - x_0)(x - x_1)\cdots(x - x_n)\underset{x_0,\,x_1,\,\ldots,\,x_n}{\triangle^{n+1}} f(x)$$

$$= \pi(x)\underset{x_0,\,x_1,\,\ldots,\,x_n}{\triangle^{n+1}} f(x) \qquad (5.16)$$

The remainder term vanishes at the points of collocation and also vanishes if $f(x)$ is a polynomial of degree n or less. The remainder term bears a close relationship to the truncation error of the collocation polynomial given by formula (1.12). This relationship will be explored in more detail in Section 5.5.

It should be noted that formula (5.15) can be written down using Sheppard's rules as presented in Section 3.6 applied down the top diagonal. The only exception is that rule 5 is not applied, that is, the kth differences in the formula are not divided by $k!h^k$ for $k = 1, 2, \ldots, n$. The fact that the terms $k!h^k$ do not appear in the denominators of divided difference interpolation formulas is offset by the denominators used in constructing the divided difference table.

The general proof of Sheppard's rules involves the derivation of formula (5.15) together with an application of the symmetrical property. Permuting the values of x merely results in different paths through the difference table. The explanation for rule 5 of Sheppard's rules in an ordinary difference interpolation formula can be found in the relationship demonstrated in formula (5.14).

Generalized versions of the central difference formulas based on a divided difference table can be derived. In all cases these formulas follow the same paths as described in Chapter 4.

Although any path through a difference table can be used with Sheppard's rules, paths are often selected in practice which will minimize computation. This is particularly effective if the difference table contains a number of zeros.

Example 5.4. Rework Example 5.1 with divided differences using the Gauss backward formula starting at $f(5)$.

The difference table is given in Table 5.2. Figure 5.1 illustrates the path taken by the Gauss backward formula. This is not the most desirable path

FIGURE 5.1

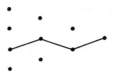

from the standpoint of minimizing computation, but it is illustrative of Sheppard's rules. The result is

$$f(3) = 125 + (3 - 5)(39) + (3 - 5)(3 - 2)(14) + (3 - 5)(3 - 2)(3 - 7)(1)$$
$$= 27$$

the anticipated answer.

5.5. RELATIONSHIPS WITH DERIVATIVES

It is useful to develop certain relationships between differences and derivatives. However, it is not our intention in this section to develop techniques of approximate differentiation, as these will be covered in Chapter 7.

Formula (1.12) gives the truncation error of the collocation polynomial as

$$f(x) - p(x) = \frac{f^{(n+1)}(\xi)\pi(x)}{(n+1)!} \tag{1.12}$$

where ξ lies in the range of collocation. Formula (5.16) gives the remainder term in the divided difference interpolation formula as

$$R_{n+1}(x) = \pi(x) \underset{x_0, x_1, \ldots, x_n}{\triangle^{n+1}} f(x) \tag{5.16}$$

Since these two must be equal, we have the following relationship between a difference and a derivative

$$\underset{x_0, x_1, \ldots, x_n}{\triangle^{n+1}} f(x) = \frac{f^{(n+1)}(\xi)}{(n+1)!} \tag{5.17}$$

A derivative can also be derived from the definition of a divided difference

$$\underset{x_0}{\triangle} f(x) = \frac{f(x) - f(x_0)}{x - x_0}$$

so that

$$\lim_{x \to x_0} \underset{x_0}{\triangle} f(x) = \lim_{x \to x_0} \frac{f(x) - f(x_0)}{x - x_0} = f'(x_0) \tag{5.18}$$

Formula (5.18) can be verified by substituting $x = x_0$ and $n = 0$ into formula (5.17) and noting that $\xi = x_0$, since it must lie in the range of collocation. The expression $\lim\limits_{x \to x_0} \underset{x_0}{\triangle} f(x)$ is often written as $\underset{x_0}{\triangle} f(x_0)$ for simplicity.

Formula (5.17) can also be used to develop relationships between higher-order derivatives and differences in which the points of collocation are all equal, namely

$$\underset{x_0, x_0, \ldots, x_0}{\triangle^{n+1}} f(x_0) = \frac{f^{(n+1)}(x_0)}{(n+1)!} \tag{5.19}$$

If formula (5.19) is applied to Newton's divided difference formula (5.15) in which all the points of collocation are equal to x_0, then we have

$$f(x) = f(x_0) + (x - x_0)\underset{x_0}{\triangle} f(x_0) + (x - x_0)^2 \underset{x_0, x_0}{\triangle^2} f(x_0)$$

$$+ \cdots + (x - x_0)^n \underset{x_0, x_0, \ldots, x_0}{\triangle^n} f(x_0) + R_{n+1}(x)$$

$$= f(x_0) + (x - x_0)f^{(1)}(x_0) + \frac{(x - x_0)^2}{2!} f^{(2)}(x_0)$$

$$+ \cdots + \frac{(x - x_0)^n}{n!} f^{(n)}(x_0) + R_{n+1}(x) \tag{5.20}$$

This is the familiar *Taylor series* from calculus which appeared in Example 1.4 with $x_0 = 0$.

The remainder term can be derived by applying the equality between formulas (1.12) and (5.16) to give

$$R_{n+1}(x) = \pi(x) \underset{x_0, x_0, \ldots, x_0}{\triangle^{n+1}} f(x)$$

$$= \frac{(x - x_0)^{n+1}}{(n+1)!} f^{(n+1)}(\xi) \tag{5.21}$$

where ξ lies between x and x_0. This is the standard formula for the remainder term in the Taylor series in calculus. The close relationship between the truncation error of the collocation polynomial and the truncation error of the Taylor series is apparent.

The Taylor series can be used as an interpolation formula in its own right.[2] In this case the interpolation is based on only one point of collocation and successive derivatives at that point. Interpolation formulas involving derivatives at the points of collocation are often called *osculatory* and will be discussed in Chapter 15.

---

[2]Precise terminology might suggest "extrapolation formula" as a more accurate descriptive phrase.

Example 5.5. *Rework Example 5.1 using the Taylor series starting at $f(7)$.*

Chapter 7 will discuss efficient algorithms for finding derivatives from the difference table. However, this example can be worked from the basic divided difference table by an interesting application of the techniques of constructing the table.

The appropriate difference table is given by Table 5.2 augmented with three additional values of $x = 7$ and with constant third differences extended:

x	$f(x)$	$\triangle f(x)$	$\triangle^2 f(x)$	$\triangle^3 f(x)$
1	1			
		7		
2	8		8	
		39		1
5	125		14	
		109		1
7	343		$\triangle^2 f(5)$	1
		$\triangle f(7)$	$_{7,7}$	
7		$_7$	$\triangle^2 f(7)$	1
			$_{7,7}$	1
7				
7				

The path taken by the Taylor series is indicated in the difference table. It is impossible to work from lower to higher differences through the difference table, since

$$\triangle_7 f(7) = \frac{f(7) - f(7)}{7 - 7}$$

is undefined. However, it is possible to work backward through the table using the definition of a divided difference as follows:

$$\frac{\triangle^2_{7,7} f(5) - 14}{7 - 2} = 1 \quad \text{or} \quad \triangle^2_{7,7} f(5) = 19$$

$$\frac{\triangle^2_{7,7} f(7) - 19}{7 - 5} = 1 \quad \text{or} \quad \triangle^2_{7,7} f(7) = 21$$

$$\frac{\triangle_7 f(7) - 109}{7 - 5} = 19 \quad \text{or} \quad \triangle_7 f(7) = 147$$

Now applying the Taylor series we have

$$f(3) = 343 + (3 - 7)(147) + (3 - 7)^2(21) + (3 - 7)^3(1)$$
$$= 27$$

the anticipated answer.

The student should note that if it is known that $f(x) = x^3$, then formula (5.19) is satisfied, since

$$147 = \underset{7}{\triangle} f(7) = \frac{f^{(1)}(7)}{1!} = \frac{3 \cdot 7^2}{1} = 147$$

$$21 = \underset{7,7}{\triangle^2} f(7) = \frac{f^{(2)}(7)}{2!} = \frac{3 \cdot 2 \cdot 7}{2} = 21$$

$$1 = \underset{7,7,7}{\triangle^3} f(7) = \frac{f^{(3)}(7)}{3!} = \frac{3 \cdot 2 \cdot 1}{6} = 1$$

5.6.   A DETERMINANT METHOD

A *determinant method*[3] for finding the collocation polynomial is given by the determinant equation

$$\begin{vmatrix} f(x) & 1 & x & x^2 & \cdots & x^n \\ f(x_0) & 1 & x_0 & x_0^2 & \cdots & x_0^n \\ f(x_1) & 1 & x_1 & x_1^2 & \cdots & x_1^n \\ \vdots & \vdots & \vdots & \vdots & & \vdots \\ f(x_n) & 1 & x_n & x_n^2 & \cdots & x_n^n \end{vmatrix} = 0 \qquad (5.22)[4]$$

To see that the above determinant gives the collocation polynomial, we can expand with minors along the first row. This produces a polynomial of degree n or less. Substituting $x = x_i$ for $i = 0, 1, 2, \ldots, n$ makes two rows equal, so that the determinant is equal to zero. Thus, $f(x) = f(x_i)$ if $x = x_i$ for $i = 0, 1, 2, \ldots, n$. This proves that $f(x)$ is the collocation polynomial, since the collocation polynomial is uniquely determined.

Example 5.6.   Rework Example 5.1 using the determinant method given in formula (5.22).

[3]Students needing a review of determinants are referred to Chapter 12.
[4]The determinant given in formula (5.22) is closely related to the well-known *Vandermonde determinant*, which is everything in formula (5.22) excluding the first row and first column.

The determinant is

$$
\begin{vmatrix}
f(3) & 1 & 3 & 9 & 27 \\
1 & 1 & 1 & 1 & 1 \\
8 & 1 & 2 & 4 & 8 \\
125 & 1 & 5 & 25 & 125 \\
343 & 1 & 7 & 49 & 343
\end{vmatrix} = 0
$$

The determinant will equal zero if the first and last columns are equal. Thus, $f(3) = 27$, the anticipated answer.

In this case the determinant method is easy to apply, since $f(x) = x^3$ and the result is immediate. However, in most cases considerably more computational effort than this is required.

5.7. THE AITKEN-NEVILLE TECHNIQUE

The *Aitken-Neville technique* finds the collocation polynomial of degree n based on the $n + 1$ points of collocation $x_0, x_1, x_2, \ldots, x_n$ by passing through a sequence of lower-degree collocation polynomials based on various subsets of the values of x. The collocation polynomial of degree n is generated by a successive application of the same technique.[5] Neither interpolation coefficients nor differences appear explicitly.

More specifically, a series of straight lines is generated by linear interpolation based on successive pairs of points. Then a series of quadratics is generated by linear interpolation based on successive pairs of the straight lines. Similarly, a series of cubics is generated by linear interpolation based on successive pairs of the quadratics. This process is continued until the required degree of accuracy in the answer is obtained.

It is not immediately clear that a linear interpolation between two polynomials of a certain degree generates a polynomial of the next higher degree. To see this let $p^{k-1}_{x_0, x_1, \ldots, x_{k-1}}(x)$ and $p^{k-1}_{x_1, x_2, \ldots, x_k}(x)$ be two successive collocation polynomials of degree $k - 1$ based on the listed points of collocation. Then the collocation polynomial of degree k based on these two lower-degree collocation polynomials is given by

$$
p^k_{x_0, x_1, \ldots, x_k}(x) = \frac{(x - x_0)p^{k-1}_{x_1, x_2, \ldots, x_k}(x) - (x - x_k)p^{k-1}_{x_0, x_1, \ldots, x_{k-1}}(x)}{x_k - x_0} \tag{5.23}
$$

[5] Aitken and Neville individually arrived at this method with slight variations in the order in which functional values are entered into the process. The approach adopted here is not identical to either one, but is a blend of the two. Since the two approaches are so similar, credit should be given to both in referring to the process.

Clearly, $p_{x_0, x_1, \ldots, x_k}^k(x)$ is a polynomial of degree k or less. The student should verify that it does collocate for $x_0, x_1, \ldots, x_{k-1}, x_k$ and thus is the required collocation polynomial.

Formula (5.23) is often written in a determinant form for simplicity

$$p_{x_0, x_1, \ldots, x_k}^k(x) = \frac{1}{x_k - x_0} \begin{vmatrix} x - x_0 & p_{x_0, x_1, \ldots, x_{k-1}}^{k-1}(x) \\ x - x_k & p_{x_1, x_2, \ldots, x_k}^{k-1}(x) \end{vmatrix} \qquad (5.24)$$

Expansion of the determinant in formula (5.24) directly yields formula (5.23).

The process used in the Aitken-Neville technique is to generate the first-degree polynomials, then generate the second-degree polynomials, and so forth. The work is usually arranged in a format similar to that used in a difference table. Table 5.3 is the standard table for the Aitken-Neville technique as far as third-degree polynomials. All entries in the table are computed using formula (5.23) or (5.24). It should be noted that $p_{x_i}^0(x) = f(x_i)$.

TABLE 5.3

x_i	$f(x_i)$	$p^1(x)$	$p^2(x)$	$p^3(x)$
x_0	$f(x_0)$			
		$p_{x_0, x_1}^1(x)$		
x_1	$f(x_1)$		$p_{x_0, x_1, x_2}^2(x)$	
		$p_{x_1, x_2}^1(x)$		$p_{x_0, x_1, x_2, x_3}^3(x)$
x_2	$f(x_2)$		$p_{x_1, x_2, x_3}^2(x)$	
		$p_{x_2, x_3}^1(x)$		
x_3	$f(x_3)$			

One interesting feature of the table is that the denominator is always the difference of the two values of x at the end points of the range of values of x brought into the process. The equivalence of this denominator with the denominator used in constructing a divided difference table is apparent and should be an aid in learning the process.

Example 5.7. ***Rework Example 5.1 using the Aitken-Neville technique.***

The standard table is as follows:

x_i	$f(x_i)$	$p^1(3)$	$p^2(3)$	$p^3(3)$
1	1			
		15		
2	8		31	
		47		27
5	125		19	
		-93		
7	343			

The entries are computed as follows:

$$p^1_{1,2}(3) = \frac{1}{2-1}\begin{vmatrix} 3-1 & 1 \\ 3-2 & 8 \end{vmatrix} = \begin{vmatrix} 2 & 1 \\ 1 & 8 \end{vmatrix} = 15$$

$$p^1_{2,5}(3) = \frac{1}{5-2}\begin{vmatrix} 3-2 & 8 \\ 3-5 & 125 \end{vmatrix} = \frac{1}{3}\begin{vmatrix} 1 & 8 \\ -2 & 125 \end{vmatrix} = 47$$

$$p^1_{5,7}(3) = \frac{1}{7-5}\begin{vmatrix} 3-5 & 125 \\ 3-7 & 343 \end{vmatrix} = \frac{1}{2}\begin{vmatrix} -2 & 125 \\ -4 & 343 \end{vmatrix} = -93$$

$$p^2_{1,2,5}(3) = \frac{1}{5-1}\begin{vmatrix} 3-1 & 15 \\ 3-5 & 47 \end{vmatrix} = \frac{1}{4}\begin{vmatrix} 2 & 15 \\ -2 & 47 \end{vmatrix} = 31$$

$$p^2_{2,5,7}(3) = \frac{1}{7-2}\begin{vmatrix} 3-2 & 47 \\ 3-7 & -93 \end{vmatrix} = \frac{1}{5}\begin{vmatrix} 1 & 47 \\ -4 & -93 \end{vmatrix} = 19$$

$$p^3_{1,2,5,7}(3) = \frac{1}{7-1}\begin{vmatrix} 3-1 & 31 \\ 3-7 & 19 \end{vmatrix} = \frac{1}{6}\begin{vmatrix} 2 & 31 \\ -4 & 19 \end{vmatrix} = 27$$

It is seen that $p^3_{1,2,5,7}(3) = 27$, the anticipated answer.

5.8. INVERSE INTERPOLATION

The problem of interpolation as discussed thus far could be called *direct interpolation*, since a value of $y = f(x)$ is determined for some intermediate value of x. *Inverse interpolation* reverses the question and attempts to find a value of x based on some intermediate value of y.

There are two basic approaches to inverse interpolation. The first and most straightforward is simply to reverse the roles of x and y, and then perform a direct interpolation. Any of the four methods of interpolation described previously in this chapter can be used.

This first approach is based on a substantially different assumption than direct interpolation. For example, if carried to third differences, it assumes that x is a third-degree polynomial in y rather than vice versa. Whether or not this is an appropriate assumption depends on the data in the problem.

The first approach should only be used for one-to-one or monotonic functions over the range of collocation. If the function is not monotonic, then more than one value of x may be associated with a given value of y, and inverse interpolation cannot be used.

The second approach is most consistent with direct interpolation in that y is assumed to be a polynomial in x. An expression for y is written using any of the standard finite difference interpolation formulas, and then the polynomial is solved for the required value of x.

In solving for x, alternative approaches exist. If the formula is carried only as far as second differences, a quadratic results and the direct solution is straightforward. If the formula is carried to third or higher differences, then a direct solution is more difficult or even impossible by use of a classical closed-form method. In these cases the usual approach is to use *successive approximation* or *iteration*. Example 5.8 illustrates an iterative approach based on a difference formula. More powerful techniques of iteration which could be applied to problems of this type are discussed in Chapter 11.

Example 5.8.   Estimate a value of x corresponding to $y = f(x) = 1.5$ given that $f(1) = 1.000$, $f(2) = 1.414$, $f(3) = 1.732$, and $f(4) = 2.000$. Use both approaches developed in this section.

The four given values of y agree to three decimal places with the function $y = \sqrt{x}$. Thus, we hope our procedures will produce an answer close to $1.5^2 = 2.25$.

The first approach assumes x is a polynomial in y. Setting up a divided difference table we obtain the following:

y	x	$\triangle x$	$\triangle^2 x$	$\triangle^3 x$
1.000	1			
		2.415		
1.414	2		.997	
		3.145		.003
1.732	3		1.000	
		3.731		
2.000	4			

The interpolated value using the top diagonal in the difference table is
$$x = 1 + (1.5 - 1)(2.415) + (1.5 - 1)(1.5 - 1.414)(.997)$$
$$+ (1.5 - 1)(1.5 - 1.414)(1.5 - 1.732)(.003)$$
$$= 2.250$$
the correct answer.

The second approach assumes y is a polynomial in x. Setting up an ordinary difference table, we obtain the following:

$x \longrightarrow$	y	z	Δz	$\Delta^2 z$	$\Delta^3 z$
1 $\longrightarrow$	-1	1.000			
			.414		
2 $\longrightarrow$	0	1.414		$-.096$	
			.318		.046
3 $\longrightarrow$	1	1.732		$-.050$	
			.268		
4 $\longrightarrow$	2	2.000			

Since third differences are not zero, a cubic results which is inconvenient to solve. Thus, we shall adopt an iterative approach. The successively approximated values of x will be denoted by $x_1, x_2, \ldots$. Several interpolation formulas could be used, but we shall use Bessel's formula which is advantageous in bisecting intervals.

Taking Bessel's formula to first differences, we have

$$1.5 = \tfrac{1}{2}(1.732 + 1.414) + \left(x_1 - \tfrac{1}{2}\right)(.318)$$

Solving for the first approximation, we have $x_1 = .27$.

Now taking Bessel's formula to second differences and using the value of x_1 in the second difference term, we have

$$1.5 = \tfrac{1}{2}(1.732 + 1.414) + \left(x_2 - \tfrac{1}{2}\right)(.318) + \frac{(.27)(.27 - 1)}{2} \cdot \tfrac{1}{2}(-.050 - .096)$$

Solving for the second approximation, we have $x_2 = .248$.

Now taking Bessel's formula to third differences and using the value of x_2 in the second and third difference terms, we have

$$1.5 = \tfrac{1}{2}(1.732 + 1.414) + \left(x_3 - \tfrac{1}{2}\right)(.318) + \frac{(.248)(.248 - 1)}{2} \cdot \tfrac{1}{2}(-.050 - .096)$$

$$+ \frac{(.248 - .5)(.248)(.248 - 1)}{6}(.046)$$

Solving for the third approximation, we have $x_3 = .248$. Since $x_2 = x_3$, further applications of the formula will produce the same answer, so the iteration is ended. Since $2.248 \rightarrow .248$ under the change in origin, the answer is 2.248 which is in error by only .002.

It is interesting that both approaches, which have such completely different assumptions, are so successful in producing reasonable answers. Unfortunately, inverse interpolation is not always this successful.

5.9. FUNCTIONS OF MORE THAN ONE VARIABLE

Interpolations with unequal intervals can be extended to functions of more than one variable. The working process is similar to that described in Section 3.7 for equal intervals. The process can best be illustrated by example.

Example 5.9. Estimate $f(4, 2)$ ***given that*** $f(0, 0) = 0$, $f(5, 0) = 25$, $f(7, 0) = 49$, $f(0, 1) = 1$, $f(5, 1) = 21$, ***and*** $f(0, 3) = 9$.

The data arranged in tabular form are as follows:

x \ y	0	1	3
0	0	1	9
5	25	21	
7	49		

In this case the functional values are arranged in triangular form. Also, they agree with the function $f(x, y) = x^2 - xy + y^2$, so that we hope our procedures will produce the answer $4^2 - 4 \cdot 2 + 2^2 = 12$.

The interpolation formula is

$$f(x, y) = \left[1 + x\triangle_x + x(x - 5)\triangle_x^2 + y\triangle_y + xy\triangle_y\triangle_x + y(y - 1)\triangle_y^2 \right] f(0, 0)$$
$$\quad\quad\quad\,\,\, 5 \quad\quad\quad\, 5,7 \quad\quad\quad\, 1 \quad\quad 1\,\,5 \quad\quad\quad 1,3$$

Calculating the differences for $f(x, 0)$, we have the following:

x	$f(x, 0)$	$\triangle f(x, 0)$	$\triangle^2 f(x, 0)$
0	0		
		5	
5	25		1
		12	
7	49		

Calculating the differences for $f(0, y)$, we have the following:

y	$f(0, y)$	$\triangle f(0, y)$	$\triangle^2 f(0, y)$
0	0		
		1	
1	1		1
		4	
3	9		

Finally

$$\triangle_y\triangle_x f(0, 0) = \frac{21 - 25 - 1 + 0}{5} = -1$$
$$\,\, 1 \,\,\, 5$$

The interpolation formula gives

$$f(4, 2) = 0 + (4)(5) + (4)(-1)(1) + (2)(1) + (4)(2)(-1) + (2)(1)(1)$$
$$= 12$$

the anticipated answer.

5.10. COMPARISON OF METHODS

This chapter has considered the technical details of four methods of interpolation. As demonstrated in the examples, all of the methods are capable of producing the same answer, since they all generate the same collocation polynomial.

Some consideration should be given to the advantages and disadvantages of the various methods as well as to the technical details. A method which may be superior to the others in a particular situation may not be superior in another situation.

The discussion of the divided difference method implicitly includes the interpolation formulas considered in Chapters 3 and 4. All finite difference interpolation methods have essentially the same advantages and disadvantages.

The determinant method is probably the least favorite method in practical applications. The computation involved in evaluating determinants is generally greater than the computation involved in the other methods. This method is of considerable theoretical interest, however.

Both the finite difference methods and the Aitken-Neville technique have the advantage over the Lagrange method that additional points of collocation can be brought into the formula merely by adding another term at the end. By contrast, if it is desired to add another point of collocation with Lagrange's formula, all the coefficients must be recomputed.

Finite difference methods have the advantage that the difference table gives a clear indication of what order of differences should be included in the interpolation. By examining the pattern of the successive orders of differences, we can tell whether the differences have yet become relatively constant or perhaps have become spurious because we have differenced too far and all that is left is roundoff error. Thus, in error analysis finite difference methods have considerable advantages.

Finite difference methods also have the advantage of offering several available formulas, whereas the other methods generally offer only one approach. As previously discussed, the Newton formulas, the central difference formulas, and divided difference interpolation all offer advantages under certain conditions.

Lagrange's formula has the advantage that the functional value at one point is directly expressed as a linear combination of functional values at other points. Thus, no intermediate step involving the construction of a difference table is involved, so that the programming is quite straightforward.

Lagrange's formula also has the advantage that if several interpolations of the same form are to be performed, the coefficients can be tabulated once and used as many times as needed.

The Aitken-Neville technique is useful in programming, since it involves only repeated linear interpolations. All of these linear interpolations are of the same form. This technique is also popular for inverse interpolation.

EXERCISES

5.1.   Introduction;   5.2.   Lagrange's formula

1. (a) Show that

$$\frac{f(x)}{\pi(x)} = \sum_{i=0}^{n} \frac{f(x_i)}{F_i(x_i)} \cdot \frac{1}{x - x_i} = \sum_{i=0}^{n} \frac{f(x_i)}{\pi'(x_i)} \cdot \frac{1}{x - x_i}$$

This result shows that Lagrange's formula is equivalent to expressing $f(x)/\pi(x)$ in partial fractions.

   (b) Find the coefficient of $\dfrac{1}{x-3}$, if

$$\frac{2x^2 - 5x + 4}{x^3 - 8x^2 + 15x}$$

is expressed in partial fractions.

2. Show that

$$L_0(x) = 1 + \frac{x - x_0}{x_0 - x_1} + \frac{(x - x_0)(x - x_1)}{(x_0 - x_1)(x_0 - x_2)} + \cdots$$

$$+ \frac{(x - x_0)(x - x_1) \cdots (x - x_{n-1})}{(x_0 - x_1)(x_0 - x_2) \cdots (x_0 - x_n)}$$

Similar expressions exist for $L_i(x)$, where $i = 1, 2, \ldots, n$.

3. Estimate $f(1)$ given that $f(0) = 3$, $f(3) = 18$, and $f(5) = 48$ using Lagrange's formula.

4. Rework Exercise 8 in Chapter 3 using Lagrange's formula.

5. Find $L_0(-1)$, that is, the coefficient of $f(0)$ using Lagrange's formula to estimate $f(-1)$, assuming that the points of collocation are $x = 0, 1, 2, \ldots, n$.

6. An interpolation for $f(4)$ is based upon the following values:

$$f(0) = 5 \qquad f(3) = 7 \qquad f(5) = 10 \qquad f(15) = 50$$

After the computations have been performed, it is discovered that $f(5)$ should have been 12. Find the error in the interpolated value of $f(4)$.

5.3. Divided differences

7. Show that $\Delta^n x^{n+1}$ for the $n+1$ points of collocation $a_0, a_1, a_2, \ldots, a_{n-1}$, a_n is equal to $a_0 + a_1 + a_2 + \cdots + a_{n-1} + a_n$. *Hint:* Proceed along the following lines:

(a) For the $n+1$ points of collocation $0, 1, 2, \ldots, n-1, n$ show that

$$\Delta^n x^{n+1} = [0 + 1 + 2 + \cdots + (n-1) + n]n!$$

(b) From the result in (a) show that for the same points of collocation

$$\underset{}{\Delta}{}^n x^{n+1} = 0 + 1 + 2 + \cdots + (n-1) + n$$

(c) Using the result that $\Delta^{n+1} x^{n+1} = 1$, show that

$$\underset{c_0,\,c_1,\,\ldots,\,c_{n-1},\,c_{n+1}}{\Delta^n} x^{n+1} = (c_{n+1} - c_n) + \underset{c_0,\,c_1,\,\ldots,\,c_{n-1},\,c_n}{\Delta^n} x^{n+1}$$

(d) Apply the result in (c) successively $n+1$ times replacing 0 by a_0, 1 by $a_1, \ldots, n$ by a_n, which produces the required result.

8. Show that

$$\underset{y,z}{\Delta}{}^2 \frac{1}{x^2} = \frac{xy + yz + xz}{x^2 y^2 z^2}$$

9. Find $\underset{3,\,5,\,7,\,9}{\Delta}{}^3 x_2^{(5)}$.

10. Express

$$\frac{\underset{b,c,d}{\Delta}{}^3 u_a - \underset{a,d,e}{\Delta}{}^3 u_c}{\underset{a,b,c}{\Delta}{}^3 u_e - \underset{a,d,e}{\Delta}{}^3 u_b}$$

in terms of a, b, c, d, e.

11. If third differences are constant and $u_{-3} = -50$, $u_0 = 55$, $u_1 = 50$, $u_3 = 52$, $u_4 = 83$, find $\underset{1,\,3}{\Delta}{}^2 u_{-2}$.

12. A divided difference table is formed through fourth differences using the values of x in the order 1, 2, 4, 7, 11. A new table is formed using the order 7, 4, 2, 11, 1. How many differences are necessarily common to both tables?

13. Using the data given in Exercise 11, evaluate $\delta^3 u_{3/2}$.

5.4. Interpolation with divided differences

14. Rework Exercise 3 using divided differences.

15. If $u_x = u_2 + A \underset{5}{\Delta} u_2 + B \underset{5,\,9}{\Delta}{}^2 u_2 + C \underset{3,\,5,\,9}{\Delta}{}^3 u_2 + D \underset{2,\,3,\,5,\,9}{\Delta}{}^4 u_{-1}$, compute $\dfrac{AD}{BC}$.

16. Given the following divided difference tables for two third-degree polynomials u_x and v_x and that $mu_x + nv_x = 3x^2 - 3x + 4$, find m and n.

x	u_x	$\triangle u_x$	$\triangle^2 u_x$	$\triangle^3 u_x$
a	7			
		-3		
b	1		2	
		-1		1
c	2		6	
		11		
d	35			

x	v_x	$\triangle v_x$	$\triangle^2 v_x$	$\triangle^3 v_x$
a	10			
		-3		
b	4		1	
		-2		2
c	6		9	
		16		
d	54			

17. Given $u_8 = u_a + 4\underset{b}{\triangle}u_a + 32\underset{b,c}{\triangle^2}u_a + 224\underset{b,c,d}{\triangle^3}u_a + 224\underset{b,c,d,e}{\triangle^4}u_a$, find $a + b + c + d$.

18. Two expressions u_x and v_x are obtained using divided difference interpolation with the paths indicated below:

x	$f(x)$	$\triangle f(x)$	$\triangle^2 f(x)$	$\triangle^3 f(x)$	$\triangle^4 f(x)$
j	J				
k	K	P	T $\;u_x$		
l	L	Q	U	X	Z
m	M	R $\;v_x$	V	Y	
n	N	S			

Find $v_x - u_x$.

19. Given $u_1 = 1$, $\underset{2}{\triangle}u_1 = 2$, $\underset{2,5}{\triangle^2}u_1 = 0$, $\underset{2,5,7}{\triangle^3}u_1 = 3$, and that higher-order differences are zero, find u_4.

5.5. Relationships with derivatives

20. Find $\underset{x_0,\,x_0}{\triangle^2}u_{x_1}$ if $u_x = x^3$.

21. Show that $D\underset{x,\,x}{\triangle^2}u_a = 2\underset{x,x,x}{\triangle^3}u_a$.

22. Show that
$$f(x) = \frac{(x_1 - x)(x + x_1 - 2x_0)}{(x_1 - x_0)^2} f(x_0)$$
$$+ \frac{(x - x_0)(x_1 - x)}{x_1 - x_0} f'(x_0) + \frac{(x - x_0)^2}{(x_1 - x_0)^2} f(x_1)$$

Hint: Consider the limit as $\epsilon \to 0$ of a three-point Lagrange formula based on points of collocation x_0, $x_0 + \epsilon$, x_1.

5.6. A determinant method

23. Rework Exercise 3 using the determinant method developed in Section 5.6.

24. (a) Show that

$$\underset{x_1}{\triangle} f(x_0) = \frac{\begin{vmatrix} 1 & f(x_0) \\ 1 & f(x_1) \end{vmatrix}}{\begin{vmatrix} 1 & x_0 \\ 1 & x_1 \end{vmatrix}}$$

(b) Show that

$$\underset{x_1, x_2}{\triangle^2} f(x_0) = \frac{\begin{vmatrix} 1 & x_0 & f(x_0) \\ 1 & x_1 & f(x_1) \\ 1 & x_2 & f(x_2) \end{vmatrix}}{\begin{vmatrix} 1 & x_0 & x_0^2 \\ 1 & x_1 & x_1^2 \\ 1 & x_2 & x_2^2 \end{vmatrix}}$$

5.7. The Aitken-Neville technique

25. Rework Exercise 3 using the Aitken-Neville technique.
26. Given that $p_{0,1}^1(2) = -1$, $p_{3,6}^1(2) = -39$, and $p_{0,1,3,6}^3(2) = 5$, find $p_{1,3}^1(2)$.

5.8. Inverse interpolation

27. If $u_x = -9/8$, $u_{-1} = 20$, $u_0 = 2$, $u_1 = 0$, $u_2 = 14$, and third differences are zero, find x.
28. If $u_{-1} = 26$, $u_0 = 8$, and $u_1 = 0$, for what value of x in the interval $0 < x < 1$ does $u_x = 2$? Assume u_x is a polynomial in x.
29. Rework Exercise 28 assuming x is a polynomial in u_x.

5.9. Functions of more than one variable; 5.10. Comparison of methods

30. Estimate $f(2, 3)$ given that $f(1, 2) = 4$, $f(1, 4) = 8$, $f(4, 2) = 1$, and $f(4, 4) = 11$ using divided difference interpolation as illustrated in Example 5.9.
31. Rework Exercise 30 using the following adaptation of Lagrange's formula

$$f(x, y) = f(a_0, b_0) \frac{(x - a_1)(y - b_1)}{(a_0 - a_1)(b_0 - b_1)} + f(a_0, b_1) \frac{(x - a_1)(y - b_0)}{(a_0 - a_1)(b_1 - b_0)}$$

$$+ f(a_1, b_0) \frac{(x - a_0)(y - b_1)}{(a_1 - a_0)(b_0 - b_1)} + f(a_1, b_1) \frac{(x - a_0)(y - b_0)}{(a_1 - a_0)(b_1 - b_0)}$$

Miscellaneous problems

32. Given points of collocation a, b, c, d, e, f, find expressions for the following divided difference interpolation formulas accurate to fourth differences:
    (a) Gauss forward formula.
    (b) Gauss backward formula.
    (c) Stirling's formula.
    (d) Bessel's formula.

33. Find $\underset{-2,\,6,\,7}{\triangle^3}\,f(-1)$ if

$$f(x) = 3 + 5(x + 1) - 6(x + 1)(x - 4) + 2(x + 1)(x - 4)(x - 7)$$
$$- 4(x + 2)(x + 1)(x - 4)(x - 7)$$

34. Find the polynomial through the points $(-1, -2)$, $(0, 0)$, and $(1, 2)$ for which the third divided differences are constant and equal to one.

35. Find the coefficient A in the following interpolation formula:

$$u_x = \tfrac{1}{2}(u_0 + u_1) + (x - \tfrac{1}{2})\underset{1}{\triangle}u_0 + \tfrac{1}{2}x(x - 1)\left(\underset{0,\,1}{\triangle^2 u_{-1}} + \underset{1,\,4}{\triangle^2 u_0}\right) + A\,\underset{0,\,1,\,4}{\triangle^3}\,u_{-1}$$

36. Use Lagrange's formula based on the points of collocation $x = 0, 1, 2, \ldots,$ $n - 1$ to show that

$$u_x = \frac{x!}{(x - n)!\,(n - 1)!}\left[\frac{u_{n-1}}{x - n + 1} - \binom{n-1}{1}\frac{u_{n-2}}{x - n + 2}\right.$$
$$\left. + \binom{n-1}{2}\frac{u_{n-3}}{x - n + 3} - \cdots + (-1)^{n-1}\binom{n-1}{n-1}\frac{u_0}{x}\right]$$

Assume that x is a positive integer greater than $n - 1$.

37. If it is known that

$$u_x = u_0 + (\tfrac{3}{4}x + A)\underset{1}{\triangle}u_0 + (\tfrac{1}{4}x - A)\underset{0}{\triangle}u_{-3}$$

is correct to second differences, express A as a function of x.

38. If it is known that

$$u_x = \tfrac{4}{3}u_1 - \tfrac{1}{3}u_4 + x\underset{4}{\triangle}u_1 + \tfrac{1}{6}x(x - 1)(x - 4)\underset{4,\,6}{\triangle^2 u_1} - A\underset{1,\,4}{\triangle^2 u_0}$$

is correct to third differences, express A as a function of x.

39. Find $\underset{b,\,c,\,d}{\triangle^3}\,u_a$ if

$$u_x = (x - a)(x - b)(x - c) + (x - a)(x - b)(x - d)$$
$$+ (x - a)(x - c)(x - d) + (x - b)(x - c)(x - d)$$

6

Summation

6.1. INTRODUCTION

IN CHAPTER 2 the analogy between differencing in numerical analysis and differentiating in calculus was apparent. The student would expect that in numerical analysis there is some analogue to integration in calculus. That analogue is the familiar process of *summation*.

We shall discover that many of the properties of integrals that are encountered in calculus also have their analogues in numerical analysis. For example, the *fundamental theorem of calculus* states that under certain conditions differentiation and integration are inverse processes. Analogous statements can be made about differencing and summation.

In order to integrate in calculus it is necessary to find *antiderivatives*. In numerical analysis it is necessary to find *antidifferences*. Thus, in evaluating $\sum_{x=a}^{b} f(x)$, we attempt to find a function $F(x)$ such that

$$\Delta F(x) = f(x) \qquad (6.1)$$

Then we have

$$
\begin{aligned}
f(a) &= F(a+1) - F(a) \\
f(a+1) &= F(a+2) - F(a+1) \\
f(a+2) &= F(a+3) - F(a+2) \\
&\;\;\vdots \\
f(b-1) &= F(b) \qquad\;\; - F(b-1) \\
f(b) &= F(b+1) - F(b)
\end{aligned}
$$

125

Summing both sides we have

$$\sum_{x=a}^{b} f(x) = \sum_{x=a}^{b} \Delta F(x) = F(x)\Big|_{a}^{b+1} = F(b+1) - F(a) \qquad (6.2)$$

It is important to note that the upper limit in formula (6.2) is $b+1$, not b.

The summation contained in formula (6.2) has limits and thus could be called a *definite summation*. It is also possible to define an *indefinite summation* as a summation without limits.

In working problems with indefinite summations a constant of summation will appear. Theoretically, the additive term may not really be a constant. For example, the first difference of the function $f(x) = x + \sin 2\pi x$ would appear to be constant when tabulated with an interval of differencing $h = 1$. This point is of only minor practical significance, since it requires a periodic function. Most of the functions of primary interest to us are polynomials or their ratios and exponentials, none of which is periodic.

Indefinite summation can be looked upon from the viewpoint of operators, so that Σ is often called the *summation operator*. From formula (6.2) it is seen that $\Sigma \Delta F(x) = F(x)$ for appropriate limits of summation. In operators for this case $\Sigma \Delta \equiv 1$, so that

$$\Delta^{-1} \equiv \Sigma \qquad (6.3)$$

Thus, in a certain sense, summation and differencing are inverse processes.

However, there are subtle errors that can be made when working with Σ as an operator. As mentioned in Section 2.2, linear operators do not always obey the commutative law of multiplication. It is easy to construct examples in which $\Delta \Sigma f(x)$ is not equal to $\Sigma \Delta f(x)$. Thus,

$$\Delta \Sigma \not\equiv \Sigma \Delta \qquad (6.4)$$

in general. Exercises 4 and 5 are examples demonstrating this result.

In practice, care should be taken in using Σ to avoid these subtle errors. Limits should be stated whenever possible to eliminate the chance of ambiguities.

6.2. SUMMATION OF POLYNOMIALS AND EXPONENTIALS

Since polynomials and exponentials occur so frequently in numerical analysis, it is important to consider approaches to use in the summation of such functions.

Polynomials can be summed by converting them to factorial notation. In factorial notation the antidifferences are immediate. For upper factorial notation, we have

$$\sum x^{(m)} = \frac{1}{m+1} x^{(m+1)} + c \qquad (6.5)$$

and for lower factorial notation, we have

$$\sum \binom{x}{m} = \binom{x}{m+1} + c \tag{6.6}$$

Formulas (6.5) and (6.6) follow from formulas (2.15) and (2.23), respectively.

It is also possible to work with negative factorial notation. From formula (2.19), we have

$$\sum x^{(-m)} = -\frac{1}{m-1} x^{(-m+1)} + c \tag{6.7}$$

This formula is useful in summing the reciprocals of certain special polynomials.

Summing exponentials can be accomplished by using formula (2.12) to give

$$\sum a^x = \frac{a^x}{a-1} + c \tag{6.8}$$

The student should verify formulas (6.5) through (6.8) by differencing the sums shown.

It is also instructive to consider formula (6.8) as a definite summation. Summing for $x = 0$ through $n - 1$, we have

$$\sum_{x=0}^{n-1} a^x = \frac{a^x}{a-1}\bigg|_0^n = \frac{a^n - 1}{a-1} \tag{6.9}$$

which is the conventional form for the sum of a geometric progression.

Example 6.1. **Find the sum of the cubes of the first n positive integers.**

Converting x^3 to factorial notation, we have $x^3 = x^{(3)} + 3x^{(2)} + x^{(1)}$. Thus,

$$\sum_{x=1}^{n} x^3 = \sum_{x=1}^{n} [x^{(3)} + 3x^{(2)} + x^{(1)}]$$

$$= \tfrac{1}{4}x^{(4)} + x^{(3)} + \tfrac{1}{2}x^{(2)} \bigg|_1^{n+1}$$

$$= \tfrac{1}{4}(n+1)n(n-1)(n-2) + (n+1)n(n-1) + \tfrac{1}{2}(n+1)n$$

$$= (n+1)n[\tfrac{1}{4}(n-1)(n-2) + (n-1) + \tfrac{1}{2}]$$

$$= \tfrac{1}{4}(n+1)n[n^2 - 3n + 2 + 4n - 4 + 2]$$

$$= \tfrac{1}{4}(n+1)n(n^2 + n)$$

$$= \left[\frac{(n+1)n}{2}\right]^2$$

Sums of powers of integers can conveniently be found for any power in this manner.

Example 6.2.   Find $\displaystyle\sum_{x=1}^{n} \frac{1}{(x+1)(x+2)}$.

We have

$$\sum_{x=1}^{n} \frac{1}{(x+1)(x+2)} = \sum_{x=1}^{n} x^{(-2)}$$

$$= -x^{(-1)} \Big|_{1}^{n+1}$$

$$= \frac{1}{2} - \frac{1}{n+2}$$

This answer can also be obtained by a direct algebraic method using partial fractions. If the summand is expressed in partial fractions, we have

$$\sum_{x=1}^{n} \frac{1}{(x+1)(x+2)} = \sum_{x=1}^{n} \left(\frac{1}{x+1} - \frac{1}{x+2} \right)$$

$$= \left(\frac{1}{2} + \frac{1}{3} + \cdots + \frac{1}{n+1} \right) - \left(\frac{1}{3} + \cdots + \frac{1}{n+1} + \frac{1}{n+2} \right)$$

$$= \frac{1}{2} - \frac{1}{n+2}$$

6.3.   SUMMATION OF FINITE DIFFERENCE FORMULAS

Summation can easily be performed upon any of the standard finite difference interpolation formulas, since each is merely a variation of the collocation polynomial. As demonstrated in Section 6.2, any polynomial can be summed. In fact, several of the standard ordinary difference formulas have coefficients already expressed in factorial notation, so that the summation is facilitated.

For example, Newton's advancing difference formula can be summed between the limits $x = 0$ and $x = n - 1$ using formula (3.1b) to give

$$\sum_{x=0}^{n-1} u_x = \sum_{x=0}^{n-1} \left[\binom{x}{0} u_0 + \binom{x}{1} \Delta u_0 + \binom{x}{2} \Delta^2 u_0 + \cdots \right]$$

$$= \left[\binom{x}{1} u_0 + \binom{x}{2} \Delta u_0 + \binom{x}{3} \Delta^2 u_0 + \cdots \right]_0^n$$

$$= \binom{n}{1} u_0 + \binom{n}{2} \Delta u_0 + \binom{n}{3} \Delta^2 u_0 + \cdots \qquad (6.10)$$

Similar expressions could be derived for Newton's backward difference formula or for any of the central difference formulas. It is also possible to derive expressions for limits other than $x = 0$ through $n - 1$.

One particularly useful formula can be developed from Stirling's formula. Consider the problem of adding u_x and u_{-x} which are symmetrically spaced about a central value u_0. From formula (4.8b), we have

$$u_x = u_0 + x\mu\delta u_0 + \frac{x^2}{2!}\delta^2 u_0 + \frac{x(x^2 - 1)}{3!}\mu\delta^3 u_0 + \frac{x^2(x^2 - 1)}{4!}\delta^4 u_0 + \cdots$$

and

$$u_{-x} = u_0 - x\mu\delta u_0 + \frac{x^2}{2!}\delta^2 u_0 - \frac{x(x^2 - 1)}{3!}\mu\delta^3 u_0 + \frac{x^2(x^2 - 1)}{4!}\delta^4 u_0 + \cdots$$

Now adding we have

$$u_x + u_{-x} = 2u_0 + x^2\delta^2 u_0 + \frac{x^2(x^2 - 1)}{12}\delta^4 u_0 + \cdots \qquad (6.11a)$$

or in terms of operators on u_0

$$E^x + E^{-x} \equiv 2 + x^2\delta^2 + \frac{x^2(x^2 - 1)}{12}\delta^4 + \cdots \qquad (6.11b)$$

This formula is useful in adding values symmetrically spaced on either side of some central value. It is particularly useful since odd difference terms vanish. Thus, third-degree accuracy can be obtained with only two terms.

This particular approach using Stirling's formula was the basis of the *summation method of graduation*[1] developed by English actuaries. This method of graduation is primarily of historical interest today. However, since the literature contains references to it, a brief discussion of the operator [n], read "summation n," and its properties is contained in Appendix D.

Example 6.3.   Find the sum to 20 terms of the following series: 1, 0, 3, 10, 21, ...

There is nothing in this example to indicate the values of x that are associated with the given functional values. However, the choice of x makes no difference in the answer obtained, since a change in values of x will be exactly offset by a change in the limits on the summation. It is usual in numerical problems of this type to start the summation at $x = 0$.

---

[1] *Graduation* is a process of smoothing data discussed in Section 15.1.

Before the summation can be performed, it is necessary to determine the form of the function. Setting up a difference table, we have the following:

x	u_x	Δu_x	$\Delta^2 u_x$
0	1		
		-1	
1	0		4
		3	
2	3		4
		7	
3	10		4
		11	
4	21		

Assuming a continuation of constant second differences equal to four, Newton's advancing difference formula gives

$$u_x = 1 - x + 2x^{(2)}$$

The sum of 20 terms has limits $x = 0$ through 19, so that

$$\sum_{x=0}^{19} u_x = \left[x - \tfrac{1}{2}x^{(2)} + \tfrac{2}{3}x^{(3)} \right]_0^{20}$$

$$= 20 - \tfrac{1}{2}(20)(19) + \tfrac{2}{3}(20)(19)(18)$$

$$= 4390$$

Upper factorial notation is used in this example, although lower factorial notation is used in formula (6.10). The student should be able to work examples in either upper or lower factorial notation. The student should also rework this example using labels of x starting at $x = 1$ and verify that the same answer is obtained.

Example 6.4.   Express $u_{-2} + 2u_{-1} + 3u_0 + 2u_1 + u_2$ **in terms of** u_0 **and central differences of** u_0.

We have

$$u_{-2} + 2u_{-1} + 3u_0 + 2u_1 + u_2 = (u_{-2} + u_2) + 2(u_{-1} + u_1) + 3u_0$$
$$= [(2 + 4\delta^2 + \delta^4) + 2(2 + \delta^2) + 3]u_0$$
$$= 9u_0 + 6\delta^2 u_0 + \delta^4 u_0$$

The student can verify this answer by expanding the differences into functional values.

6.4. SUMMATION BY PARTS

It is possible to derive a formula for *summation by parts* which is analogous to the calculus formula for integration by parts. Formula (2.10) gives

$$\Delta(u_x v_x) = u_x \Delta v_x + v_{x+1}\Delta u_x \tag{2.10}$$

which can be written as

$$u_x \Delta v_x = \Delta(u_x v_x) - v_{x+1}\Delta u_x$$

The standard summation by parts formula is obtained by summing this expression between the limits $x = 0$ and $x = n - 1$ to obtain

$$\sum_{x=0}^{n-1} (u_x \Delta v_x) = u_x v_x \Big|_0^n - \sum_{x=0}^{n-1}(v_{x+1}\Delta u_x) \tag{6.12}$$

The label "$x + 1$" on the v in the summation on the right-hand side should be noted. Also it should be noted that u_x and v_x are interchangeable.

As would be expected, summation by parts is useful in summing the product of two functions of x. One type of problem commonly encountered in practice is the sum of the product of an exponential and a polynomial. If the polynomial is of degree n, then n applications of formula (6.12) will have to be made, which is inconvenient.

A general formula for the summation of the product of an exponential and a polynomial can be obtained as a special case of a more general summation by parts formula now to be derived.

Let Δ, E, and Σ be operators applying to the product function $u_x v_x$; let Δ_u, E_u, and Σ_u be operators applying to u_x alone; and let Δ_v, E_v, and Σ_v be operators applying to v_x alone. Then we have

$$\begin{aligned}
\Sigma u_x v_x &= \Delta^{-1}u_x v_x \\
&= (E - 1)^{-1}u_x v_x \\
&= (E_u E_v - 1)^{-1}u_x v_x \\
&= (\Delta_u + \Delta_v + \Delta_u \Delta_v)^{-1}u_x v_x \\
&= [\Delta_v + \Delta_u(1 + \Delta_v)]^{-1}u_x v_x \\
&= \Delta_v^{-1}[1 + \Delta_u \Delta_v^{-1}(1 + \Delta_v)]^{-1}u_x v_x \\
&= \Delta_v^{-1}[1 - \Delta_u \Delta_v^{-1}(1 + \Delta_v) + \Delta_u^2 \Delta_v^{-2}(1 + \Delta_v)^2 - \cdots]u_x v_x \\
&= [\Delta_v^{-1} - \Delta_u \Delta_v^{-2}E_v + \Delta_u^2 \Delta_v^{-3}E_v^2 - \cdots]u_x v_x \\
&= (u_x)(\Sigma v_x) - (\Delta u_x)(\Sigma^2 v_{x+1}) + (\Delta^2 u_x)(\Sigma^3 v_{x+2}) - \cdots \tag{6.13}^2
\end{aligned}$$

---

[2]The properties of Σ^n for $n > 1$ are subtle and are illustrated in Exercise 19.

In Exercise 18 the student will show that a slight rearrangement of the above derivation gives the following formula:

$$\Sigma\, u_x v_x = (u_{x-1})(\Sigma\, v_x) - (\Delta u_{x-2})(\Sigma^2\, v_x) + (\Delta^2 u_{x-3})(\Sigma^3\, v_x) - \cdots \quad (6.14)$$

Because of the pattern of differences of u_x, formula (6.13) is sometimes called the *generalized forward difference summation by parts formula* and formula (6.14) the *generalized backward difference summation by parts formula*.

The required formula for summing the product of an exponential and a polynomial can be obtained from formula (6.13) by letting u_x be the polynomial and setting $v_x = a^x$ as the exponential. Then, we have

$$\Sigma\, v_x = \Sigma\, a^x = \frac{a^x}{a-1}$$

$$\Sigma^2\, v_{x+1} = \Sigma^2\, a^{x+1} = \frac{a^{x+1}}{(a-1)^2}$$

$$\vdots$$

Formula (6.13) now gives

$$\sum_{x=0}^{n-1} a^x u_x = \left[\frac{a^x}{a-1} u_x - \frac{a^{x+1}}{(a-1)^2}\Delta u_x + \frac{a^{x+2}}{(a-1)^3}\Delta^2 u_x - \cdots\right]_0^n$$

$$= \left[\frac{a^x}{a-1}\left\{1 - \frac{a\Delta}{a-1} + \frac{a^2\Delta^2}{(a-1)^2} - \cdots\right\} u_x\right]_0^n \quad (6.15)$$

The apparent infinite series in formula (6.15) has only a finite number of terms as long as u_x is a polynomial.

Example 6.5. *Compute* $\displaystyle\sum_{x=1}^{10} 2^x x^2$ *using*—

(1)　*Formula (6.12).*
(2)　*Formula (6.15).*

(1)　One application of formula (6.12) with $u_x = x^2$ and $\Delta v_x = 2^x$ gives

$$\sum_{x=1}^{10} 2^x x^2 = 2^x x^2 \Big|_1^{11} - \sum_{x=1}^{10} 2^{x+1}(2x+1)$$

Another application of formula (6.12) with $u_x = 2x+1$ and $\Delta v_x = 2^{x+1}$ gives

$$\sum_{x=1}^{10} 2^x x^2 = 2^x x^2 \Big|_1^{11} - 2^{x+1}(2x+1)\Big|_1^{11} + \sum_{x=1}^{10} 2^{x+2}\cdot 2$$

Now the last term can be summed with formula (6.9) to give

$$\sum_{x=1}^{10} 2^x x^2 = 2^x x^2 \Big|_1^{11} - 2^{x+1}(2x+1)\Big|_1^{11} + 2^{x+3}\Big|_1^{11}$$
$$= 2^{11} \cdot 11^2 - 2 - 2^{12} \cdot 23 + 4 \cdot 3 + 2^{14} - 2^4$$
$$= 169{,}978$$

(2) Formula (6.15) gives immediately

$$\sum_{x=1}^{10} 2^x x^2 = 2^x\left[x^2 - 2(2x+1) + 4 \cdot 2 \right]_1^{11}$$
$$= 2^{11}\left(11^2 - 2 \cdot 23 + 8\right) - 2(1 - 2 \cdot 3 + 8)$$
$$= 169{,}978$$

The simpler working process with formula (6.15) is apparent. As the degree of the polynomial increases, the advantages become even more pronounced.

6.5. MISCELLANEOUS SUMMATION EXAMPLES

Summation problems can be a source of frustration to the student, since a variety of techniques are required for the different types of problems encountered. In some cases considerable ingenuity may be necessary to sum a series. Again, this situation has its analogue in calculus, where the antiderivative of a function is often not readily apparent.

The approach developed in this chapter is based on finite differences. Finding the sum of a given series is contingent upon determining the antidifference of the summand. As described, this is relatively straightforward for polynomials and exponentials. It may also be possible to sum certain other series by finding antidifferences by trial and error methods.

Another approach which finds occasional use is summation by separation of symbols. This method is particularly effective when the operator expression representing the series can be summed by an algebraic technique.

Many series can be summed using results of classical mathematical analysis without using finite differences at all. As an illustration, in Example 1.5 the series

$$1 - \tfrac{1}{2} + \tfrac{1}{3} - \tfrac{1}{4} + \cdots$$

appears. This series cannot be directly summed using finite differences, since no antidifference exists. However, we know from the series expansion of $\log_e (1 + x)$ that this series is equal to $\log_e 2$. The student should always be alert to the possibility of summation by standard mathematical methods other than finite difference techniques.

One final approach to summation, which is often necessary, is to do no mathematical analysis at all and to merely sum the series by direct computation. The availability of high-speed digital computers has made the straightforward computational approach a much more practical possibility than it was previously.

In summing convergent infinite series by direct computation, the number of terms required to keep truncation error below a certain level is determined, if possible. This type of determination is illustrated in Example 1.4 and 1.5.

In computation we are concerned with how rapidly the series converges. A rapidly converging series is ideal for computation. However, a slowly converging series requires more computer time and is subject to sizable roundoff error. Example 1.4 shows that eight terms are required to produce three-decimal accuracy for computing e^x, $-1 \le x \le 1$, using the standard series expansion. This is rapid enough convergence for computation.

However, Example 1.5 shows that 2000 terms will be required to produce three-decimal accuracy for computing $\log_e 2$ using the series expansion mentioned above. This is unacceptably slow convergence for computation.

Advanced summation theory considers algorithms for speeding up the convergence of slowly converging series. These algorithms are called *acceleration methods*. Generally these methods involve the rearrangement or replacement of a given series by an equivalent series which converges more rapidly. Exercise 8 in Chapter 1 develops an acceleration method for Example 1.5. Another example of a slowly converging series and a different type of acceleration method is given in Example 6.11.

The following miscellaneous examples are illustrative of the variety of approaches which can be used in summation.

Example 6.6. **Find** $\displaystyle\sum_{x=1}^{n} \log \frac{x+3}{x+1}$.

Finite difference methods could be adapted to this example, since

$$\log \frac{x+3}{x+1} = \log(x+3) - \log(x+1) = \underset{2}{\Delta} \log(x+1)$$

The interval of differencing $h = 2$ introduces a complexity. However, the finite difference approach is not required at all, since directly from the properties of logarithms we have

$$\sum_{x=1}^{n} \log \frac{x+3}{x+1} = \log \frac{4}{2} + \log \frac{5}{3} + \cdots + \log \frac{n+2}{n} + \log \frac{n+3}{n+1}$$

$$= \log \frac{4}{2} \cdot \frac{5}{3} \cdot \cdots \cdot \frac{n+2}{n} \cdot \frac{n+3}{n+1}$$

$$= \log \frac{(n+2)(n+3)}{2 \cdot 3}$$

Example 6.7. **Find** $\displaystyle\sum_{x=1}^{n}\left[\dfrac{x^2 4^x}{(x+1)(x+2)}\right].$

In this example, finding the antidifference of the summand requires skillful manipulation (and perhaps a little luck). We have

$$\frac{x^2}{(x+1)(x+2)} = \frac{(x^2+3x+2)-(3x+2)}{(x+1)(x+2)} = 1 - \frac{3x+2}{(x+1)(x+2)}$$

If partial fractions are used, we have

$$1 - \frac{3x+2}{(x+1)(x+2)} = 1 - \frac{4}{x+2} + \frac{1}{x+1}$$

Now the summation becomes

$$\sum_{x=1}^{n}\left[\frac{x^2 4^x}{(x+1)(x+2)}\right] = \sum_{x=1}^{n}\left[1 - \frac{4}{x+2} + \frac{1}{x+1}\right]4^x$$

$$= \sum_{x=1}^{n}\left[4^x - \Delta\frac{4^x}{x+1}\right]$$

$$= \left[\frac{4^x}{3} - \frac{4^x}{x+1}\right]_{1}^{n+1}$$

$$= \frac{2}{3} + \frac{4^{n+1}}{3}\left(\frac{n-1}{n+2}\right)$$

upon simplification.

Example 6.8. **Find** $\displaystyle\sum_{r=0}^{n}(-1)^r\dfrac{\dbinom{n}{r}}{x+r}.$

In this case separation of symbols provides a workable approach. We have

$$\sum_{r=0}^{n}(-1)^r\frac{\dbinom{n}{r}}{x+r} = \frac{\dbinom{n}{0}}{x} - \frac{\dbinom{n}{1}}{x+1} + \frac{\dbinom{n}{2}}{x+2} - \cdots + (-1)^n\frac{\dbinom{n}{n}}{x+n}$$

$$= \left[\binom{n}{0} - \binom{n}{1}E + \binom{n}{2}E^2 - \cdots + (-1)^n\binom{n}{n}E^n\right]\frac{1}{x}$$

$$= (1-E)^n\frac{1}{x}$$

$$= (-1)^n\Delta^n(x-1)^{(-1)}$$

$$= (-1)^n(-1)^n n!(x-1)^{(-n-1)}$$

$$= \frac{n!}{x(x+1)(x+2)\cdots(x+n)}$$

Example 6.9. **Estimate** $f(3)$ **given that** $f(1) = \sum\limits_{x=1}^{1} f(x) = 1$, $\sum\limits_{x=2}^{4} f(x) = 23$, **and** $\sum\limits_{x=5}^{9} f(x) = 225$.

Since three values of $\sum f(x)$ are given, we assume that $f(x)$ is a second-degree polynomial, that is,

$$f(x) = ax^{(2)} + bx^{(1)} + c$$

Thus, we have

$$\sum_{x=1}^{1} f(x) = \sum_{x=1}^{1} \left(ax^{(2)} + bx^{(1)} + c\right) = \qquad b + \ c = \ 1$$

$$\sum_{x=2}^{4} f(x) = \sum_{x=2}^{4} \left(ax^{(2)} + bx^{(1)} + c\right) = \ 20a + \ 9b + 3c = \ 23$$

$$\sum_{x=5}^{9} f(x) = \sum_{x=5}^{9} \left(ax^{(2)} + bx^{(1)} + c\right) = 220a + 35b + 5c = 225$$

Solving these three equations in three unknowns, we obtain $a = 1$, $b = 0$, $c = 1$. Thus,

$$f(x) = x^{(2)} + 1$$

and

$$f(3) = 3^{(2)} + 1 = 7$$

Example 6.10. **It is given that** $f(x) = x(x + 3)(x + 6)(x + 9)$. **Find the following:**

(1) $f(1) + f(4) + f(7) + \cdots + f(22) + f(25)$.

(2) $f(1) + f(2) + f(3) + \cdots + f(24) + f(25)$.

This example analyzes the summation of polynomials in factorial notation for intervals of differencing other than one. The summand can be written as $f(x) = (x + 9)_3^{(4)}$, that is, $(x + 9)^{(4)}$ with interval of differencing $h = 3$. The two questions ask that a summation be performed in which the interval of differencing in the summation is first $h = 3$, and then $h = 1$.

In (1) the intervals of differencing in the factorial and in the summation are both equal to three. From formula (2.16)

$$\underset{h}{\Delta} x_h^{(m)} = hm x_h^{(m-1)} \tag{2.16}$$

we can obtain

$$\sum_h x_h^{(m)} = \frac{1}{h} \cdot \frac{1}{m+1} x_h^{(m+1)} \tag{6.16}$$

Thus,

$$\sum_{x=1}^{25}{}_{3}(x+9)_{3}^{(4)} = \tfrac{1}{3} \cdot \tfrac{1}{5}(x+9)_{3}^{(5)}\Big|_{1}^{28}$$

$$= \tfrac{1}{15}(x+9)(x+6)(x+3)x(x-3)\Big|_{1}^{28}$$

Note that the upper limit in evaluating the answer is one interval of differencing, that is, three, above the upper limit on the summation.

In (2) the interval of differencing in the factorial is equal to three, while the interval of differencing in the summation is equal to one. The answer can be expressed in the convenient form

$$\sum_{x=1}^{25}(x+9)_{3}^{(4)} = \tfrac{1}{15}(x+9)(x+6)(x+3)x(x-3)\Big|_{1,\,2,\,3}^{26,\,27,\,28}$$

in which the three upper limits are evaluated and the results added, and similarly with the results at the three lower limits. The sum of the results at the three lower limits is then subtracted from the sum of the results at the three upper limits. This intriguing result can be seen by arranging the summation as follows:

Values of x	*Limits*
1, 4, 7, 10, 13, 16, 19, 22, 25	1, 28
2, 5, 8, 11, 14, 17, 20, 23	2, 26
3, 6, 9, 12, 15, 18, 21, 24	3, 27

Thus, the summation can be broken down into three summations each of which is performed as in (1). This result easily generalizes to other problems.

Example 6.11.   Compute $1 - \tfrac{1}{3} + \tfrac{1}{5} - \tfrac{1}{7} + \cdots$ *to four-decimal accuracy.*

The sum of this infinite series is equal to $\pi/4 = .7854$ to four decimal places. This can be seen from

$$\frac{1}{1+t^2} = 1 - t^2 + t^4 - t^6 + \cdots$$

and integrating

$$\int_{0}^{x} \frac{1}{1+t^2}\, dt = \arctan x = x - \frac{x^3}{3} + \frac{x^5}{5} - \frac{x^7}{7} + \cdots$$

Evaluated at $x = 1$, we have

$$\arctan 1 = \frac{\pi}{4} = 1 - \frac{1}{3} + \frac{1}{5} - \frac{1}{7} + \cdots$$

If S is the true sum and S_n is the sum to n terms, then as in Example 1.5

$$|S - S_n| < |a_{n+1}|$$

$$= \left|\frac{1}{2n + 1}\right|$$

$$< .00005$$

Thus, $2n + 1 > \dfrac{1}{.00005}$ or $n > 9999.5$, so that an incredible 10,000 terms must be computed to guarantee only four-decimal accuracy! This is another example of a slowly converging series.

An acceleration method for this problem can be developed using the *Euler transformation*. This transformation is useful in summing series with alternating signs as in this example. The transformation is given by

$$1 - E + E^2 - E^3 + \cdots \equiv \frac{1}{1 + E} \equiv \frac{1}{2}\left[1 - \frac{\Delta}{2} + \frac{\Delta^2}{4} - \frac{\Delta^3}{8} + \cdots\right] \quad (6.17)$$

The derivation of formula (6.17) is left as Exercise 23.

The best procedure in finding a sum of this type is to compute a few terms directly and then apply the Euler transformation to the rest of the series. The first 10 terms produce

$$1 - \tfrac{1}{3} + \tfrac{1}{5} - \cdots - \tfrac{1}{19} = .760460$$

The next five terms together with their difference table are as follows:

	Δ	Δ^2	Δ^3	Δ^4
$\frac{1}{21} = .047619$				
	$-.004141$			
$\frac{1}{23} = .043478$		$.000663$		
	$-.003478$		$-.000148$	
$\frac{1}{25} = .040000$		$.000515$		$.000042$
	$-.002963$		$-.000106$	
$\frac{1}{27} = .037037$		$.000409$		
	$-.002554$			
$\frac{1}{29} = .034483$				

The Euler transformation gives

$$\tfrac{1}{2}(.047619) - \tfrac{1}{4}(-.004141) + \tfrac{1}{8}(.000663)$$

$$- \tfrac{1}{16}(-.000148) + \tfrac{1}{32}(.000042) = .024938$$

The answer is $.760460 + .024938 = .7854$ to four decimals places, which is correct. This process uses only 15 terms instead of 10,000 terms.

6.6. THE EULER-MACLAURIN FORMULA

The *Euler-Maclaurin formula* is a formula which relates the sum of a function over a certain range to the integral of the function over the same range together with adjustment terms. The adjustment terms involve derivatives at the end points of the range involved.

As background for the derivation of the Euler-Maclaurin formula, we need the Taylor series as given by formula (5.20) with $x_0 = 0$, that is,

$$f(x) = f(0) + xf^{(1)}(0) + \frac{x^2}{2!} f^{(2)}(0) + \frac{x^3}{3!} f^{(3)}(0) + \cdots$$

If this formula is expressed in terms of operators on $f(0)$, we have

$$E^x \equiv 1 + xD + \frac{x^2}{2!} D^2 + \frac{x^3}{3!} D^3 + \cdots$$

$$\equiv e^{xD}$$

Thus,

$$E \equiv 1 + \Delta \equiv e^D \tag{6.18}$$

and

$$D \equiv \log_e E \equiv \log_e (1 + \Delta) \tag{6.19}$$

Formulas (6.18) and (6.19) give a relationship between the operators Δ and D which will be quite useful in developing approximate differentiation formulas in Chapter 7.

Consider the indefinite summation of some function $f(x)$, for which we have

$$\begin{aligned}
\Sigma f(x) &= \Delta^{-1} f(x) \\
&= (e^D - 1)^{-1} f(x) &&\text{(formula (6.18))} \\
&= \left(D + \frac{D^2}{2!} + \frac{D^3}{3!} + \frac{D^4}{4!} + \cdots \right)^{-1} f(x) &&\text{(series expansion)} \\
&= \left(D^{-1} - \frac{1}{2} + \frac{D}{12} - \frac{D}{720} + \cdots \right) f(x) &&\text{(long division)} \\
&= \int f(x)\, dx - \tfrac{1}{2} f(x) + \tfrac{1}{12} f^{(1)}(x) - \tfrac{1}{720} f^{(3)}(x) + \cdots {}^3
\end{aligned}$$

---

[3] The coefficients in the Euler-Maclaurin formula are closely related to *Bernoulli numbers* defined in Exercise 56.

Now placing the limits $x = 0$ to $x = n - 1$ on the summation and evaluating the right-hand side between $x = 0$ and $x = n$, we have

$$\sum_{x=0}^{n-1} f(x) = \int_0^n f(x)\, dx - \tfrac{1}{2}[f(n) - f(0)]$$
$$+ \tfrac{1}{12}[f^{(1)}(n) - f^{(1)}(0)] - \tfrac{1}{720}[f^{(3)}(n) - f^{(3)}(0)] + \cdots$$

The Euler-Maclaurin formula is normally written in a slightly different form by adding $f(n)$ to both sides to give

$$\sum_{x=0}^{n} f(x) = \int_0^n f(x)\, dx + \tfrac{1}{2}[f(n) + f(0)] + \tfrac{1}{12}[f^{(1)}(n) - f^{(1)}(0)]$$
$$- \tfrac{1}{720}[f^{(3)}(n) - f^{(3)}(0)] + \cdots \quad (6.20)$$

Formula (6.20) is exact for fourth-degree polynomials, since all derivatives of even degree two and higher vanish.

The Euler-Maclaurin formula has several uses in numerical analysis as follows:

1. It can serve as a summation formula in its own right. This requires that the integral and the derivatives can be evaluated.
2. It is useful in deriving summation formulas for subdivision of intervals. This topic will be discussed in Section 6.7.
3. It can occasionally serve as an effective acceleration method. The concept of an acceleration method is discussed in Section 6.5.
4. It can serve as an approximate integration formula. This use reverses the perspective of the formula as a summation formula. In this case the sum and derivatives would have to be evaluated in order to approximate the integral. Approximate integration is discussed in Chapter 8.

Example 6.12.   Apply the Euler-Maclaurin formula to $\sum\limits_{x=1}^{100} \log_e x$ ***and compare the result with that obtained in Example 1.7.***

We have for $f(x) = \log_e x$,

$$\int_1^{100} f(x)\, dx = \left[x \log_e x - x \right]_1^{100} = 100(4.60517) - 100 + 1 = 361.517$$
$$f^{(1)}(x) = x^{-1}$$

and

$$f^{(3)}(x) = 2x^{-3}$$

Applying the Euler-Maclaurin formula gives

$$\sum_{x=1}^{100} f(x) = \int_1^{100} f(x)\, dx + \tfrac{1}{2}[f(100) + f(1)] + \tfrac{1}{12}[f^{(1)}(100) - f^{(1)}(1)]$$

$$- \tfrac{1}{720}[f^{(3)}(100) - f^{(3)}(1)] + \cdots$$

$$= 361.517 + \tfrac{1}{2}(4.60517) + \tfrac{1}{12}(.01 - 1) - \tfrac{1}{720}(.000002 - 2)$$

$$= 363.740$$

The exact answer to three decimal places given in Example 1.7 is 363.739. Thus, the error in the Euler-Maclaurin formula in this example is only .001.

6.7. SUBDIVISION OF INTERVALS

On occasion it is important to be able to find a summation at intervals which are smaller than those at which functional values are tabulated. If the missing intermediate functional values can be computed without an undue volume of computation being required, then no problem is created. For example, the results contained in Example 6.10 can easily be generalized for the summation of polynomials at various intervals of differencing.

However, in many cases this exact approach is not justifiable if the volume of computation becomes too overwhelming. Furthermore, for some types of data it is either impractical or impossible to tabulate functional values at all these intermediate points, since an underlying mathematical law either cannot readily be determined or does not even exist.

One straightforward approach to the problem is to interpolate for the missing values and then perform the summation. However, this would often be excessively time-consuming. Two algorithms which accomplish this result in an efficient manner can be developed. These are *Lubbock's formula* and *Woolhouse's formula*.

Consider a situation in which functional values are tabulated at integral points $x = 0, 1, \ldots, n - 1, n$. Each one of the n intervals lying between two successive points is now subdivided into m subintervals. Thus, there is a total of mn subintervals. Figure 6.1 illustrates the situation.

FIGURE 6.1

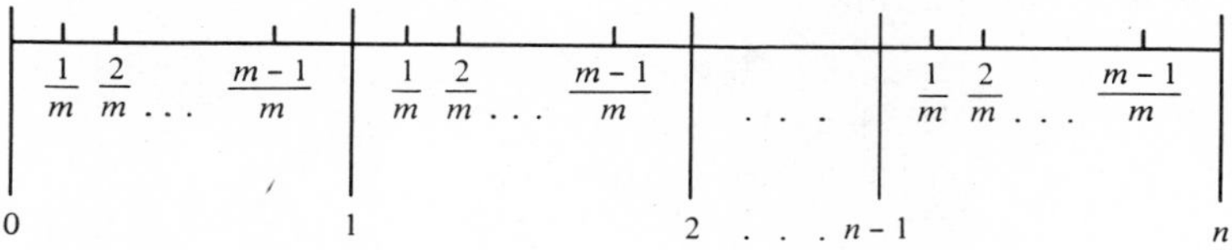

Both Lubbock's formula and Woolhouse's formula are similar in that they express the sum at the $m$thly points as approximately equal to m times the sum at integral points with adjustment terms. This seems reasonable, since the sum at $m$thly points has m times as many terms as the sum at integral points. However, the details of the two formulas are quite different.

Lubbock's formula is derived first. The required sum is

$$u_0 + u_{\frac{1}{m}} + u_{\frac{2}{m}} + \cdots + u_{n - \frac{1}{m}} = \left(1 + E^{\frac{1}{m}} + E^{\frac{2}{m}} + \cdots + E^{n - \frac{1}{m}}\right)u_0$$

$$= \frac{E^n - 1}{E^{\frac{1}{m}} - 1}u_0$$

$$= \frac{\Delta}{(1 + \Delta)^{\frac{1}{m}} - 1} \cdot \frac{E^n - 1}{\Delta}u_0$$

$$= \Delta\left[\frac{1}{m}\Delta + \frac{\left(\frac{1}{m}\right)\left(\frac{1}{m} - 1\right)}{2}\Delta^2\right.$$

$$\left. + \frac{\left(\frac{1}{m}\right)\left(\frac{1}{m} - 1\right)\left(\frac{1}{m} - 2\right)}{6}\Delta^3 + \cdots\right]^{-1}\frac{E^n - 1}{\Delta}u_0$$

$$= \Delta\left[m\Delta^{-1} + \frac{m - 1}{2} - \frac{m^2 - 1}{12m}\Delta + \frac{m^2 - 1}{24m}\Delta^2 - \cdots\right]\frac{E^n - 1}{\Delta}u_0$$

$$= \left[m\frac{E^n - 1}{E - 1} + \left\{\frac{m - 1}{2} - \frac{m^2 - 1}{12m}\Delta + \frac{m^2 - 1}{24m}\Delta^2 - \cdots\right\}(E^n - 1)\right]u_0$$

$$= m(u_0 + u_1 + u_2 + \cdots + u_{n-1}) + \frac{m - 1}{2}(u_n - u_0) - \frac{m^2 - 1}{12m}(\Delta u_n - \Delta u_0)$$

$$+ \frac{m^2 - 1}{24m}(\Delta^2 u_n - \Delta^2 u_0) - \cdots \tag{6.21}$$

Formula (6.21) is the standard version of Lubbock's formula.

Woolhouse's formula is derived from the Euler-Maclaurin formula, which for interval of differencing $h = 1$ is

$$\int_0^n u_x \, dx = (u_0 + u_1 + u_2 + \cdots + u_n) - \tfrac{1}{2}(u_n + u_0)$$

$$- \tfrac{1}{12}(u_n^{(1)} - u_0^{(1)}) + \tfrac{1}{720}(u_n^{(3)} - u_0^{(3)}) - \cdots$$

The Euler-Maclaurin formula for interval of differencing $h = 1/m$ is

$$m \int_0^n u_x \, dx = \left(u_0 + u_{\frac{1}{m}} + u_{\frac{2}{m}} + \cdots + u_n \right)$$

$$- \frac{1}{2}(u_n + u_0) - \frac{1}{12m}(u_n^{(1)} - u_0^{(1)}) + \frac{1}{720m^3}(u_n^{(3)} - u_0^{(3)}) - \cdots$$

If the first expression above is multiplied by m and equated to the second expression, we have

$$u_0 + u_{\frac{1}{m}} + u_{\frac{2}{m}} + \cdots + u_n = m(u_0 + u_1 + u_2 + \cdots + u_n) - \frac{m-1}{2}(u_n + u_0)$$

$$- \frac{m^2 - 1}{12m}\left(u_n^{(1)} - u_0^{(1)} \right) + \frac{m^4 - 1}{720m^3}(u_n^{(3)} - u_0^{(3)}) - \cdots$$

$$(6.22)$$

Formula (6.22) is the standard version of Woolhouse's formula.

In comparing Lubbock's formula and Woolhouse's formula the following should be noted:

1. In Lubbock's formula the *m*thly summation includes *mn* terms and the integral summation includes *n* terms. In Woolhouse's formula the *m*thly summation includes *mn* + 1 terms and the integral summation includes *n* + 1 terms.
2. The adjustments in Lubbock's formula involve differences, while the adjustments in Woolhouse's formula involve derivatives. Thus, Lubbock's formula has an advantage over Woolhouse's formula, if only tabular data are given with no underlying mathematical law present.
3. Lubbock's formula involves data outside the interval $0 \le x \le n$, because of the differences which appear. No similar problem exists for Woolhouse's formula. This is a disadvantage of Lubbock's formula, since these functional values may not be available.
4. Woolhouse's formula has missing terms of even degree two and higher, while Lubbock's formula has no missing terms. Thus, if higher degree accuracy is required, the computations with Lubbock's formula are greater than with Woolhouse's formula. Also, the higher degree terms in Lubbock's formula are more cumbersome to use and harder to derive than in Woolhouse's formula. Finally, convergence of the series in Lubbock's formula is slower than in Woolhouse's formula.

In summary, Woolhouse's formula is usually superior to Lubbock's formula, if the derivatives can be evaluated. Lubbock's formula is used primarily in those situations in which the derivatives cannot be evaluated.

EXERCISES

6.1.   Introduction

1. Find the sum of the first differences of the function $u_x = 2^x - x^3 + x^2 - 1$ from $x = 0$ through $x = 7$.

2. Find $\sum_{x=1}^{9} \Delta^2 f(x)$ if $f(x) = x^4$.

3. Given that $u_x = \delta v_x$, express $\sum_{x=0}^{10} u_x$ in terms of the function v_x.

4. Show that if $\Delta[F(x) + c] = f(x)$, then $F(x) + c = \Sigma \Delta F(x)$. In operators, this result states that $\Sigma \Delta \neq 1$ in this case because of the constant of summation.

5. (a)   In calculus the following result holds

$$D \int_a^x f(t)\, dt = f(x)$$

Derive the analogous numerical analysis result

$$\Delta \sum_{t=a}^{x} f(t) = f(x+1)$$

In operators, this result states that for this special type of summation $\Delta \Sigma \equiv E \neq 1$.

(b)   In calculus the following result holds

$$D \int_x^a f(t)\, dt = -f(x)$$

Derive the analogous numerical analysis result

$$\Delta \sum_{t=x}^{a} f(t) = -f(x)$$

In operators, this result states that for this special type of summation $\Delta \Sigma \equiv -1 \neq 1$.

6.2.   Summation of polynomials and exponentials

6. Find

$$\sum_{x=4}^{10} \left[\binom{x}{3} + x^{(3)} \right]$$

7. Show that

$$\sum_{x=1}^{n} x^2 = \frac{n(n+1)(2n+1)}{6}$$

8. Express $\sum_{x=0}^{100} (x+1)(x+2)(x+3)(x+8)$ in factorial notation.

9. Sum the following series to n terms

$$1 \cdot 2 \cdot 3 + 2 \cdot 3 \cdot 5 + 3 \cdot 4 \cdot 7 + 4 \cdot 5 \cdot 9 + \cdots.$$

10. It is known that $\sum_{x=0}^{n-1} f(x) = n^3 - n$. Find $f(x)$.

11. Find

$$\sum_{x=1}^{n} \frac{1}{(3x - 2)(3x + 1)(3x + 4)}$$

12. Find

$$\sum_{x=1}^{10} 3^{x^2}(9^x - \tfrac{1}{3})$$

6.3. Summation of finite difference formulas

13. Derive an expression for $\sum_{x=1}^{n} u_x$ if u_x is expressed to fourth differences using the Gauss backward formula.

14. Sum the following series to 20 terms: 2, 1, 2, 5, ...

15. Find $\sum_{x=0}^{n} u_x$, if $u_0 = 0$, $u_1 = 0$, $u_2 = 2$, $u_3 = 8$, $u_4 = 22$, $u_5 = 52$.

16. Assuming fourth and higher differences are zero, express $u_{-5} + u_{-3} + u_{-1} + u_1 + u_3 + u_5$ in terms of u_0 and $\delta^2 u_0$.

17. If u_x is a second-degree polynomial, express $\Delta^2 u_0$ in terms of u_0, u_1, and $\sum_{x=0}^{7} u_x$.

6.4. Summation by parts

18. Derive formula (6.14).

19. Find $\sum_{x=1}^{10}{}^3 \, x$ by direct summation of a polynomial three times. Verify the answer with a four-column work sheet in which the column headings are x, $\Sigma \, x$, $\Sigma^2 \, x$, $\Sigma^3 \, x$, and values of x from 1 through 10 are arranged vertically. Form each column as a successive downward summation of the previous column. Assume all constants of summation are equal to zero.

20. Find $\sum_{x=1}^{10} (3x + 1)4^x$.

21. Find $\sum_{x=0}^{\infty} x^2 4^{-x}$.

22. Find $\sum_{x=0}^{n-1} x^{(2)} x^{(6)}$.

6.5.   Miscellaneous summation examples

23.   Derive formula (6.17).

24.   Show that $\sum_{x=1}^{n} (x^2 + 1)x! = n(n + 1)!$.

25.   Find $\sum_{x=1}^{n} (x + 3)(x + 5)$.

26.   Find

$$\sum_{x=1}^{\infty} \frac{1}{(x + 2)(x + 4)}$$

27.   Sum the following series: $12^7 - \binom{6}{1}13^7 + \binom{6}{2}14^7 - \cdots + \binom{6}{6}18^7$.

28.   If $u_0 = 1$, $u_1 + u_2 = 12$, and $u_3 + u_4 + u_5 = 58$, find u_4.

29.   Find

$$1 + \frac{1}{n + 1} + \frac{1 \cdot 2}{(n + 1)(n + 2)} + \frac{1 \cdot 2 \cdot 3}{(n + 1)(n + 2)(n + 3)} + \cdots$$

where $n > 1$.

30.   Find $\Delta x^{(m)} - 2\Delta^2 x^{(m)} + 3\Delta^3 x^{(m)} - 4\Delta^4 x^{(m)} + \cdots$, where m is a positive integer.

31.   Find

$$\Delta^{-1} \log_b \left(1 + \frac{b}{bx + c}\right)$$

32.   Find

$$\sum_{x=0}^{n-1} 3^x \frac{2x + 1}{(x + 1)(x + 2)}$$

33.   Rework Example 1.5 to find

$$\log_e 2 = 1 - \tfrac{1}{2} + \tfrac{1}{3} - \cdots$$

by computing the first 10 terms directly and then using the Euler transformation on the next 5 terms as was done in Example 6.11. This algorithm uses 15 terms of the series as compared with 32 terms in Exercise 8 in Chapter 1 and 2000 terms in Example 1.5.

6.6.   The Euler-Maclaurin formula

34.   Use the Euler-Maclaurin formula to find $\sum_{x=0}^{n} f(x)$, given that

$$\int_0^n f(x)\, dx = 100, \quad f(0) = 2, \quad f(n) = 20, \quad f'(0) = 1, \quad f'(n) = 25.$$

Ignore third and higher derivatives.

35. Use the Euler-Maclaurin formula to develop an approximation for $\sum_{x=1}^{100} \sqrt{x}$ to two decimal places. Ignore third and higher derivatives.

36. Use the Euler-Maclaurin formula to develop an approximation for $\int_{1}^{2} u_x \, dx$ to four decimal places based on the following data:

x	1.0	1.2	1.4	1.6	1.8	2.0	*Total*
u_x	.5000	.4545	.4167	.3846	.3571	.3333	2.4462
u_x'	−.2500					−.1111	

Ignore third and higher derivatives. Note that formula (6.20) must be modified to reflect the fact that $h \neq 1$.

37. Use the Euler-Maclaurin formula to find an exact expression for $\sum_{x=1}^{n} x^4$. The Euler-Maclaurin formula is an efficient algorithm for finding sums of powers of the first n positive integers.

38. Find $\sum_{x=1}^{\infty} \dfrac{1}{x^3}$ to seven decimal places by computing the first nine terms directly and then by using the Euler-Maclaurin formula as an acceleration method for the balance of the series.

6.7. Subdivision of intervals

39. Example 1.7 contains the result that $\sum_{x=1}^{100} \log_e x = 363.739$ to three decimal places. Use Lubbock's formula to second differences to estimate

$$\log_e 1 + \log_e 1.1 + \log_e 1.2 + \cdots + \log_e 99.9 + \log_e 100$$

to two decimal places. For ease in computation the following values are given:

x	$\log_e x$
1	.0000
2	.6931
3	1.0986
100	4.6052
101	4.6151
102	4.6250

40. Rework Exercise 38 using Woolhouse's formula to third derivatives.

41. Show that Lubbock's formula to second differences can be obtained from Woolhouse's formula by substituting

$$D \equiv \log_e (1 + \Delta) \equiv \Delta - \frac{\Delta^2}{2} + \cdots$$

in Woolhouse's formula for first derivatives and by ignoring third and higher derivatives.

Miscellaneous problems

42. Given that $\sum_{x=0}^{n-1} f(x) = 2n^{(3)} - 5n^{(2)} + 3n^{(1)}$, find $\sum_{x=0}^{n-1} (x+1)f(x)$.

43. Assuming that $\Delta\Sigma \equiv \Sigma\Delta \equiv 1$, find

$$[1 - \Sigma + \Sigma^2 - \Sigma^3 + \cdots]x^{-1}$$

44. Find $\sum_{x=1}^{n} \sum_{y=1}^{x} xy$.

45. If it is known that

$$\sum_{x=1}^{5} u_x = \tfrac{5}{3}(u_1 + 2u_4)$$

for all second-degree polynomials, show that the following are also true:

(a) $\sum_{x=1}^{5} u_x = \tfrac{5}{3}(2u_2 + u_5)$.

(b) $\sum_{x=1}^{10} u_x = \tfrac{5}{3}(u_1 + 2u_4 + 2u_7 + u_{10})$.

(c) $\sum_{x=1}^{10} u_x = \tfrac{5}{3}(2u_2 + u_5 + u_6 + 2u_9)$.

46. Sum the following series to 10 terms

$$\frac{1^2}{1} + \frac{1^2 + 2^2}{2} + \frac{1^2 + 2^2 + 3^2}{3} + \cdots$$

47. Find $\displaystyle\int_0^{10} u_x \, dx$ if $\sum_{x=0}^{n-1} u_x = n^4$ for all positive integers n.

48. Find

$$\sum_{x=0}^{n-1} \left(\frac{\Delta^2}{E} \sum_{t=1}^{x} u_t \right)$$

49. The number 200 is divided into 16 parts which sum to 200. The differences between successive parts form an arithmetic progression with common difference one, and the last part is double the first. Find the 10th part.

50. Show that $\sum_{x=1}^{n} \sum_{y=1}^{x} f(x, y) = \sum_{y=1}^{n} \sum_{x=y}^{n} f(x, y)$.[4]

51. Sum the following series

$$\frac{1}{x+3} + \frac{1}{(x+3)(x+4)} + \frac{2}{(x+3)(x+4)(x+5)}$$

$$+ \frac{6}{(x+3)(x+4)(x+5)(x+6)} + \cdots$$

---

[4]This result is analogous to a well-known result in calculus for reversing the order of integration.

52. Sum the following series

$$x^2 + \frac{(x+1)^2}{3} + \frac{(x+2)^2}{9} + \frac{(x+3)^2}{27} + \cdots$$

53. The operator M is defined by

$$M \equiv \int_{-1/2}^{1/2} E^t \, dt$$

Show that $DM \equiv \delta$ in terms of operators.[5]

54. Find the function whose first difference is $ax^3 + bx^2 + cx + d$.

55. Find

$$\Delta^{-1}\left[2^x x \frac{x!}{(2x+1)!}\right]$$

56. The *Bernoulli numbers* are defined as the coefficients, B_i, in the series expansion of

$$\frac{x}{e^x - 1} = \sum_{i=0}^{\infty} B_i \frac{x^i}{i!}$$

Find the first five Bernoulli numbers B_0, B_1, B_2, B_3, B_4. The coefficients in the Euler-Maclaurin formula are given by $\frac{B_i}{i!}$.

---

[5]The operator M appears in advanced graduation theory.

7

Approximate
differentiation

7.1. INTRODUCTION

APPROXIMATE METHODS of differentiation are occasionally required in practical work. For example, such methods are needed in estimating derivatives for data in which an underlying mathematical law either does not exist or cannot readily be determined.

Numerical analysis offers several approaches to approximate the required derivatives. The primary device used in this chapter is again the collocation polynomial in which the derivatives of the collocation polynomial serve as estimates of the derivatives of the approximated function.

Techniques of approximate differentiation have not proven to be completely satisfactory in practice. Approximate differentiation tends to be an unstable process which should be avoided, if possible. This is an exception to the analogy that we have drawn between differencing and differentiating. Broadly speaking, the problem is that although the collocation polynomial may intersect the given functional values at the points of collocation, the slopes of the two curves may differ considerably at any given point. More technically oriented reasons for the instability of approximate differentiation are discussed in Section 7.5.

Approximate differentiation formulas based on least-squares polynomials have proven to be superior to formulas based on collocation polynomials in certain cases. This alternative approach will be examined in Chapter 15.

7.2. RELATIONSHIPS BETWEEN DIFFERENCES AND DERIVATIVES

Relationships between differences and derivatives have been touched upon in Sections 5.5 and 6.6. These relationships can be formalized to produce some useful formulas for approximate differentiation.

Formulas (6.18) and (6.19), which are derived from the Taylor series given by formula (5.20), state the important relationship between the operators D and Δ, that is,

$$E \equiv 1 + \Delta \equiv e^D \tag{6.18}$$

and

$$D \equiv \log_e E \equiv \log_e (1 + \Delta) \tag{6.19}$$

These two formulas assume that the interval of differencing in Δ is equal to one. It is important to generalize these results to an interval of differencing h. Also, these results can be expressed as series expansions which are important in applying the formulas to problems. The more general results are

$$\underset{h}{E} \equiv 1 + \underset{h}{\Delta} \equiv e^{hD} \tag{7.1a}$$

$$\equiv 1 + (hD) + \frac{(hD)^2}{2!} + \frac{(hD)^3}{3!} + \cdots \tag{7.1b}$$

and

$$hD \equiv \log_e \underset{h}{E} \equiv \log_e \left(1 + \underset{h}{\Delta}\right) \tag{7.2a}$$

$$\equiv \underset{h}{\Delta} - \tfrac{1}{2}\underset{h}{\Delta}^2 + \tfrac{1}{3}\underset{h}{\Delta}^3 - \cdots \tag{7.2b[1]}$$

It is instructive to derive formula (7.2b) for the case $h = 1$ by an alternative procedure which is used in derivations of other approximate differentiation formulas. Using Newton's advancing difference formula, we have

$$u_x = u_0 + x\Delta u_0 + \frac{x(x-1)}{2!}\Delta^2 u_0 + \frac{x(x-1)(x-2)}{3!}\Delta^3 u_0 + \cdots$$

$$= u_0 + x\Delta u_0 + \frac{x^2 - x}{2}\Delta^2 u_0 + \frac{x^3 - 3x^2 + 2x}{6}\Delta^3 u_0 + \cdots$$

---

[1] From this point on the "h" is dropped from the difference operators in approximate differentiation identities. This simplifies the notation and should cause no confusion, since the interval of differencing is clear from the rest of the identity.

Now differentiating with respect to x gives

$$Du_x = \Delta u_0 + \frac{2x - 1}{2} \Delta^2 u_0 + \frac{3x^2 - 6x + 2}{6} \Delta^3 u_0 + \cdots$$

If $x = 0$, then

$$Du_0 = \Delta u_0 - \tfrac{1}{2}\Delta^2 u_0 + \tfrac{1}{3}\Delta^3 u_0 - \cdots$$

or in terms of operators

$$D \equiv \Delta - \tfrac{1}{2}\Delta^2 + \tfrac{1}{3}\Delta^3 - \cdots \tag{7.2c}$$

Higher-order derivatives can be obtained by multiplying the operator expressions together and grouping the appropriate cross products. To fourth differences the results are

$$hD \equiv \Delta - \tfrac{1}{2}\Delta^2 + \tfrac{1}{3}\Delta^3 - \tfrac{1}{4}\Delta^4 + \cdots \tag{7.2b}$$

$$h^2 D^2 \equiv \quad\quad \Delta^2 - \Delta^3 + \tfrac{11}{12}\Delta^4 - \cdots \tag{7.3a}$$

$$h^3 D^3 \equiv \quad\quad\quad\quad \Delta^3 - \tfrac{3}{2}\Delta^4 + \cdots \tag{7.3b}$$

$$h^4 D^4 \equiv \quad\quad\quad\quad\quad\quad \Delta^4 - \cdots \tag{7.3c}$$

The derivations of these results are left as Exercise 1.

It is also useful to develop the relationship between the operators D and δ. The result is

$$hD \equiv \delta - \tfrac{1}{24}\delta^3 + \tfrac{3}{640}\delta^5 - \cdots \tag{7.4}$$

This formula can be derived using Bessel's formula in a fashion quite similar to the derivation of formula (7.2c) using Newton's advancing difference formula. The details are left as Exercise 2.

Higher-order derivatives can again be obtained from formula (7.4). The results are

$$hD \equiv \delta \quad - \tfrac{1}{24}\delta^3 \quad\quad + \tfrac{3}{640}\delta^5 \quad\quad - \cdots \tag{7.4}$$

$$h^2 D^2 \equiv \delta^2 \quad - \tfrac{1}{12}\delta^4 \quad\quad + \tfrac{1}{90}\delta^6 \quad - \cdots \tag{7.5a}$$

$$h^3 D^3 \equiv \quad\quad \delta^3 \quad\quad - \tfrac{1}{8}\delta^5 \quad\quad + \cdots \tag{7.5b}$$

$$h^4 D^4 \equiv \quad\quad\quad\quad \delta^4 \quad\quad\quad - \tfrac{1}{6}\delta^6 + \cdots \tag{7.5c}$$

The derivations of these results are left as Exercise 3.

Finally, it is possible to develop the relationship between the operators D and ∇ using Newton's backward difference formula. The result is

$$hD \equiv \nabla + \tfrac{1}{2}\nabla^2 + \tfrac{1}{3}\nabla^3 + \tfrac{1}{4}\nabla^4 + \cdots \tag{7.6}$$

The derivation of formula (7.6) is left as Exercise 4. The higher-order derivatives are

$$hD \equiv \nabla + \tfrac{1}{2}\nabla^2 + \tfrac{1}{3}\nabla^3 + \tfrac{1}{4}\nabla^4 + \cdots \tag{7.6}$$

$$h^2 D^2 \equiv \quad\quad \nabla^2 + \nabla^3 + \tfrac{11}{12}\nabla^4 + \cdots \tag{7.7a}$$

$$h^3 D^3 \equiv \quad\quad\quad\quad \nabla^3 + \tfrac{3}{2}\nabla^4 + \cdots \tag{7.7b}$$

$$h^4 D^4 \equiv \quad\quad\quad\quad\quad\quad \nabla^4 + \cdots \tag{7.7c}$$

The derivations of these results are left as Exercise 5. It is interesting to compare the formulas involving ∇ with the formulas involving Δ. The formulas have the same form, except that all the terms are positive for the formulas involving ∇.

Example 7.1. **Estimate** $f''(4)$ **given that** $f(0) = 0$, $f(2) = 16$, $f(4) = 256$, $f(6) = 1296$, **and** $f(8) = 4096$.

The five given functional values agree with the polynomial $f(x) = x^4$. Thus, we hope our procedures will produce the answer $f''(4) = 4 \cdot 3 \cdot 4^2 = 192$. Setting up the difference table, we have the following:

x	$f(x)$	$\Delta f(x)$	$\Delta^2 f(x)$	$\Delta^3 f(x)$	$\Delta^4 f(x)$
0	0				384
		16		192	
2	16		224		384
		240		576	
4	256		800		384
		1040		960	
6	1296		1760		384
		2800		1344	
8	4096				384

If it is assumed that fourth differences can be extended as a constant 384, then some additional differences can be obtained. These appear above the top diagonal and below the bottom diagonal.

If formula (7.3a) is applied, we have

$$f''(4) = \frac{1}{h^2}[\Delta^2 - \Delta^3 + \tfrac{11}{12}\Delta^4]f(4)$$

$$= \tfrac{1}{4}[1760 - 1344 + \tfrac{11}{12}(384)]$$

$$= 192$$

the anticipated answer.

Alternatively, formula (7.5a) can be applied to give

$$f''(4) = \frac{1}{h^2}[\delta^2 - \tfrac{1}{12}\delta^4]f(4)$$

$$= \tfrac{1}{4}[800 - \tfrac{1}{12}(384)]$$

$$= 192$$

Finally, formula (7.7a) can be applied to give

$$f''(4) = \frac{1}{h^2}[\nabla^2 + \nabla^3 + \tfrac{11}{12}\nabla^4]f(4)$$

$$= \tfrac{1}{4}[224 + 192 + \tfrac{11}{12}(384)]$$

$$= 192$$

7.3. OTHER APPROXIMATE DIFFERENTIATION FORMULAS

The formulas given in Section 7.2 offer a variety of possible approximate differentiation formulas. However, there are some other formulas which are also useful in practice.

The set of formulas giving the relationship between the operators D and δ have the property that every other term vanishes and the coefficients decrease fairly rapidly in absolute value. Thus, the convergence of these formulas is generally superior to either of the other two sets of formulas, that is, those relating D and Δ or D and ∇.

However, there is a problem in using these central difference formulas to find the odd derivatives at the tabulated points of collocation, since the required differences do not appear in the difference table. Thus, there is no problem in finding $f''(4)$ in Example 7.1 using central differences with formula (7.5a), but formula (7.4) would not have worked for $f'(4)$.

An alternative to formula (7.4) can be derived using Stirling's formula instead of Bessel's formula, which gives

$$hD \equiv \mu\delta - \tfrac{1}{6}\mu\delta^3 + \tfrac{1}{30}\mu\delta^5 - \cdots \tag{7.8}$$

The derivation of formula (7.8) is similar to the derivations in Section 7.2 and is left as Exercise 13. The convergence of formula (7.8) is not as rapid as formula (7.4). Since third derivatives with formula (7.5b) would also be a problem at the points of collocation, formulas (7.8) and (7.5a) can be applied to yield the following alternative formula

$$h^3 D^3 \equiv \mu\delta^3 - \tfrac{1}{4}\mu\delta^5 + \cdots \tag{7.9}$$

The derivation is left as Exercise 14. It is interesting to note that formulas derived for even derivatives from Stirling's formula would present the same problem as the formulas for odd derivatives from Bessel's formula, that is, the required differences would not appear in the difference table.

The student is cautioned against trying to memorize too many of the above results. Formulas for first derivatives—that is, formulas (7.2b), (7.4), (7.6), and (7.8)—should be learned. However, formulas for higher-order derivatives can be derived as needed.

All of the above formulas are expressed in terms of differences. Any of these formulas can be expanded into the functional values upon which the differences are based. Two results of this type deserve special mention.

If one term of formula (7.8) is used, we have

$$hD \equiv \mu\delta$$
$$\equiv \tfrac{1}{2}[E^{1/2} + E^{-1/2}][E^{1/2} - E^{-1/2}]$$
$$\equiv \tfrac{1}{2}[E - E^{-1}] \tag{7.10a}$$

or in terms of functional values

$$Df(x) = \frac{f(x + h) - f(x - h)}{2h} \tag{7.10b}$$

This simple approximate differentiation formula is accurate for second-degree polynomials and is widely used in practice. In fact, it appears in most elementary calculus books.

Formula (7.10b) has an interesting geometrical interpretation as illustrated in Figure 7.1.

FIGURE 7.1

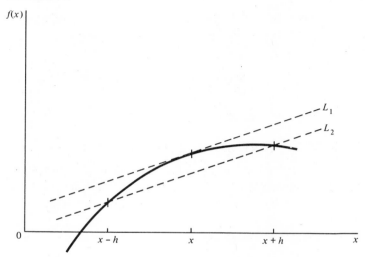

The left-hand side of formula (7.10b) gives the slope of L_1, the tangent line at $f(x)$. The right-hand side gives the slope of L_2, the secant line through $f(x + h)$ and $f(x - h)$. Geometrically, it would appear that the two slopes will often be approximately equal. In fact, by the *mean value theorem* there is a point $x + c$, where $-h < c < h$, such that the slopes are exactly equal. However, the value of c may not be exactly zero.

If two terms of formula (7.8) are used, we have

$$hD \equiv \mu\delta - \tfrac{1}{6}\mu\delta^3$$
$$\equiv \tfrac{1}{2}[E - E^{-1}] - \tfrac{1}{6} \cdot \tfrac{1}{2}[E - E^{-1}][E - 2 + E^{-1}]$$
$$\equiv \tfrac{1}{12}[8(E - E^{-1}) - (E^2 - E^{-2})] \tag{7.11a}$$

or in terms of functional values

$$Df(x) = \frac{8[f(x + h) - f(x - h)] - [f(x + 2h) - f(x - 2h)]}{12h} \tag{7.11b}$$

Formula (7.11b) constitutes a refinement of formula (7.10b) and is accurate for fourth-degree polynomials.

Example 7.2.   Estimate $f'(4)$ *using the data given in Example 7.1. Use formulas (7.8) and (7.11b)*.

We hope our procedures will produce the answer $f'(4) = 4 \cdot 4^3 = 256$.
Using the difference table given in Example 7.1, formula (7.8) produces

$$f'(4) = \frac{1}{h} \left[\mu\delta - \tfrac{1}{6}\mu\delta^3\right] f(4)$$
$$= \tfrac{1}{2}[\tfrac{1}{2}(1040 + 240) - \tfrac{1}{6} \cdot \tfrac{1}{2}(960 + 576)]$$
$$= 256$$

the anticipated answer.

Formula (7.11b) works directly with functional values and gives

$$f'(4) = \frac{8[f(6) - f(2)] - [f(8) - f(0)]}{12 \cdot 2}$$
$$= \frac{8(1296 - 16) - (4096 - 0)}{24}$$
$$= 256$$

7.4.   UNEQUAL INTERVALS OF DIFFERENCING

The formulas derived in the preceding sections assume functional values are tabulated at equal intervals. It is possible to develop approximate differentiation formulas which can be applied when functional values are tabulated at unequal intervals.

There are three common approaches used in practice.

The first approach is to use divided differences. A divided difference formula for $f(x)$ is written down using Sheppard's rules and then differentiated. This method is popular in finding values of derivatives at specific points.

The second approach is to use Lagrange's formula which directly expresses $f(x)$ as a linear combination of functional values. The formula is then differentiated and the answer results. The only problem with this method is that the coefficients become unwieldy if very many terms are used. This method is popular in deriving general formulas for derivatives.

The third approach is called the *method of undetermined coefficients* and is also popular in deriving general formulas for derivatives. Denote the $n + 1$ points of collocation by $x_0, x_1, x_2, \ldots, x_n$. The answer is of the form

$$Df(x) = \sum_{i=0}^{n} H_i f(x_i) \tag{7.12}$$

which is a linear combination of functional values. Thus, the problem reduces to finding the $n + 1$ coefficients $H_0, H_1, H_2, \ldots, H_n$. The simplest method of determining these $n + 1$ unknowns is to apply the formula to the $n + 1$ independent functions $f(x) = 1, x, x^2, \ldots, x^{n-1}, x^n$, respectively. Thus, the equations become

$$
\left.
\begin{aligned}
D1 &= 0 &&= \sum_{i=0}^{n} H_i \\[6pt]
Dx &= 1 &&= \sum_{i=0}^{n} H_i x_i \\[6pt]
Dx^2 &= 2x &&= \sum_{i=0}^{n} H_i x_i^2 \\[6pt]
&\;\;\vdots && \\[2pt]
Dx^n &= nx^{n-1} &&= \sum_{i=0}^{n} H_i x_i^n
\end{aligned}
\right\} \tag{7.13}
$$

For any given set of points of collocation $x_0, x_1, x_2, \ldots, x_n$ and for any given x, the set of formulas (7.13) consists of $n + 1$ linear equations in the $n + 1$ unknown H_i's.

One interesting observation from the first equation in (7.13) is that the sum of the coefficients is equal to zero. This provides a quick check on any approximate differentiation formula expressed strictly in terms of functional values. The student should note that this property holds for both formulas (7.10b) and (7.11b).

Example 7.3. *Use divided differences to find* Du_3 *if* $u_1 = 7$, $u_4 = 16$, *and* $u_8 = 112$.

Setting up a divided difference table, we have the following:

$\dot{x}$	u_x	$\triangle u_x$	$\triangle^2 u_x$
1	7		
		3	
4	16		3
		24	
8	112		

Now $u_x = 7 + 3(x-1) + 3(x-1)(x-4) = 3x^2 - 12x + 16$, and $Du_x = 6x - 12$, so that $Du_3 = 18 - 12 = 6$.

Example 7.4. Develop an approximate differentiation formula expressing $f'(1)$ in terms of $f(0)$, $f(1)$, and $f(3)$. Use both Lagrange's formula and the method of undetermined coefficients.

Using Lagrange's formula, we have

$$f(x) = \frac{(x-1)(x-3)}{(0-1)(0-3)}f(0) + \frac{(x-0)(x-3)}{(1-0)(1-3)}f(1) + \frac{(x-0)(x-1)}{(3-0)(3-1)}f(3)$$

$$= \left(\tfrac{1}{3}x^2 - \tfrac{4}{3}x + 1\right)f(0) + \left(-\tfrac{1}{2}x^2 + \tfrac{3}{2}x\right)f(1) + \left(\tfrac{1}{6}x^2 - \tfrac{1}{6}x\right)f(3)$$

so that

$$f'(x) = \left(\tfrac{2}{3}x - \tfrac{4}{3}\right)f(0) + \left(-x + \tfrac{3}{2}\right)f(1) + \left(\tfrac{1}{3}x - \tfrac{1}{6}\right)f(3)$$

and

$$f'(1) = -\tfrac{2}{3}f(0) + \tfrac{1}{2}f(1) + \tfrac{1}{6}f(3)$$

Note that the sum of the coefficients is equal to zero.

Using the method of undetermined coefficients, we have[2]

$$f'(1) = H_0 f(0) + H_1 f(1) + H_3 f(3)$$

and

$$
\begin{aligned}
f(x) = 1: &\quad f'(1) = 0 = H_0 + H_1 + H_3\\
f(x) = x: &\quad f'(1) = 1 = H_1 + 3H_3\\
f(x) = x^2: &\quad f'(1) = 2 = H_1 + 9H_3
\end{aligned}
$$

Solving these three equations in three unknowns gives $H_0 = -\tfrac{2}{3}$, $H_1 = \tfrac{1}{2}$, and $H_3 = \tfrac{1}{6}$, which agrees with the result from Lagrange's formula.

---

[2]Actually this formulation of the problem does not agree with formula (7.12) in the labels on the H_i's. If formula (7.12) is changed to $Df(x) = \sum_{i=0}^{n} H_{xi} f(x_i)$ and $x_0 = 0$, $x_1 = 1$, and $x_2 = 3$, then the formulation would be completely consistent. However, the above approach avoids this unnecessary transformation and is simpler to use.

7.5. SUBDIVISION OF INTERVALS

In many applications of numerical analysis subdivision of intervals is a useful procedure to improve the accuracy of some approximating technique. However, subdivision of intervals in approximate differentiation does not always improve the approximation and may actually worsen it in some cases.

This is illustrative of the inherent instability of approximate differentiation. Basically there is a trade-off between truncation error and roundoff error. As the interval of differencing is decreased, truncation error is decreased while roundoff error is increased, and conversely.

Thus, if the interval of differencing is too large, the approximating functional values are too far removed from the value for which the derivative is being approximated to be reliable estimates of the slope at that point. For example, in Figure 7.1 as h increases, $f(x - h)$ and $f(x + h)$ move further and further away from $f(x)$, so that the slope of the secant line through these two points generally becomes a poorer estimate of the slope of the tangent line at $f(x)$.

On the other hand, if the interval of differencing is too small, roundoff error may become severe. For example, in formula (7.10b) as h is decreased the numerator becomes the subtraction of two nearly equal numbers. As discussed in Chapter 1, the subtraction of two nearly equal numbers often results in the loss of significant digits. Furthermore, the denominators of all the approximate differentiation formulas developed contain h. As h is decreased, the shrinking denominator magnifies all the other algorithm errors present.

In many situations as the interval of differencing is decreased, results are temporarily improved, reach their optimum value, and then worsen again. In some cases it is possible to reach a crude estimate of this optimum interval of differencing. To do this we find a value of h which minimizes the sum of the absolute values of maximum truncation error and maximum roundoff error.

Expressions for maximum truncation error in approximate differentiation can be obtained from formula (1.12)

$$f(x) - p(x) = \frac{f^{(n+1)}(\xi)\pi(x)}{(n + 1)!} \tag{1.12}$$

where $\pi(x) = (x - x_0)(x - x_1) \cdots (x - x_n)$ and ξ is some value of x in the range of collocation.

If both sides are differentiated, we have

$$f'(x) - p'(x) = \frac{f^{(n+1)}(\xi)\pi'(x)}{(n + 1)!} \tag{7.14^3}$$

----

[3] This step assumes that ξ is independent of x, which is not strictly correct. A more thorough treatment of the truncation error in approximate differentiation is contained in Appendix E.

Now assume that we evaluate this expression at x_i, one of the $n + 1$ points of collocation $x_0, x_1, \ldots, x_n$. Then from Section 5.2 we know that $\pi'(x_i) = F_i(x_i)$, where $F_i(x_i) = (x_i - x_0) \cdots (x_i - x_{i-1})(x_i - x_{i+1}) \cdots (x_i - x_n)$, so that formula (7.14) becomes

$$f'(x_i) - p'(x_i) = \frac{f^{(n+1)}(\xi)F_i(x_i)}{(n+1)!} \tag{7.15a}$$

A value of ξ can be chosen to maximize $f^{(n+1)}(\xi)$ over the range of collocation. The maximum truncation error is then given by

$$\text{maximum truncation error} = \left| \frac{f^{(n+1)}(\xi)F_i(x_i)}{(n+1)!} \right| \tag{7.15b}$$

For example, to determine the maximum truncation error in formula (7.10b), we have three points of collocation $x - h, x, x + h$ with the derivative being sought at x. Thus, $n = 2$ and $F_i(x_i) = [x - (x - h)][x - (x + h)] = -h^2$, so that formula (7.15b) gives

$$\text{maximum truncation error} = \frac{h^2}{6} \left| f^{(3)}(\xi) \right| \tag{7.16}$$

In a similar fashion the maximum truncation error of formula (7.11b) is given by

$$\text{maximum truncation error} = \frac{h^4}{30} \left| f^{(5)}(\xi) \right| \tag{7.17}$$

The derivation of formula (7.17) is similar to formula (7.16) and is left as Exercise 24.

Maximum truncation errors for other approximate differentiation formulas can be found in a similar fashion from formula (7.15b).

It should be noted that it is usually quite difficult in practice to estimate $f^{(n+1)}(\xi)$ in formula (7.15b). To do this we would have to be able to find higher-order derivatives than we are approximating, an unlikely possibility! However, in some cases it may be possible to find upper bounds for these higher-order derivatives independently of the approximate differentiation process.

Another approach is often used, if it is impossible to get upper bounds on these higher-order derivatives. This approach uses the first omitted term in the finite difference expansion as an estimate of truncation error. Since this term is based strictly on functional values, it is readily available.

For example, formula (7.10a) is based on one term of formula (7.8). The first omitted term is $-\dfrac{1}{6h}\mu\delta^3$. We know from formula (7.9) that $\mu\delta^3 \equiv h^3 D^3$ to third differences, so that we have

$$\text{truncation error estimate} = \frac{h^2}{6}\left|f^{(3)}(x)\right| \qquad (7.18)$$

The similarity of formula (7.18) to formula (7.16) is strikingly apparent.

Similarly, formula (7.11a) is based on two terms of formula (7.8). The first omitted term is $\dfrac{1}{30h}\mu\delta^5$. We know that $\mu\delta^5 \equiv h^5 D^5$ to fifth differences, so that we have

$$\text{truncation error estimate} = \frac{h^4}{30}\left|f^{(5)}(x)\right| \qquad (7.19)$$

Formula (7.19) is seen to be similar to formula (7.17).

This alternative approach is quite simple to apply and can be used for any of the approximate differentiation formulas developed in this chapter.

Expressions for maximum roundoff error can be derived by expressing the approximate differentiation formula in terms of functional values and using formula (1.5) for maximum roundoff error.[4]

If each functional value has a maximum roundoff error of E,[5] for $E > 0$, then the maximum roundoff error in formula (7.10b) is given by

$$\text{maximum roundoff error} = E\,\frac{1+1}{2h} = \frac{E}{h} \qquad (7.20)$$

and the maximum roundoff error in formula (7.11b) is given by

$$\text{maximum roundoff error} = E\,\frac{8+8+1+1}{12h} = \frac{3E}{2h} \qquad (7.21)$$

The optimum interval of differencing for formula (7.10b) can be found by minimizing

$$E(h) = \frac{h^2}{6}\left|f^{(3)}(\xi)\right| + \frac{E}{h}$$

Now

$$E'(h) = \frac{h}{3}\left|f^{(3)}(\xi)\right| - \frac{E}{h^2} = 0$$

---

[4]An alternative would be to use estimated roundoff error given by formula (1.8) instead of maximum roundoff error given by formula (1.5).

[5]The student should be careful not to confuse this error E with the operator E.

or

$$h = \left| \frac{3E}{f^{(3)}(\xi)} \right|^{1/3} \tag{7.22}$$

Similarly, the optimum interval of differencing for formula (7.11b) can be found by minimizing

$$E(h) = \frac{h^4}{30} \left| f^{(5)}(\xi) \right| + \frac{3E}{2h}$$

Now

$$E'(h) = \frac{2h^3}{15} \left| f^{(5)}(\xi) \right| - \frac{3E}{2h^2} = 0$$

or

$$h = \left| \frac{45E}{4 f^{(5)}(\xi)} \right|^{1/5} \tag{7.23}$$

Example 7.5.   *Compute $D^2 f(1)$ to three decimal places for $f(x) = \sqrt{x}$, where approximate differentiation is used based on three functional values evaluated to five decimal places which are symmetrically spaced about $f(1)$. Formula (7.5a) is to be used at three intervals of differencing: $h = .20$, $.10$, and $.05$.*

The exact value is $D^2 f(x) = \left(\frac{1}{2}\right)\left(-\frac{1}{2}\right)x^{-3/2}$, which evaluated at $x = 1$ gives $D^2 f(1) = -.25$. The approximate differentiation formula is

$$D^2 \equiv \frac{\delta^2}{h^2}$$

The three difference tables are:

	$h = .20$				$h = .10$		
x	$\sqrt{x}$	$\delta\sqrt{x}$	$\delta^2\sqrt{x}$	x	$\sqrt{x}$	$\delta\sqrt{x}$	$\delta^2\sqrt{x}$
.80	.89443			.90	.94868		
		.10557				.05132	
1.00	1.00000		−.01012	1.00	1.00000		−.00251
		.09545				.04881	
1.20	1.09545			1.10	1.04881		

	$h = .05$		
x	$\sqrt{x}$	$\delta\sqrt{x}$	$\delta^2\sqrt{x}$
.95	.97468		
		.02532	
1.00	1.00000		−.00062
		.02470	
1.05	1.02470		

The respective answers are

$$\frac{-.01012}{(.2)^2} = -.253 \qquad \frac{-.00251}{(.1)^2} = -.251 \qquad \frac{-.00062}{(.05)^2} = -.248$$

and the respective absolute errors are .003, .001, and .002. This example demonstrates that up to a certain point reducing the interval of differencing improves accuracy, but thereafter worsens it. Apparently, .10 is to be preferred over either .20 or .05 as the interval of differencing. It should also be noted that none of the intervals of differencing can produce even three-decimal accuracy despite the carrying of five decimal places and the tightly packed functional values.

Example 7.6.   **Estimate the optimum interval of differencing for Example 7.5.**

The approximate differentiation formula expressed in terms of functional values is

$$D^2f(1) = \frac{f(1 + h) - 2f(1) + f(1 - h)}{h^2}$$

The truncation error estimate is developed as the first omitted term in formula (7.5a), that is,

$$-\frac{1}{12h^2}\,\delta^4 f(x)$$

Now $\delta^4 \equiv h^4 D^4$ to fourth differences, so that the truncation error estimate becomes

$$\frac{h^2}{12}\left|f^{(4)}(x)\right|$$

Since $f(x) = x^{1/2}$, $f^{(4)}(x) = \left(\frac{1}{2}\right)\left(-\frac{1}{2}\right)\left(-\frac{3}{2}\right)\left(-\frac{5}{2}\right)x^{-7/2}$ which equals approximately $\left(\frac{1}{2}\right)\left(-\frac{1}{2}\right)\left(-\frac{3}{2}\right)\left(-\frac{5}{2}\right) = -1.875$ in the vicinity of $x = 1.$[6] Thus, maximum truncation error is approximately equal to

$$\frac{h^2}{12}(1.875) = .15625h^2$$

Maximum roundoff error with five decimal places being carried is

$$\frac{1 + 2 + 1}{h^2}(.000005) = .00002h^{-2}$$

---

[6]Complete precision would require using a value of ξ to maximize this expression instead of using $x = 1$. However, at this stage in the problem, this refinement is often not made. The effect on the already highly approximate answer would be slight.

Total maximum error is

$$E(h) = .15625h^2 + .00002h^{-2}$$

We seek to minimize $E(h)$. Differentiating, we have

$$E'(h) = .3125h - .00004h^{-3} = 0$$

so that

$$h = \left(\frac{.00004}{.3125}\right)^{1/4} = .11$$

to two decimal places. This answer appears to confirm the results of Example 7.5, since $h = .10$ produces better results than either $h = .05$ or $h = .20$.

7.6. FUNCTIONS OF MORE THAN ONE VARIABLE

It is possible to generalize the results of this chapter to be applicable to partial derivatives. The manipulations necessary are quite similar to those required for interpolation with functions of more than one variable as discussed in Section 3.7.

For example, with two variables we have

$$\frac{\partial^2}{\partial x\, \partial y} f(x, y) = D_x D_y f(x, y)$$

$$= \frac{1}{h_x}(\Delta_x - \tfrac{1}{2}\Delta_x^2 + \cdots)\frac{1}{h_y}(\Delta_y - \tfrac{1}{2}\Delta_y^2 + \cdots)f(x, y) \quad (7.24)$$

In formula (7.24) all appropriate cross products would be taken. Other partial derivatives can be worked out analogously.

Example 7.7.   Using the data in Example 3.9, estimate $\dfrac{\partial^2}{\partial x\, \partial y} f(1, 2)$.

The given functional values are as follows:

x / y	1	3	5
2	2	18	50
5	5	45	125

These functional values agree with $f(x, y) = x^2 y$. Since

$$\frac{\partial^2}{\partial x\, \partial y} x^2 y = 2x,$$

we hope our procedures will produce the answer $2 \cdot 1 = 2$.

Applying formula (7.24),

$$\frac{\partial^2}{\partial x\, \partial y} f(1, 2) = \tfrac{1}{2}(\Delta_x - \tfrac{1}{2}\Delta_x^2)\tfrac{1}{3}(\Delta_y)f(1, 2)$$

$$= \left(\tfrac{1}{6}\Delta_x \Delta_y - \tfrac{1}{12}\Delta_x^2 \Delta_y\right)f(1, 2)$$

$$= \tfrac{1}{6}(24) - \tfrac{1}{12}(24)$$

$$= 2$$

the anticipated answer. The required differences are evaluated in Example 3.9.

EXERCISES

7.1. Introduction; 7.2. Relationships between differences and derivatives

1. Derive formulas (7.3a), (7.3b), and (7.3c) from (7.2b).
2. Derive formula (7.4) for $h = 1$ using Bessel's formula (see Exercise 19 in Chapter 4).
3. Derive formulas (7.5a), (7.5b), and (7.5c) from (7.4).
4. Derive formula (7.6) for $h = 1$ using Newton's backward difference formula.
5. Derive formulas (7.7a), (7.7b), and (7.7c) from (7.6).
6. Find $e^{2D}(x^2)$.
7. If $u_0 = 7$, $u_1 = 9$, $u_2 = 14$, $u_3 = 24$, and $u_4 = 41$, find $D^2 u_0$.
8. If $f(8) = -3$, $f(10) = -1$, $f(12) = 2$, $f(14) = 4$, and $f(16) = 5$, find $Df(8)$.
9. At a certain point on a third-degree polynomial the values of the first three derivatives are equal to 1, 2, and 3, respectively. Find the value of the first advancing difference at this same point.
10. Find $\Delta^2 f(0)$ given that $f(0) = 3$, $Df(0) = 5$, $D^2 f(0) = -1$, $D^3 f(0) = 7$, and $D^4 f(0) = 12$.
11. If third differences are constant, $Df(3)$ can be expressed as

$$[a\Delta + b\Delta^2 + c\Delta^3]f(2)$$

Find a, b, and c.
12. Express ∇ as a series expansion in D.

7.3. Other approximate differentiation formulas

13. Derive formula (7.8) for $h = 1$ using Stirling's formula.
14. Derive formula (7.9) from formulas (7.8) and (7.5a).
15. Derive a formula to approximate first derivatives, which is correct to third differences, using the Gauss forward formula.

16. Derive a formula to approximate first derivatives, which is correct to third differences, using the Gauss backward formula.

17. If $u_{-1} = 3$, $u_0 = 5$, $u_1 = 9$, $u_2 = 12$, and $u_3 = 11$, find Du_1 by—
    (a) Using formula (7.8).
    (b) Using formula (7.11b).

18. Using the data given in Exercise 17, find $Du_{1/2}$. What formula is most convenient for this problem?

7.4. Unequal intervals of differencing

19. Use the method of undetermined coefficients to develop an approximate differentiation formula which expresses $f'(0)$ in terms of $f(-1)$, $f(0)$, and $f(1)$. Show that this formula is equivalent to formula (7.10b).

20. If $u_0 = 14$, $u_1 = 15$, and $u_4 = 30$, find Du_2.

21. If $f(0) = 10$, $f(2) = -126$, $f(4) = -46$, and $f(5) = 135$, estimate the value of x for which $f(x)$ has a relative minimum.

22. If u_x is a fourth-degree polynomial and $\triangle^4_{-1,1,3,4} u_{-2} = \frac{1}{2}$, find $D^4 u_0$.

23. Develop an approximate differentiation formula expressing $f'(2)$ in terms of $f(0)$, $f(2)$, and $f(5)$.

7.5. Subdivision of intervals

24. Derive formula (7.17).

25. Use formula (7.10b) to approximate the first derivative of e^{-x} for values of x in the interval $1 \leq x \leq 2$. If values accurate to four decimal places are used in the computations, find the optimum interval of differencing (to the nearest .005).

26. The following tables of values of $f(x) = (1 + x^2)^{-1}$ are given:

TABLE 7.1		TABLE 7.2		TABLE 7.3	
x	$f(x)$	x	$f(x)$	x	$f(x)$
.00	1.00000	.30	.91743	.46	.82535
.25	.94118	.40	.86207	.48	.81274
.50	.80000	.50	.80000	.50	.80000
.75	.64000	.60	.73529	.52	.78715
1.00	.50000	.70	.67114	.54	.77423

Use formula (7.5a) to approximate $f''(.5)$. Carry the answers to four decimal places.
(a) Find the exact value of $f''(.5)$.
(b) Find the approximate value of $f''(.5)$ using data in Table 7.1.
(c) Find the approximate value of $f''(.5)$ using data in Table 7.2.
(d) Find the approximate value of $f''(.5)$ using data in Table 7.3.

7.6. Functions of more than one variable

27. Two sets of data are given below. Choose the appropriate one and estimate $\dfrac{\partial}{\partial y} f(x, y)$ at the point $(1, 1)$. Assume third differences in each set are zero.

x	$f(x, 1)$		y	$f(1, y)$
1.00	35		1.00	35
1.25	37		1.50	41
1.50	41		2.00	45

28. Derive a formula accurate to fourth differences for $\dfrac{\partial^2}{\partial x\, \partial y} f(x, y)$ in terms of backward difference operators.

29. Express $\dfrac{\partial^2}{\partial x\, \partial y} u_{1:1}$ in terms of the nine functional values in the following array:

$$
\begin{array}{ccc}
u_{0:0} & u_{0:1} & u_{0:2} \\
u_{1:0} & u_{1:1} & u_{1:2} \\
u_{2:0} & u_{2:1} & u_{2:2}
\end{array}
$$

Miscellaneous problems

30. What does $Du_{x+\frac{1}{4}}$ equal in terms of advancing differences of u_x, if fourth and higher differences are ignored?

31. Find $\Delta^2 u_0$ if $u_0 = 5$, $Du_0 = 16$, and $Du_1 = 20$.

32. If $f(x)$ is a second-degree polynomial and $Df(3) = a\Delta f(4) + b\Delta^2 f(5)$, find a and b.

33. Show that
$$
Du_x = \frac{1}{h}\left[(u_{x+h} - u_{x-h}) - \tfrac{1}{2}(u_{x+2h} - u_{x-2h}) + \tfrac{1}{3}(u_{x+3h} - u_{x-3h}) - \cdots\right]
$$

34. If $u_0 = 4$, $u_1 = 10$, $Du_0 = 3$, and $Du_2 = 23$, find $\Delta^2 u_1$.

35. (a) The hyperbolic sine of x is defined by
$$
\sinh x = \frac{e^x - e^{-x}}{2}
$$
Show that $\delta = 2 \sinh \tfrac{1}{2}D$.
    (b) Express δ as a series expansion in D.

36. (a) The hyperbolic cosine of x is defined by
$$
\cosh x = \frac{e^x + e^{-x}}{2}
$$
Show that $\mu = \cosh \tfrac{1}{2}D$.
    (b) Express μ as a series expansion in D.

8

Approximate integration

8.1. INTRODUCTION

APPROXIMATE METHODS of integration are often required in practical work. As the student recalls from calculus, many apparently simple functions do not possess antiderivatives. Classical mathematics can be rather capricious in this regard. For example, in Section 1.1 it is pointed out that e^{x^2} cannot be integrated analytically, whereas functions such as e^x, xe^x, and xe^{x^2}, which are similar in form to e^{x^2}, can be.

Approximate integration is a technique of finding numerical values for definite integrals for functions such as e^{x^2} which cannot be integrated analytically, or for functions such as $x^{27} \sin^{42} x$ for which the analytical solution is quite cumbersome. Furthermore, approximate integration can also be applied to a broad range of problems in which only tabular data are available and in which no underlying mathematical law exists.

One approach in approximating $\int_a^b f(x)\, dx$ can be developed from the definition of the definite integral, which is based on subdividing the interval $a \le x \le b$ into n subintervals of equal length.[1] Let the n subintervals be

---

[1] The subintervals do not have to be of equal length as long as they all tend toward zero. However, the development here is expedited by assuming them equal. The more general case is considered in standard textbooks in calculus.

determined by the $n + 1$ points $x_0, x_1, x_2, \ldots, x_{n-1}, x_n$, such that $x_0 = a$, $x_n = b$, and $x_1 - x_0 = x_2 - x_1 = \cdots = x_n - x_{n-1} = \Delta x_i$. Then the definite integral is defined by

$$\int_a^b f(x) \, dx = \lim_{\Delta x_i \to 0} \sum_{i=0}^{n-1} f(x_i) \Delta x_i \tag{8.1}$$

The student should note that as $\Delta x_i \to 0$, $n \to \infty$, that is, the number of points at which $f(x)$ is tabulated increases without bound.

Formula (8.1) forms the basis for an elementary method of approximate integration known as the *rectangular rule*. In this method a value of n is chosen and the summation is performed. For example, one application of the rectangular rule over the entire interval is given by

$$\int_a^b f(x) \, dx = (b - a)f(a) \tag{8.2}$$

If the interval is subdivided into n subintervals, then the rectangular rule gives

$$\int_a^b f(x) \, dx = \frac{b - a}{n} \sum_{i=0}^{n-1} f(x_i) \tag{8.3}$$

since

$$\Delta x_i = \frac{b - a}{n}.$$

Figure 8.1 is a geometrical interpretation of the rectangular rule. In Figure 8.1 the area under the series of rectangles is used as an approximation for the area under $f(x)$.

FIGURE 8.1

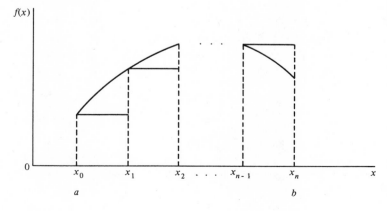

The rectangular rule as given by formulas (8.2) and (8.3) and illustrated in Figure 8.1 is based on the left-hand end point in each subinterval. If right-hand end points are used instead, formulas (8.2) and (8.3) become[2]

$$\int_a^b f(x)\, dx = (b-a)f(b) \tag{8.4}$$

and

$$\int_a^b f(x)\, dx = \frac{b-a}{n} \sum_{i=1}^n f(x_i) \tag{8.5}$$

Figure 8.2 illustrates this approximation.

FIGURE 8.2

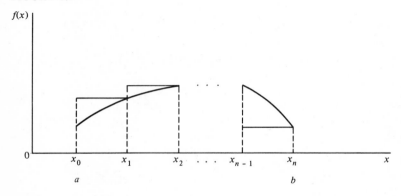

It would appear from Figures 8.1 and 8.2 that improved approximations might result if the midpoint in each subinterval is used instead of either end point. This is often the case, and an approximate integration formula based on this approach is called the *midpoint rule*. Formulas (8.2) and (8.3) for the midpoint rule become

$$\int_a^b f(x)\, dx = (b-a)f\left(\frac{a+b}{2}\right) \tag{8.6}$$

and

$$\int_a^b f(x)\, dx = \frac{b-a}{n} \sum_{i=0}^{n-1} f\left(\frac{x_i + x_{i+1}}{2}\right) \tag{8.7}$$

Figure 8.3 illustrates the midpoint rule.

---

[2]Formulas (8.3) and (8.5) are mixtures of *upper* and *lower sums* as defined in calculus and are identical to neither type of sum.

FIGURE 8.3

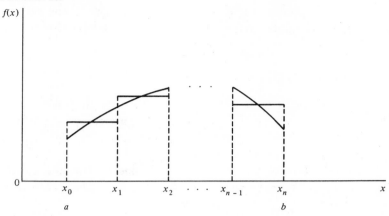

Although the rectangular and midpoint rules are directly obtained from the definition of the definite integral, they are not widely used in practice. The most commonly used approximate integration formulas to be developed in subsequent sections of this chapter are based on the collocation polynomial. Thus, the integral of the collocation polynomial serves as an estimate of the integral of the function being approximated.

Approximate integration has proven to be a very successful technique in practice. In general, it does not exhibit the instability of approximate differentiation. This can be interpreted geometrically by noting that the area under the collocation polynomial is usually a good approximation to the area under the given function, although the slopes of the two curves may differ considerably at any given point.

Example 8.1.   ***Evaluate*** $\displaystyle\int_0^1 e^{x^2}\, dx$ ***to five decimal places using one application of***—

   (1)   *The rectangular rule.*
   (2)   *The midpoint rule.*

(1)   Using the left-hand end point, formula (8.2) gives

$$e^0 = 1.00000$$

Alternatively, using the right-hand end point, formula (8.4) gives

$$e^1 = 2.71828$$

(2)   Using the midpoint rule, formula (8.6) gives

$$e^{1/4} = 1.28403$$

In subsequent sections of this chapter, which introduce algorithms for obtaining better approximations, we shall see just how crude the results produced by the rectangular rule and the midpoint rule really are.

8.2. NEWTON-COTES FORMULAS

Newton-Cotes formulas are a family of approximate integration formulas which are widely used in practice. These formulas are based on $n + 1$ values of x, for $n = 1, 2, 3, \ldots$, tabulated at equal intervals, such that the range of integration is equal to the range of collocation.

Derivations of Newton-Cotes formulas are based on integration of Newton's advancing difference formula. For now, we assume that values of x are tabulated at integral points $0, 1, 2, \ldots, n$.

The lowest-order Newton-Cotes formula is based on two points ($n = 1$) and is derived by carrying Newton's advancing difference formula to two terms

$$
\begin{aligned}
\int_0^1 f(x)\, dx &= \int_0^1 [f(0) + x \Delta f(0)]\, dx \\
&= f(0) + \tfrac{1}{2} \Delta f(0) \\
&= \tfrac{1}{2}[f(0) + f(1)]
\end{aligned}
\tag{8.8}
$$

Formula (8.8) is commonly called the *trapezoidal rule*. The name reflects the fact that the area under $f(x)$ from $x = 0$ to 1 is approximated by a trapezoid rather than a rectangle, as in Section 8.1. Figure 8.4 is similar to Figures 8.1, 8.2, 8.3 and illustrates this approach, assuming that the trapezoidal rule is applied over each subinterval separately.

FIGURE 8.4

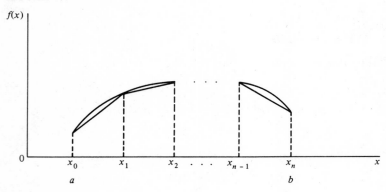

The Newton-Cotes formula based on three points ($n = 2$) is called *Simpson's rule*[3] and is derived by carrying Newton's advancing difference formula to three terms

$$\int_0^2 f(x)\,dx = \int_0^2 \left[f(0) + x\Delta f(0) + \binom{x}{2}\Delta^2 f(0) \right] dx$$

$$= 2f(0) + 2\Delta f(0) + \tfrac{1}{3}\Delta^2 f(0)$$

$$= \tfrac{1}{3}[f(0) + 4f(1) + f(2)] \tag{8.9}$$

The Newton-Cotes formula based on four points ($n = 3$) is called the *three-eighths rule* and is derived by carrying Newton's advancing difference formula to four terms

$$\int_0^3 f(x)\,dx = \int_0^3 \left[f(0) + x\Delta f(0) + \binom{x}{2}\Delta^2 f(0) + \binom{x}{3}\Delta^3 f(0) \right] dx$$

$$= 3f(0) + \tfrac{9}{2}\Delta f(0) + \tfrac{9}{4}\Delta^2 f(0) + \tfrac{3}{8}\Delta^3 f(0)$$

$$= \tfrac{3}{8}[f(0) + 3f(1) + 3f(2) + f(3)] \tag{8.10}$$

It is possible to continue in this fashion and generate Newton-Cotes formulas for $n > 3$. However, these higher-order Newton-Cotes formulas are not widely used in practice. Alternative approaches of increasing accuracy rather than using higher-degree collocation polynomials are generally more efficient and are discussed later in this chapter.

The above versions of the first three Newton-Cotes formulas assume that the interval of differencing $h = 1$ and that the range of integration is 0 to n. It is necessary to remove these restrictions, so that Newton-Cotes formulas can be applied to the general problem of evaluating $\int_a^b f(x)\,dx$ for any real numbers a and b such that $a < b$. There are two tools for this purpose: (1) *change in origin and scale*, and (2) *composite formulas*.

A *change in origin* involves a linear shift along the x-axis. For example, it may be required to apply the three-eighths rule to $\int_3^6 f(x)\,dx$. Then the change of variable $f(x) = g(y)$ in which $x = y + 3$ can be made, which gives

$$\int_3^6 f(x)\,dx = \int_0^3 g(y)\,dy$$

$$= \tfrac{3}{8}[g(0) + 3g(1) + 3g(2) + g(3)]$$

$$= \tfrac{3}{8}[f(3) + 3f(4) + 3f(5) + f(6)]$$

The latter formula is clearly the result which would be expected.

---

[3]Simpson's rule is sometimes referred to as the *one-third rule*.

A *change in scale* involves a stretching or shrinking of the x-axis to reflect a general interval of differencing $h \neq 1$. For example, it may be required to apply Simpson's rule for $\int_0^{2h} f(x)\,dx$. Then the change of variable $f(x) = g(y)$ in which $x = hy$ can be made. Now $dx = h\,dy$, so that we have

$$\int_0^{2h} f(x)\,dx = h \int_0^2 g(y)\,dy$$

$$= \frac{h}{3}[g(0) + 4g(1) + g(2)]$$

$$= \frac{h}{3}[f(0) + 4f(h) + f(2h)]$$

The latter formula is seen to be based on formula (8.9) multiplied by the interval of differencing h.

A *composite formula* is an extended version of any of the above formulas applied more than once to successive intervals.[4] For example, the trapezoidal rule applied first to the interval $0 \leq x \leq 1$, then to the interval $1 \leq x \leq 2$, and so forth until the interval $n - 1 \leq x \leq n$, becomes

$$\int_0^n f(x)\,dx = \tfrac{1}{2}[f(0) + 2f(1) + 2f(2) + \cdots + 2f(n-1) + f(n)] \quad (8.11)$$

Similarly, Simpson's rule applied over n intervals, each of length two becomes

$$\int_0^{2n} f(x)\,dx = \tfrac{1}{3}[f(0) + 4f(1) + 2f(2) + 4f(3) + \cdots + 4f(2n-1) + f(2n)]$$

$$(8.12)$$

Finally, the three-eighths rule applied over n intervals, each of length three, becomes

$$\int_0^{3n} f(x)\,dx = \tfrac{3}{8}[f(0) + 3f(1) + 3f(2)$$

$$+ 2f(3) + 3f(4) + \cdots + 3f(3n-1) + f(3n)] \quad (8.13)$$

An approximate integration formula can be tested to find the highest degree polynomial which it reproduces exactly. If a formula reproduces $\int_a^b f(x)\,dx$ for $f(x) = 1, x, x^2, \ldots, x^n$ exactly and fails for $f(x) = x^{n+1}$, then the formula is exact for polynomials of degree n.

---

[4]The term *piecewise-polynomial* is sometimes used in describing composite formulas, since the collocation polynomial assumed in the approximate integration formula changes from interval to interval. See Section 4.8 for a further discussion of piecewise-polynomials.

If the trapezoidal rule is tested by this technique, it is found to reproduce first-degree polynomials. Similarly, the three-eighths rule is found to reproduce third-degree polynomials. These results are consistent with formulas (8.8) and (8.10), since these formulas are derived by carrying Newton's advancing difference formula to first and third differences, respectively.

However, if Simpson's rule is tested, it is found to reproduce third-degree polynomials, despite the fact that the derivation is carried only as far as second differences. Thus, Simpson's rule is a very popular approximate integration formula in practice, since it gains this additional degree of accuracy without the use of additional points.

Actually it can be shown that any symmetrical approximate integration formula involving an odd number of terms gains an extra degree of accuracy. This applies to both Newton-Cotes formulas and other symmetrical formulas as well. Thus, the symmetrical midpoint rule reproduces a first-degree polynomial, whereas the nonsymmetrical rectangular rule reproduces only a polynomial of degree zero (that is, a constant).

The student should note that all of the formulas developed so far have the property that the sum of the coefficients is equal to the range of integration. This result holds for any approximate integration formula expressed strictly in terms of functional values. A proof of this result is contained in Section 8.4.

Example 8.2.   Rework Example 8.1 using one application of—
(1)   The trapezoidal rule.
(2)   Simpson's rule.
(3)   The three-eighths rule.

(1)   The trapezoidal rule gives

$$\tfrac{1}{2}(e^0 + e^1) = 1.85914$$

(2)   Simpson's rule applied with $h = \tfrac{1}{2}$ gives

$$\tfrac{1}{2} \cdot \tfrac{1}{3}(e^0 + 4e^{1/4} + e^1) = 1.47573$$

(3)   The three-eighths rule applied with $h = \tfrac{1}{3}$ gives

$$\tfrac{1}{3} \cdot \tfrac{3}{8}(e^0 + 3e^{1/9} + 3e^{4/9} + e^1) = 1.46871$$

8.3.   TRUNCATION AND ROUNDOFF ERROR

The total algorithm error in using approximate integration arises from truncation error and roundoff error. In this section both types of error are analyzed in approximating $\int_a^b f(x)\,dx$.

Expressions for maximum truncation error in approximate integration can be obtained from formula (1.12)

$$f(x) - p(x) = \frac{f^{(n+1)}(\xi)\pi(x)}{(n+1)!} \tag{1.12}$$

where $\pi(x) = (x - x_0)(x - x_1) \cdots (x - x_n)$ and ξ is some value of x in the range of collocation. If both sides are integrated, we have

$$\int f(x) \, dx - \int p(x) \, dx = \frac{f^{(n+1)}(\xi)}{(n+1)!} \int \pi(x) \, dx \tag{8.14}[5]$$

If the trapezoidal rule is applied to points of collocation 0 and h, then $\pi(x) = x(x - h) = x^2 - hx$ and $n = 1$. Thus, formula (8.14) gives

$$\frac{f^{(2)}(\xi)}{2} \int_0^h (x^2 - hx) \, dx = -\frac{h^3}{12} f^{(2)}(\xi) \tag{8.15}$$

Simpson's rule can be handled in a similar fashion if we remember that $n = 3$, since it reproduces third-degree polynomials. If points of collocation 0, h, $2h$, $3h$ are chosen, then $\pi(x) = x(x - h)(x - 2h)(x - 3h) = x^4 - 6x^3h + 11x^2h^2 - 6xh^3$. Thus, formula (8.14) gives

$$\frac{f^{(4)}(\xi)}{24} \int_0^{2h} (x^4 - 6x^3h + 11x^2h^2 - 6xh^3) \, dx = -\frac{h^5}{90} f^{(4)}(\xi) \tag{8.16}$$

Finally, the truncation error for the three-eighths rule is given by

$$\frac{f^{(4)}(\xi)}{24} \int_0^{3h} (x^4 - 6x^3h + 11x^2h^2 - 6xh^3) \, dx = -\frac{3h^5}{80} f^{(4)}(\xi) \tag{8.17}$$

Formulas (8.15), (8.16), and (8.17) each assume one application of the approximate integration formula. It is necessary to generalize the above results to composite formulas which cover the entire range of integration $a \le x \le b$. Assume that the points of collocation are $x_0, x_1, x_2, \ldots, x_{n-1}, x_n$ such that $x_0 = a$, $x_n = b$, and $x_1 - x_0 = x_2 - x_1 = \cdots = x_n - x_{n-1} = h$.

If the trapezoidal rule is applied n times to cover the interval $a \le x \le b$, then from formula (8.15) the absolute value of the truncation error becomes

$$\frac{nh^3}{12} |f^{(2)}(\xi)| = \frac{(b-a)h^2}{12} |f^{(2)}(\xi)| \tag{8.18}$$

since $nh = b - a$. Note that $a \le \xi \le b$.[6]

---

[5]This step assumes that ξ is independent of x, which is not strictly correct. A more thorough treatment of the truncation error in approximate integration is contained in Appendix E.

[6]A rigorous proof of this result depends upon continuous second derivatives and an application of the intermediate value theorem. The details are given in Appendix E.

Similarly, if Simpson's rule is applied $n/2$ times to cover the interval $a \leq x \leq b$, then from formula (8.16) the absolute value of the truncation error becomes

$$\frac{n}{2} \cdot \frac{h^5}{90} \left| f^{(4)}(\xi) \right| = \frac{(b-a)h^4}{180} \left| f^{(4)}(\xi) \right| \tag{8.19}$$

Finally, if the three-eighths rule is applied $n/3$ times to cover the interval $a \leq x \leq b$, then from formula (8.17) the absolute value of the truncation error becomes

$$\frac{n}{3} \cdot \frac{3h^5}{80} \left| f^{(4)}(\xi) \right| = \frac{(b-a)h^4}{80} \left| f^{(4)}(\xi) \right| \tag{8.20}$$

Expressions for maximum truncation error are obtained in all of the above formulas by choosing a value of ξ which maximizes $\left| f^{(n+1)}(\xi) \right|$ over the range of collocation.

Expressions for maximum roundoff error can be derived by expressing the approximate integration formula in terms of functional values and using formula (1.5) for maximum roundoff error.[7]

Let the maximum roundoff error in each functional value be E, for $E > 0$.[8] As mentioned in Section 8.2 the sum of the coefficients in any approximate integration formula expressed strictly in terms of functional values is equal to the range of integration, that is, $b - a$. Thus, for any of the approximate integration formulas considered so far, we have

$$\text{maximum roundoff error} = (b-a)E \tag{8.21}$$

Formula (8.21) shows the inherent stability of approximate integration in contrast with approximate differentiation. Maximum roundoff error for approximate integration is not sensitive to subdivision of intervals, that is, decreasing h, as approximate differentiation is. The student should note, however, that *estimated* roundoff error (as contrasted with *maximum* roundoff error; see formulas (1.5) and (1.8)) does increase somewhat with increasing numbers of functional values used, that is, decreasing values of h.

The student should also note that formula (8.21) is valid only if the coefficients in the approximate integration formula are all positive, as they are for the trapezoidal rule, Simpson's rule, and the three-eighths rule. If any of the coefficients is negative, then the sum of the absolute values of the coefficients exceeds $b - a$ and must be used instead of $b - a$ in formula (8.21). Negative coefficients do appear in higher-order Newton-Cotes formulas.

---

[7]An alternative would be to use estimated roundoff error given by formula (1.8) instead of maximum roundoff error given by formula (1.5).

[8]The student should be careful not to confuse this error E with the operator E.

Example 8.3.   **The trapezoidal rule is to be applied 20 times to compute** $\int_0^1 e^{x^2}\, dx$. **Functional values are computed to three decimal places.**

(1)   **Find the maximum truncation error.**
(2)   **Find the maximum roundoff error.**

(1)   We have

$$f(x) = e^{x^2}$$
$$f'(x) = 2xe^{x^2}$$
$$f''(x) = 2e^{x^2} + 4x^2 e^{x^2} = (4x^2 + 2)e^{x^2}$$

$f''(x)$ is maximized for $0 \leq x \leq 1$ at $x = 1$, and its value is $6e = 16.30969$. The interval of differencing $h = .05$. Thus, formula (8.18) gives

$$\frac{(1 - 0)(.05)^2}{12}\,(16.30969) = .0034$$

(2)   Since functional values are computed to three decimal places, $E = .0005$. Thus, formula (8.21) gives

$$(1 - 0)(.0005) = .0005$$

8.4.   UNEQUAL INTERVALS OF DIFFERENCING

Newton-Cotes formulas assume functional values are tabulated at equal intervals. It is possible to develop approximate integration formulas which can be applied when functional values are tabulated at unequal intervals.

There are three common approaches used in practice.

The first approach is to use divided differences. A divided difference formula for $f(x)$ is written down using Sheppard's rules and then integrated. This method is popular in finding approximate integrals when numerical functional values are given.

The second approach is to use Lagrange's formula which directly expresses $f(x)$ as a linear combination of functional values. The formula is then integrated and the answer results. The only problem with this method is that the coefficients become unwieldy if very many terms are used. This method is popular in deriving general formulas for approximate integrals.

The third approach is called the *method of undetermined coefficients* and is also popular in deriving general formulas for approximate integrals. Denote the $n + 1$ points of collocation by $x_0, x_1, x_2, \ldots, x_n$. The answer is of the form

$$\int_a^b f(x)\, dx = \sum_{i=0}^n H_i f(x_i) \tag{8.22}$$

which is a linear combination of functional values. Thus, the problem reduces to finding the $n + 1$ coefficients $H_0, H_1, H_2, \ldots, H_n$. The simplest method of determining these $n + 1$ unknowns is to apply the formula to the $n + 1$ independent functions $f(x) = 1, x, x^2, \ldots, x^{n-1}, x^n$, respectively. Thus, the equations become

$$
\left.
\begin{aligned}
\int_a^b dx &= b - a &&= \sum_{i=0}^n H_i \\
\int_a^b x\, dx &= \frac{b^2 - a^2}{2} &&= \sum_{i=0}^n H_i x_i \\
\int_a^b x^2\, dx &= \frac{b^3 - a^3}{3} &&= \sum_{i=0}^n H_i x_i^2 \\
&\;\;\vdots & &\;\;\vdots \\
\int_a^b x^n\, dx &= \frac{b^{n+1} - a^{n+1}}{n + 1} &&= \sum_{i=0}^n H_i x_i^n
\end{aligned}
\right\}
\tag{8.23}
$$

For any given set of points of collocation $x_0, x_1, x_2, \ldots, x_n$, the set of formulas (8.23) consists of $n + 1$ linear equations in the $n + 1$ unknown H_i's.

Obviously, the methods developed in this section can also be applied to derive formulas for equal intervals of differencing. For example, the Newton-Cotes formulas can be derived in this manner.

The first equation in (8.23) proves the general result that the sum of the coefficients is equal to the range of integration. This provides a quick check on any approximate integration formula expressed strictly in terms of functional values.

Example 8.4. *Derive an approximate integration formula expressing* $\int_0^2 f(x)\, dx$ *in terms of* $f(0), f(1),$ *and* $f(3)$*. Use both Lagrange's formula and the method of undetermined coefficients.*

Using Lagrange's formula, we have from Example 7.4

$$
f(x) = \left(\tfrac{1}{3}x^2 - \tfrac{4}{3}x + 1\right)f(0) + \left(-\tfrac{1}{2}x^2 + \tfrac{3}{2}x\right)f(1) + \left(\tfrac{1}{6}x^2 - \tfrac{1}{6}x\right)f(3)
$$

Thus, we have

$$
\int_0^2 f(x)\, dx = \left[\tfrac{1}{9}x^3 - \tfrac{2}{3}x^2 + x\right]_0^2 f(0) + \left[-\tfrac{1}{6}x^3 + \tfrac{3}{4}x^2\right]_0^2 f(1)
$$
$$
+ \left[\tfrac{1}{18}x^3 - \tfrac{1}{12}x^2\right]_0^2 f(3)
$$
$$
= \tfrac{2}{9} f(0) + \tfrac{5}{3} f(1) + \tfrac{1}{9} f(3)
$$

Note that the sum of the coefficients is equal to two, the range of integration.

Using the method of undetermined coefficients, we have[9]

$$\int_0^2 f(x)\,dx = H_0\,f(0) + H_1\,f(1) + H_3\,f(3)$$

and

$$f(x) = 1: \qquad \int_0^2 f(x)\,dx = 2 = H_0 + H_1 + H_3$$

$$f(x) = x: \qquad \int_0^2 f(x)\,dx = 2 = \qquad H_1 + 3H_3$$

$$f(x) = x^2: \qquad \int_0^2 f(x)\,dx = \tfrac{8}{3} = \qquad H_1 + 9H_3$$

Solving these three equations in three unknowns gives $H_0 = \tfrac{2}{9}$, $H_1 = \tfrac{5}{3}$, and $H_3 = \tfrac{1}{9}$, which agrees with the result from Lagrange's formula.

Example 8.5.   Estimate $\int_0^1 e^{x^2}\,dx$ **based on values of** $x = 0, .1, .3, .6,$ **and** 1. **Compute the answer to five decimal places.**

We have

$$\int_0^1 e^{x^2}\,dx = H_0\,f(0) + H_{.1}f(.1) + H_{.3}\,f(.3) + H_{.6}\,f(.6) + H_1 f(1)$$

and

$$f(x) = 1: \qquad \int_0^1 f(x)\,dx = 1 = H_0 + \qquad H_{.1} + \qquad H_{.3} + \qquad H_{.6} + H_1$$

$$f(x) = x: \qquad \int_0^1 f(x)\,dx = \tfrac{1}{2} = \qquad .1000H_{.1} + .3000H_{.3} + .6000H_{.6} + H_1$$

$$f(x) = x^2: \qquad \int_0^1 f(x)\,dx = \tfrac{1}{3} = \qquad .0100H_{.1} + .0900H_{.3} + .3600H_{.6} + H_1$$

$$f(x) = x^3: \qquad \int_0^1 f(x)\,dx = \tfrac{1}{4} = \qquad .0010H_{.1} + .0270H_{.3} + .2160H_{.6} + H_1$$

$$f(x) = x^4: \qquad \int_0^1 f(x)\,dx = \tfrac{1}{5} = \qquad .0001H_{.1} + .0081H_{.3} + .1296H_{.6} + H_1$$

---

[9]Actually this formulation of the problem does not agree with formula (8.22) in the labels on the H_i's. This point is similar to footnote 2 in Chapter 7.

Solving these five equations in five unknowns gives

$$H_0 = -.14814815$$
$$H_{.1} = .55555556$$
$$H_{.3} = -.13227513$$
$$H_{.6} = .60185185$$
$$H_1 = .12301587$$

and evaluating $H_0 e^0 + H_{.1} e^{.01} + H_{.3} e^{.09} + H_{.6} e^{.36} + H_1 e^1$, we have

$$\int_0^1 e^{x^2} \, dx = 1.46530$$

Example 8.6. *Derive an approximate integration formula expressing* $\int_0^1 f(x) \, dx$ *in terms of* $f(0), f(1), f'(0),$ *and* $f'(1)$ *using the method of undetermined coefficients.*

We have

$$\int_0^1 f(x) \, dx = H_0 f(0) + H_1 f(1) + G_0 f'(0) + G_1 f'(1)$$

and

$$f(x) = 1: \quad \int_0^1 f(x) \, dx = 1 = H_0 + H_1$$

$$f(x) = x: \quad \int_0^1 f(x) \, dx = \tfrac{1}{2} = \quad H_1 + G_0 + G_1$$

$$f(x) = x^2: \quad \int_0^1 f(x) \, dx = \tfrac{1}{3} = \quad H_1 \qquad + 2G_1$$

$$f(x) = x^3: \quad \int_0^1 f(x) \, dx = \tfrac{1}{4} = \quad H_1 \qquad + 3G_1$$

Solving these four equations in four unknowns gives $H_0 = \tfrac{1}{2}$, $H_1 = \tfrac{1}{2}$, $G_0 = \tfrac{1}{12}$, $G_1 = -\tfrac{1}{12}$. Thus, the approximate integration formula is given by

$$\int_0^1 f(x) \, dx = \tfrac{1}{2}[f(0) + f(1)] + \tfrac{1}{12}[f'(0) - f'(1)]$$

This example is interesting, since it derives a more general approximate integration formula involving derivatives at the points of collocation as well as functional values. This particular formula is called the *corrected trapezoidal rule*, for obvious reasons.

The student should note that if derivatives are involved, then the sum of the coefficients is no longer equal to the range of integration in all cases, although it is in this example. Also, this type of formula can only be used if the derivatives are known, which may limit its practicality.

8.5.   OTHER APPROXIMATE INTEGRATION FORMULAS

In this section a variety of other approximate integration formulas which possess higher degrees of accuracy are developed. These formulas are characterized by expressions involving a linear combination of functional values together with correction terms. The correction terms involve either differences or derivatives.

One of these formulas is the *Euler-Maclaurin formula*. This formula was introduced in Chapter 6 as a summation formula. However, it can also serve as an approximate integration formula. A slight rearrangement of formula (6.20) gives

$$\int_0^n f(x)\, dx = \tfrac{1}{2}[f(0) + 2f(1) + 2f(2) + \cdots + 2f(n-1) + f(n)]$$

$$- \tfrac{1}{12}[f^{(1)}(n) - f^{(1)}(0)] + \tfrac{1}{720}[f^{(3)}(n) - f^{(3)}(0)] - \cdots \quad (8.24)$$

Written in the above form, it is apparent that the Euler-Maclaurin formula produces the trapezoidal rule with correction terms involving derivatives.

It is interesting to note that if $n = 1$ in formula (8.24), then the corrected trapezoidal rule for functional values and first derivatives as derived by the method of undetermined coefficients in Example 8.6 is obtained.

Since the derivatives in formula (8.24) may be unavailable, it is important to derive equivalent formulas in which the correction terms are expressed in differences. One of these is the *Gauss-Encke formula*. This formula is derived from the Euler-Maclaurin formula by substituting formulas (7.8) and (7.9) with $h = 1$, that is,

$$D \equiv \mu\delta - \tfrac{1}{6}\mu\delta^3 + \cdots \qquad (7.8)$$

$$D^3 \equiv \qquad \mu\delta^3 - \cdots \qquad (7.9)$$

If these substitutions are made in the Euler-Maclaurin formula, then the following formula results

$$\int_0^n f(x)\, dx = \tfrac{1}{2}[f(0) + 2f(1) + 2f(2) + \cdots + 2f(n-1) + f(n)]$$

$$- \tfrac{1}{12}\mu[\delta f(n) - \delta f(0)] + \tfrac{11}{720}\mu[\delta^3 f(n) - \delta^3 f(0)] - \cdots \quad (8.25)$$

The details of the derivation are left as Exercise 28. Formula (8.25) is the standard version of the Gauss-Encke formula. The formula can alternatively be derived using Bessel's formula.

Formula (8.25) has the disadvantage that functional values outside the range of collocation are required to evaluate the differences. *Gregory's formula* is a variation which involves only functional values in the range of collocation. This formula is derived from the Euler-Maclaurin formula by substituting formulas (7.2c), (7.3b), (7.6), and (7.7b) with $h = 1$, that is,

$$D \equiv \Delta - \tfrac{1}{2}\Delta^2 + \tfrac{1}{3}\Delta^3 - \cdots \tag{7.2c}$$

$$D^3 \equiv \Delta^3 - \cdots \tag{7.3b}$$

$$D \equiv \nabla + \tfrac{1}{2}\nabla^2 + \tfrac{1}{3}\nabla^3 + \cdots \tag{7.6}$$

$$D^3 \equiv \nabla^3 + \cdots \tag{7.7b}$$

If these substitutions are made in the Euler-Maclaurin formula, using the forward difference formulas at $f(0)$ and the backward difference formulas at $f(n)$, then the following formula results

$$\int_0^n f(x)\, dx = \tfrac{1}{2}[f(0) + 2f(1) + 2f(2) + \cdots + 2f(n-1) + f(n)]$$
$$- \tfrac{1}{12}[\nabla f(n) - \Delta f(0)] - \tfrac{1}{24}[\nabla^2 f(n) + \Delta^2 f(0)]$$
$$- \tfrac{19}{720}[\nabla^3 f(n) - \Delta^3 f(0)] - \cdots \tag{8.26}$$

The details of the derivation are left as Exercise 29. Formula (8.26) is the standard version of Gregory's formula.

It is also possible to derive a formula for *Simpson's rule with correction terms*. If Stirling's formula is carried as far as sixth differences, integrated over the interval $-1 \leq x \leq 1$, shifted by a change in origin to cover the interval $0 \leq x \leq 2$, and applied successively to develop a composite formula, then the following formula results where n is an even positive integer

$$\int_0^n f(x)\, dx = \tfrac{1}{3}[f(0) + 4f(1) + 2f(2) + 4f(3) + \cdots + 4f(n-1) + f(n)]$$
$$- \tfrac{1}{90}[\delta^4 f(1) + \delta^4 f(3) + \cdots + \delta^4 f(n-1)]$$
$$+ \tfrac{1}{756}[\delta^6 f(1) + \delta^6 f(3) + \cdots + \delta^6 f(n-1)] - \cdots \tag{8.27}$$

The details of the derivation are left as Exercise 30. Formula (8.27) is the standard composite version of Simpson's rule with correction terms.

8.6. SUBDIVISION OF INTERVALS

In many cases the accuracy of approximate integration is improved by subdivision of intervals rather than by using higher-order Newton-Cotes formulas or formulas with correction terms as developed in Section 8.5. In practice subdivision of intervals is usually applied to the trapezoidal rule. Subdivision of intervals requires that functional values can be computed for any value of x. This may not be possible for tabular data not based on a mathematical formula.

Romberg's method, sometimes called *extrapolation to the limit*, is a systematic procedure of using subdivision of intervals with the trapezoidal rule which is easily applied on high-speed digital computers. The method is based on the fact that the truncation error in the trapezoidal rule given by formula (8.18) is proportional to h^2.

The first step in Romberg's method is to define a series of trapezoidal sums $T_0, T_1, T_2, \ldots$, which result from successively bisecting the interval of differencing. Let the integral to be approximated be $\int_a^b f(x)\, dx$. Figure 8.5 illustrates the nature of these trapezoidal sums.

The first trapezoidal sum, T_0, is based on one application of the trapezoidal rule with interval of differencing h_0. Let $a = x_0$, so that $b = x_0 + h_0$. Then we have

$$T_0 = \frac{h_0}{2}\left[f(x_0) + f(x_0 + h_0)\right]$$

FIGURE 8.5

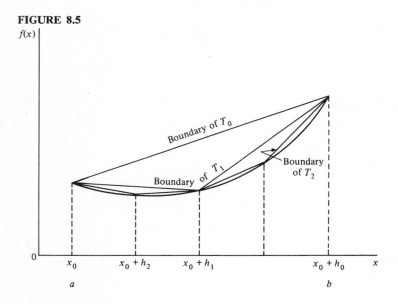

The second trapezoidal sum, T_1, is based on two applications of the trapezoidal rule with interval of differencing $h_1 = \dfrac{h_0}{2}$. Then we have

$$T_1 = \frac{h_1}{2} \left[f(x_0) + 2f(x_0 + h_1) + f(x_0 + 2h_1) \right]$$

If this process is continued $k + 1$ times, the trapezoidal sum, T_k, is based on 2^k applications of the trapezoidal rule with interval of differencing $h_k = \dfrac{h_0}{2^k}$. Then we have

$$T_k = \frac{h_k}{2} \left[f(x_0) + 2 \sum_{i=1}^{2^k-1} f(x_0 + ih_k) + f(x_0 + 2^k h_k) \right] \tag{8.28}$$

Let the true value of the integral be denoted by I and the error in T_k by E_k, that is,

$$E_k = I - T_k \tag{8.29}$$

From formula (8.18) we know that if the interval of differencing in the trapezoidal rule is halved, then the error is reduced by a factor of approximately $\frac{1}{4}$. Thus, we have

$$E_{k+1} \doteq \tfrac{1}{4} E_k \tag{8.30}$$

Now for the first two trapezoidal sums we have

$$I = T_0 + E_0$$

and

$$I = T_1 + E_1 \doteq T_1 + \tfrac{1}{4} E_0 = T_1 + \tfrac{1}{4}(I - T_0)$$

so that

$$I \doteq \frac{4T_1 - T_0}{3}$$

Denote the right-hand side of the above equation by I_1^1.

This result can be extended to any two successive trapezoidal sums to give

$$I_k^1 = \frac{4T_k - T_{k-1}}{3} \tag{8.31}$$

The subscript in I_k^1 refers to the highest order of the trapezoidal sum appearing, that is, T_k, while the superscript indicates that these results comprise the first set of approximations based directly upon the trapezoidal sums. It can be shown that the approximation given by formula (8.31) is equivalent to Simpson's rule, which the student can verify by expanding the trapezoidal sums T_{k-1} and T_k. The details are left as Exercise 37.

A second set of approximations, I_k^2, can be developed from I_k^1. Just as the error in successive terms of T_k is reduced by a factor of approximately $\frac{1}{4}$, the error in successive terms of I_k^1 is reduced by a factor of approximately $\frac{1}{16}$, since truncation error in Simpson's rule is proportional to h^4. Now using a similar derivation to formula (8.31) we have

$$I_k^2 = \frac{16I_k^1 - I_{k-1}^1}{15}$$

If this process is continued m times, we have in general

$$I_k^m = \frac{4^m I_k^{m-1} - I_{k-1}^{m-1}}{4^m - 1} \qquad \text{where } m \geq 1, k \geq m \qquad (8.32)$$

The student should note that $T_k = I_k^0$ in the context of formula (8.32). The process is continued until the required degree of accuracy is obtained and is illustrated in Example 8.7.

Example 8.7.   Evaluate $\displaystyle\int_0^1 e^{x^2}\, dx$ ***to six decimal places using Romberg's method.***

Formulas (8.28), (8.31), and (8.32) are applied to develop the value of the integral. The results to seven decimal places are displayed in Table 8.1.

TABLE 8.1

h	k	$T_k = I_k^0$	I_k^1	I_k^2	I_k^3
1.0	0	1.8591409			
.5	1	1.5715831	1.4757305		
.25	2	1.4906789	1.4637108	1.4629094	
.125	3	1.4697123	1.4627234	1.4626576	1.4626536
.0625	4	1.4644203	1.4626563	1.4626518	1.4626517
.03125	5	1.4630941	1.4626520	1.4626517	1.4626517

Convergence to the true answer 1.462652 to six decimal places is quite rapid. The student should note that computation of the values of I_5^m is not necessary to achieve the required level of accuracy.

8.7.   GAUSSIAN INTEGRATION

All of the approximate integration formulas developed in the preceding sections assume that the values of x on which the formula is based are fixed in advance. *Gaussian integration* is a technique of approximate integration which gains additional accuracy by appropriately choosing the values of x to maximize the accuracy of the approximation.

Consider an approximate integration formula based on $n + 1$ points of collocation $x_0, x_1, x_2, \ldots, x_n$. From the method of undetermined coefficients described in Section 8.4, we can select $n + 1$ coefficients $H_0, H_1, H_2, \ldots, H_n$ such that the formula is accurate for polynomials of degree n.[10]

In Gaussian integration the $n + 1$ values of x_i are selectable as well as the $n + 1$ coefficients H_i. Thus, there are $2n + 2$ selectable quantities, and the resulting approximate integration formula is accurate for polynomials of degree $2n + 1$.

Gaussian integration proceeds in three steps:

1. The integral to be approximated is standardized, so that the lower limit is -1 and the upper limit is $+1$. A change of variable is generally necessary to accomplish this.
2. The $n + 1$ values of x_i satisfy $-1 \leq x_i \leq 1$, for $i = 0, 1, 2, \ldots, n$, and are symmetrically spaced in this interval. The values of x_i are based on *orthogonal polynomials* which are subsequently developed in this section.
3. The $n + 1$ values of H_i are found by the method of undetermined coefficients.

The development of the appropriate values of x_i is based on the property of *orthogonality*. A set of polynomials $O_0(x), O_1(x), \ldots, O_n(x)$, where the subscript denotes the degree of the polynomial, is said to be *simply orthogonal* if the integral over the range of orthogonality of the product of any two such polynomials of different degree in the set is equal to zero, that is, if

$$\int_a^b O_i(x)O_j(x)\, dx = 0 \qquad \text{for } i \neq j \tag{8.33}$$

The special case of orthogonal polynomials in which we are interested is the set of *Legendre polynomials*, $O_n^L(x)$, for which the range of orthogonality is $-1 \leq x \leq 1$, that is,

$$\int_{-1}^1 O_i^L(x)O_j^L(x)\, dx = 0 \qquad \text{for } i \neq j \tag{8.34}$$

Legendre polynomials have the additional powerful property that

$$\int_{-1}^1 O_{n+1}^L(x)Q_n(x)\, dx = 0 \tag{8.35}$$

---

[10]The student should recall that if the formula is symmetrical and if the number of points of collocation is odd, then the formula gains an extra degree of accuracy and is accurate for polynomials of degree $n + 1$.

where $Q_n(x)$ is any polynomial of degree n or less.[11] As a special case of formula (8.35), if $Q_n(x) = 1$, then

$$\int_{-1}^{1} O_{n+1}^{L}(x) \, dx = 0$$

The Legendre polynomial of degree n is defined by

$$O_n^L(x) = \frac{1}{2^n n!} \, D^n(x^2 - 1)^n \tag{8.36}$$

This polynomial has n distinct, real roots in the interval $-1 \leq x \leq 1$ which are symmetrically spaced. The Legendre polynomials for $n = 1, 2, 3, 4, 5$ are given in Table 8.2.

TABLE 8.2

n	$O_n^L(x)$
1	x
2	$\frac{1}{2}(3x^2 - 1)$
3	$\frac{1}{2}(5x^3 - 3x)$
4	$\frac{1}{8}(35x^4 - 30x^2 + 3)$
5	$\frac{1}{8}(63x^5 - 70x^3 + 15x)$
6	$\frac{1}{16}(231x^6 - 315x^4 + 105x^2 - 5)$

Formulas (8.34) and (8.35) are based on the product of two polynomials of different degree. The following formulas are of interest if two polynomials are of the same degree:

$$\int_{-1}^{1} x^n O_n^L(x) \, dx = \frac{2^{n+1}(n!)^2}{(2n + 1)!} \tag{8.37}$$

and

$$\int_{-1}^{1} [O_n^L(x)]^2 \, dx = \frac{2}{2n + 1} \tag{8.38}$$

With this background in Legendre polynomials, we are now in a position to find the required values of x_i for Gaussian integration. Assume that the change of variable has been made, so that we are attempting to find $\int_{-1}^{1} f(x) \, dx$

---

[11] A complete development of this property of Legendre polynomials and several properties to be stated subsequently is not given here, since it is not required for a full mastery of Gaussian integration. The interested reader should see Francis Scheid, *Theory and Problems of Numerical Analysis* (New York: McGraw-Hill Book Co., 1968), pp. 125–38, for a more complete development of the properties of Legendre polynomials.

based on the $n + 1$ points of collocation $x_0, x_1, x_2, \ldots, x_n$. We shall now show that the objective in Gaussian integration of obtaining accuracy for polynomials of degree $2n + 1$ can be obtained if the $n + 1$ points of collocation are chosen as the $n + 1$ roots of $O_{n+1}^L(x)$.

Assume that $f(x)$ is a polynomial of degree $2n + 1$. From formulas (5.3) and (1.12) we have

$$f(x) = \sum_{i=0}^{n} L_i(x) f(x_i) + \frac{f^{(n+1)}(\xi)\pi(x)}{(n + 1)!} \tag{8.39}$$

where $L_i(x)$ is the Lagrange coefficient of $f(x_i)$. If we integrate formula (8.39) over the range $-1 \le x \le 1$, we have

$$\int_{-1}^{1} f(x)\, dx = \int_{-1}^{1} \sum_{i=0}^{n} L_i(x) f(x_i)\, dx + \int_{-1}^{1} \frac{f^{(n+1)}(\xi)\pi(x)}{(n + 1)!}\, dx$$

$$= \sum_{i=0}^{n} H_i f(x_i) + \int_{-1}^{1} \frac{f^{(n+1)}(\xi)\pi(x)}{(n + 1)!}\, dx \tag{8.40}$$

from the development of formulas for unequal intervals of differencing in Section 8.4. The required result follows if we can show that

$$\int_{-1}^{1} \frac{f^{(n+1)}(\xi)\pi(x)}{(n + 1)!}\, dx = 0 \tag{8.41}$$

We can replace $f^{(n+1)}(\xi)$ by $f^{(n+1)}(x)$ without loss of generality.

Let the $n + 1$ points of collocation $x_0, x_1, x_2, \ldots, x_n$ be the $n + 1$ roots of $O_{n+1}^L(x)$. Then from the factor theorem,[12] we have

$$O_{n+1}^L(x) = c(x - x_0)(x - x_1) \cdots (x - x_n) = c\pi(x)$$

Furthermore, if $f(x)$ is a polynomial of degree $2n + 1$, then $f^{(n+1)}(x)$ is a polynomial of degree n. Now formula (8.41) is satisfied by applying formula (8.35), since the degree of $O_{n+1}^L(x) = c\pi(x)$, that is, $n + 1$, is greater than the degree of $f^{(n+1)}(x)$, that is, n.

Once the values of x_i are determined, then the values of H_i can be determined by the method of undetermined coefficients. The values of x_i and H_i for Legendre polynomials of degree 1–6 are contained in Appendix F.

It can be shown that the truncation error in Gaussian integration is given by

$$\frac{f^{(2n+2)}(\xi)}{(2n + 2)!} \int_{-1}^{1} [\pi(x)]^2\, dx \tag{8.42}[13]$$

where $-1 \le \xi \le 1$.

---

[12]The factor theorem is discussed in Section 1.6.
[13]For a proof of formula (8.42) see Scheid, *Theory and Problems of Numerical Analysis*, p. 128.

Example 8.8.   Evaluate $\int_0^1 e^{x^2} dx$ ***using Gaussian integration based on five points of collocation.***

The first step is to make the change of variable $x = \dfrac{z+1}{2}$ so that $dx = \tfrac{1}{2} dz$ and

$$\int_0^1 e^{x^2} dx = \frac{1}{2} \int_{-1}^1 e^{\left(\frac{z+1}{2}\right)^2} dz$$

The values of z_i and H_i for a five-point Gaussian integration contained in Appendix F are reproduced in Table 8.3.

TABLE 8.3

i	z_i	H_i
0	$-.9061798459$	.2369268851
1	$-.5384693101$	.4786286705
2	.0000000000	.5688888888
3	.5384693101	.4786286705
4	.9061798459	.2369268851

The answer is

$$\frac{1}{2} \sum_{i=0}^4 H_i e^{\left(\frac{z_i+1}{2}\right)^2} = 1.462652$$

which is correct to six decimal places. The student should note that this five-point Gaussian integration is exact for polynomials of degree nine.

8.8.   SINGULAR INTEGRALS

Approximate integration techniques should not be blindly applied to *singular integrals*. These are integrals with infinite functional values or with an infinite range of integration. For example, the following integrals are singular:

$$\int_0^1 \frac{1}{\sqrt{x}} dx$$

and

$$\int_0^\infty e^{-x}\sqrt{x}\, dx$$

Approximate integration may also prove troublesome if low-order derivatives of the function are singular, even if the function itself is nonsingular. For example, the following integral has no singularity itself, but the first derivative of the function does have a singularity at $x = 0$

$$\int_0^1 \sqrt{x} \, dx$$

In some cases approximate integration can be successfully applied even in the presence of singularities in the function or in its derivatives. However, a substantial amount of subdivision of intervals may be required, and even then convergence to the true answer may be considerably slowed. Generally, the higher the order of the derivative which first possesses singularities (counting the function itself as the zero-order derivative) the less difficulty there is in applying approximate integration.

In other cases approximate integration cannot be successfully applied to a singular integral and alternative approaches are required. The following are three approaches which have proven successful in many cases:

1. *Change of variable.* In this approach a substitution or change of variable is made which removes the singularity. This is a method which is quite popular and produces excellent results when it can be applied.
2. *Series expansion.* In this approach the function is expanded as a series and integrated term by term. This method will be successful if the series has rapid enough convergence.
3. *Subtracting the singularity.* In this approach the integral is split into two pieces; one a singular piece which can be integrated analytically, and the other a nonsingular piece which can be integrated by an approximate method.

Example 8.9. *It is required to find $\int_0^1 \dfrac{\cos x}{\sqrt{x}} \, dx$. Demonstrate the three methods described in this section.*

The integral is singular, since the function is infinite at the lower limit.
(1) *Change of variable.* If the substitutions $x = t^2$ and $dx = 2t \, dt$ are made, then the integral becomes

$$2 \int_0^1 \cos t^2 \, dt$$

Neither this function nor any of its derivatives have singularities. Any of the techniques of approximate integration should now prove successful.

(2) *Series expansion.*   The series expansion for the function is given by

$$\frac{\cos x}{\sqrt{x}} = \frac{1 - \dfrac{x^2}{2!} + \dfrac{x^4}{4!} - \dfrac{x^6}{6!} + \cdots}{\sqrt{x}}$$

$$= x^{-1/2} - \frac{x^{3/2}}{2} + \frac{x^{7/2}}{24} - \frac{x^{11/2}}{720} + \cdots$$

which can be integrated term by term. Convergence is quite rapid in this case.

(3) *Subtracting the singularity.*   We can express the function as

$$\frac{\cos x}{\sqrt{x}} = \frac{1}{\sqrt{x}} + \frac{\cos x - 1}{\sqrt{x}}$$

The first term has a singularity, but can easily be integrated analytically. The second term has no singularity since

$$\lim_{x \to 0} \frac{\cos x - 1}{\sqrt{x}} = 0$$

from L'Hospital's rule.

However, the second term above does have a singularity in the first derivative, which may prove bothersome. This singularity can be moved to higher derivatives with a more refined subtraction. For example, if the function is written as

$$\frac{\cos x}{\sqrt{x}} = \frac{1 - \dfrac{x^2}{2}}{\sqrt{x}} + \frac{\cos x - 1 + \dfrac{x^2}{2}}{\sqrt{x}}$$

then the first term can again be directly integrated, and the second term has no singularity in either the function or its first derivative.

8.9. FUNCTIONS OF MORE THAN ONE VARIABLE

It is possible to generalize the results of this chapter to be applicable to multiple integrals. The generalizations are straightforward applications of an approximate integration formula separately for each variable.

For example, if we have two variables, then a generalization of the trapezoidal rule becomes

$$\int_0^1 \int_0^1 f(x, y)\, dx\, dy = \frac{1}{2} \int_0^1 [f(0, y) + f(1, y)]\, dy$$

$$= \frac{1}{4}[f(0, 0) + f(0, 1) + f(1, 0) + f(1, 1)] \quad (8.43)$$

Similarly, Simpson's rule applied to two variables becomes

$$\int_0^2 \int_0^2 f(x, y)\, dx\, dy = \frac{1}{3} \int_0^2 [f(0, y) + 4f(1, y) + f(2, y)]\, dy$$

$$= \frac{1}{9}[f(0, 0) + 4f(0, 1) + f(0, 2)$$

$$+ 4f(1, 0) + 16f(1, 1) + 4f(1, 2)$$

$$+ f(2, 0) + 4f(2, 1) + f(2, 2)] \quad (8.44)$$

Example 8.10. **Evaluate** $\int_1^3 \int_1^3 f(x, y)\, dx\, dy$ **using the double integral version of Simpson's rule based on the following data:**

x	1	2	3
y			
1	1	4	9
2	2	8	18
3	3	12	27

The given functional values agree with $f(x, y) = x^2 y$. Thus, we hope our procedures will produce the answer

$$\int_1^3 \int_1^3 x^2 y\, dx\, dy = \frac{104}{3}$$

Simpson's rule given by formula (8.44) with the appropriate change in origin gives

$$\tfrac{1}{9}[1 + (4)(4) + 9 + (4)(2) + (16)(8) + (4)(18) + 3 + (4)(12) + 27] = \tfrac{104}{3}$$

the anticipated answer.

EXERCISES

8.1. Introduction

1. Rework Example 8.1 using two applications of the rectangular rule—
    (a) Based on left-hand end points.
    (b) Based on right-hand end points.

2. Rework Example 8.1 using two applications of the midpoint rule. It is known that $e^{.0625} = 1.06449$ and $e^{.5625} = 1.75505$.

3. Find an approximation for $\int_0^n x^2 \, dx$ based on n applications of the rectangular rule using right-hand end points.

4. (a) Show that the midpoint rule reproduces a first-degree polynomial exactly.
    (b) Show that the rectangular rule does not reproduce a first-degree polynomial exactly.

8.2. Newton-Cotes formulas

5. Estimate $\int_0^{12} f(x) \, dx$ from the following data:

x	0	2	4	6	9	12
$f(x)$	4	3	5	6	7	4

    using both Simpson's rule and the three-eighths rule, as appropriate.

6. Estimate $\int_{-2}^{4} f(x) \, dx$ from the following data:

x	-2	$-\frac{1}{2}$	1	$2\frac{1}{2}$	4
$f(x)$	12	12	12	9	0

    using Simpson's rule.

7. If a single application of Simpson's rule is used to determine the area bounded by the curve $f(x) = 2x^3 - 3x^2 - x$, the x-axis, and the lines $x = 0$ and $x = 4$, by how much does the result differ from the actual area?

8. The driver of an automobile was traveling at 40 ft./sec. when he heard an ambulance siren. He came to a full stop six seconds later. Two seconds after he heard the siren his speed was 34 ft./sec., and four seconds after he heard the siren his speed was 22 ft./sec. Approximately how many feet did he travel after he heard the siren?

9. Verify that Simpson's rule reproduces any third-degree polynomial.

10. Given that $u_x = 2^{2x+1} - 1$, evaluate $\int_0^6 u_x \, dx$ using Simpson's rule with interval of differencing $h = 1$.

11. If $\int_1^4 u_x \, dx = mu_1 + nu_2 + pu_4 + qu_5$, use change in origin and scale to find expressions for—

(a) $\int_{-6}^6 u_x \, dx.$

(b) $\int_1^2 u_x \, dx.$

12. Find an approximation for $\int_0^n x^2 \, dx$ based on n applications of the trapezoidal rule.

13. Derive the Newton-Cotes formula based on five points ($n = 4$)

$$\int_0^4 f(x) \, dx = \tfrac{2}{45}[7f(0) + 32f(1) + 12f(2) + 32f(3) + 7f(4)]$$

This formula is called *Boole's rule*.

14. Derive an expression for $\int_3^4 f(x) \, dx$ based on Simpson's rule applied twice over the interval $0 \le x \le 4$ and the three-eighths rule applied once over the interval $0 \le x \le 3$.

15. Derive an expression for $\int_0^5 f(x) \, dx$ in terms of $f(0), f(1), f(2), f(3), f(4), f(5)$ by averaging the results obtained from the following two approaches:
(a) Simpson's rule for $0 \le x \le 2$; three-eighths rule for $2 \le x \le 5$.
(b) Three-eighths rule for $0 \le x \le 3$; Simpson's rule for $3 \le x \le 5$.

16. Find the error in using Simpson's rule to approximate $\int_0^2 x^4 \, dx.$

8.3. Truncation and roundoff error

17. By using the two-point trapezoidal rule, the following approximation has been determined

$$\int_2^3 \log_e x \, dx = \tfrac{1}{2}(\log_e 2 + \log_e 3)$$

Using the standard expression for the truncation error of the trapezoidal rule, find the absolute value of maximum truncation error in this approximation.

18. The value of $\int_1^2 \dfrac{dx}{x}$ is to be approximated using the trapezoidal rule. What is the largest value of h that can be used if accuracy is required to four decimal places? Ignore roundoff error.

19. Rework Exercise 18 using Simpson's rule.

20. Rework Exercise 18 using the three-eighths rule.

21. The trapezoidal rule is applied 10 times to obtain an approximation for $\int_1^3 (x^3 - x^2)\, dx$. Use the truncation error formula to find the maximum truncation error in this approximation.

22. The value of $\int_{-1}^1 (x^4 - 12x^2)\, dx$ is to be approximated using the trapezoidal rule. What is the largest value of h that can be used if accuracy is required to four decimal places? Ignore roundoff error.

23. If a value of $h = .01$ is chosen and Simpson's rule is applied to approximate $\int_0^1 f(x)\, dx$, how many applications of Simpson's rule are required?

8.4. Unequal intervals of differencing

24. Find n, if $\int_0^n u_x\, dx = 9u_0 + u_{12} + 19u_{24} - 5u_{28}$.

25. (a) If $\int_{-1}^1 f(x)\, dx = af(0) + b[f(2) + f(-2)]$, find a and b.

    (b) The formula derived in (a) is accurate for polynomials of what degree?

26. (a) Derive an approximate integration formula for $\int_0^2 u_x\, dx$ expressed as a linear combination of u_{-1}, u_0, and u_3.

    (b) If the maximum roundoff error in each functional value is equal to E, find the maximum roundoff error in the approximate integration formula developed in (a).

27. If $\int_{-2}^2 f(x)\, dx = a[f(-1) + f(1)] + bf''(0)$, find a and b.

8.5. Other approximate integration formulas

28. Derive formula (8.25).

29. Derive formula (8.26).

30. Derive formula (8.27).

31. Derive the following approximate integration formula

$$\int_{-n}^n f(x)\, dx = 2n[1 + \tfrac{1}{6}n^2\delta^2 + \tfrac{1}{360}(3n^4 - 5n^2)\delta^4 + \cdots]f(0)$$

32. The integral $\int_0^1 \dfrac{dx}{1 + x^2}$ is to be used to find $\dfrac{\pi}{4} = .785398$.

    (a) If Simpson's rule is applied twice based on values of $x = 0, .25, .50, .75,$ and 1, find the value of the approximate integral. The following table of functional values is given:

x	$f(x)$	$\delta f(x)$	$\delta^2 f(x)$	$\delta^3 f(x)$	$\delta^4 f(x)$	$\delta^5 f(x)$	$\delta^6 f(x)$
$-.25$	.94118						
		.05882					
.00	1.00000		$-.11764$				
		$-.05882$		.03528			
.25	.94118		$-.08236$		.02826		
		$-.14118$		.06354		$-.05298$	
.50	.80000		$-.01882$		$-.02472$		.04912
		$-.16000$		.03882		$-.00386$	
.75	.64000		.02000		$-.02858$		
		$-.14000$		.01024			
1.00	.50000		.03024				
		$-.10976$					
1.25	.39024						

(b) On the basis of formula (8.27) demonstrate that much of the impressive accuracy of the answer in (a) is fortuitous.

8.6. Subdivision of intervals

33. If the first three trapezoidal sums in an application of Romberg's method are $T_0 = .60$, $T_1 = .90$, and $T_2 = .99$, find I_2^3. Compute the answer to four decimal places.

34. Show that
$$I_4^3 = \tfrac{64}{63} I_4^2 - \tfrac{1}{64} I_3^3 - \tfrac{1}{64} \cdot \tfrac{1}{63} I_2^2.$$

35. The three-eighths rule is applied once to approximate an integral. If the interval of differencing is trisected and the three-eighths rule is applied again, truncation error is reduced by a factor of approximately K. Find K.

36. Let $S_0, S_1, S_2, \ldots$ be a series of sums for approximate integration using Simpson's rule which are based on successively bisecting the interval of differencing.
    (a) Derive a formula analogous to formula (8.31) for Simpson's rule.
    (b) Derive a formula analogous to formula (8.32) for Simpson's rule.

37. Show that Simpson's rule is obtained if T_{k-1} and T_k are expanded in formula (8.31).

38. Romberg's method of bisection of intervals using the trapezoidal rule is generalized so that on each application the interval of differencing is subdivided into m parts instead of two parts. If n applications of this procedure are made, then truncation error is reduced by a factor of approximately K. Find K.

8.7. Gaussian integration

39. (a) Using formula (8.36) derive the formula for $O_n^L(x)$ appearing in Table 8.2 for $n = 1$.
    (b) Find exact expressions for the values of x_i and the corresponding H_i for a Gaussian integration.

40. Rework Exercise 39 for $n = 2$.

41. Rework Exercise 39 for $n = 3$.

42. If the values of x in a Gaussian integration problem are the roots of $O_7^L(x)$, what is the maximum degree of polynomial for which the resulting approximate integration formula is accurate?

43. If the range of orthogonality of the polynomials $O_1(x) = x^2$ and $O_2(x) = x^2 - 1$ is the interval $-a \le x \le a$ for $a > 0$, find a.

44. Find an expression for the truncation error in using a two-point Gaussian integration formula.

45. (a) Find $\int_{-1}^{1} O_2^L(x) O_3^L(x) \, dx$.

    (b) Find $\int_{-1}^{1} O_3^L(x) O_3^L(x) \, dx$.

46. Evaluate $\int_{-1}^{1} [O_2^L(x) + O_1^L(x) + 1][O_2^L(x) - 1] \, dx$.

8.8 Singular integrals

47. Use the change of variable technique to remove the singularities from the following integrals:

    (a) $\int_{0}^{1} \frac{e^x}{\sqrt{x}} \, dx$.

    (b) $\int_{-1}^{1} \frac{x^2}{\sqrt{1 - x^2}} \, dx$.

48. Consider the integral $\int_{0}^{1} \frac{\log_e (1 + x)}{x^{1/3}} \, dx$.

    (a) Can the undefined lower limit be ignored and one of the approximate integration formulas used directly?

    (b) Can the singularities in the two integrals in Exercise 47 be similarly ignored?

49. Apply the change of variable technique to the integral in Exercise 48.

50. Approximate the integral in Exercise 48 using a series expansion to four terms.

8.9. Functions of more than one variable

51. Derive a formula for the three-eighths rule applied to two variables to find $\int_{0}^{3} \int_{0}^{3} f(x, y) \, dx \, dy$.

52. Apply both the trapezoidal rule and Simpson's rule together to evaluate $\int_0^4 \int_0^4 f(x, y)\, dx\, dy$ based on the following data:

x \ y	0	2	4
0	1	9	25
4	1	17	41

Miscellaneous problems

53. Which of the following formulas is better as an approximation for $\int_0^4 u_x\, dx$?

   (a) $\frac{1}{9}[5(u_0 + u_4) + 4(u_1 + u_3) + 18u_2]$.
   (b) $\frac{2}{15}[2(u_0 + u_2 + u_4) + 12(u_1 + u_3)]$.

54. In some cases approximate integration gives more accurate answers than exact integration. This can happen if the evaluation of the two limits in an exact integration results in the subtraction of two nearly equal numbers with resulting loss of significant digits. To four decimal places it is known that $\int_8^{10} \log_e x\, dx = 4.3904$. The following numerical values are given:

x	$\log_e x$
8	2.079
9	2.197
10	2.303

   (a) Find the value of $\int_8^{10} \log_e x\, dx$ by exact integration with the above values.

   (b) Find the value of $\int_8^{10} \log_e x\, dx$ using Simpson's rule with the above values.

55. *Weddle's rule* is given by

$$\int_0^6 u_x\, dx = \frac{3}{10}[(u_0 + u_6) + 5(u_1 + u_5) + (u_2 + u_4) + 6u_3]$$

Derive Weddle's rule as a seven-point Newton-Cotes formula in which a deliberate error of $-\frac{1}{140}\Delta^6 u_0$ is introduced. Thus, Weddle's rule is strictly accurate for fifth-degree polynomials, but is virtually accurate for seventh-degree polynomials, since it is symmetrical with an odd number of points.

9

Difference equations

9.1. INTRODUCTION

A DIFFERENCE EQUATION is an equation to be solved involving differences. As might be expected, difference equations are the finite difference analogue of differential equations in calculus. In fact, many difference equations can be solved by methods analogous to solution techniques for differential equations.

A general difference equation involving some function u_x and differences of u_x through order m can be expressed as

$$F(u_x, \Delta u_x, \Delta^2 u_x, \ldots, \Delta^m u_x) = 0 \tag{9.1}$$

It is assumed in this chapter that functional values are tabulated at an equal interval of differencing $h = 1$. If the interval of differencing $h \neq 1$, then a change in scale should be made so that $h = 1$. Also, generalizations to equations involving divided differences, where the intervals of differencing are unequal, are of little practical significance and are not considered here. Finally, for convenience all differences are assumed to be advancing differences. Any backward or central differences can easily be converted to advancing differences.

It is also assumed in this chapter that u_x is a function of one variable. Although it is possible to develop a theory of *partial difference equations* which is analogous to the theory of partial differential equations for functions of

200

more than one variable, such a development is beyond the scope of this book.[1]
Also, *systems of difference equations* which arise in problems involving functions of more than one variable are not considered here.

Difference equations are not only of academic interest as a further development of the analogy between finite differences and calculus, but they are of considerable practical significance as well. First of all, they do appear extensively in many applied problems encountered in practice. Secondly, they are of considerable importance in the numerical solution of differential equations. This second point will become evident in the discussion of differential equations in Chapter 10.

9.2. RECURSION FORMULAS

The general difference equation given by formula (9.1) can be written in terms of functional values by expanding the differences using formula (2.7a)

$$\Delta \equiv E - 1 \tag{2.7a}$$

The general difference equation given by formula (9.1) then can be expressed as

$$f\left(u_x, u_{x+1}, u_{x+2}, \ldots, u_{x+m}\right) = 0 \tag{9.2}$$

An equation, such as formula (9.2), which relates successive functional values is often called a *recursion formula*. Difference equations are generally expressed as recursion formulas to facilitate the development of solutions.

The *order* of a difference equation is defined as the difference between the highest and lowest values of x appearing in the recursion formula. Thus, the difference equation given by formula (9.2) is of order m, since

$$(x + m) - x = m$$

It should be noted that the order of a difference equation as just defined is generally also given by the highest-order difference appearing when the equation is written in the format of formula (9.1), that is, m.

However, formula (9.1) can be misleading in certain cases, and the order can be less than the highest-order difference appearing. For example, consider the apparent second-order difference equation

$$\Delta^2 u_x + 2\Delta u_x + u_x = 0$$

which reduces to

$$u_{x+2} = 0$$

when expressed as a recursion formula. This latter equation is of order zero, and thus ceases to be a true difference equation.

---

[1]The interested reader is referred to Carl-Erik Fröberg, *Introduction to Numerical Analysis* (Boston: Addison-Wesley Publishing Co., Inc., 1969), pp. 257–58, for a discussion of partial difference equations.

The above elementary example shows that a recursion formula may characterize the nature of a difference equation better than the original equation involving differences. Thus, it is important to define order using formula (9.2) rather than formula (9.1).

The *general solution* of a difference equation is defined as a sequence of values of u_x on a discrete set of values of x tabulated at interval of differencing $h = 1$ for which formula (9.2) is satisfied for each subset of $m + 1$ successive values of x throughout the entire sequence.

The general solution of a difference equation as just defined is generally not unique and many solutions exist. However, a first-order difference equation does have a unique solution if one *boundary value* u_0 is given. Similarly, a second-order difference equation has a unique solution if two boundary values u_0 and u_1 are given. In general, a difference equation of order m requires m boundary values to have a unique solution. Such a unique solution in which the boundary values are specified is called a *particular solution*.

The recursion formula and the boundary values together allow the successive values of the solution to be computed. For example, consider the third-order difference equation

$$f(u_x, u_{x+1}, u_{x+2}, u_{x+3}) = 0$$

for which the three boundary values u_0, u_1, u_2 are given. The above recursion formula applied for $x = 0, 1, 2, 3$ becomes one equation in one unknown which can be solved for u_3. The recursion formula can be applied again based on u_1, u_2, u_3 to find u_4. In likewise fashion, successive values of $u_5, u_6, \ldots$ can be generated. Successive values can also be generated in the opposite direction, since the recursion formula can be used to find u_{-1} from the boundary values u_0, u_1, u_2. Similarly, successive values $u_{-2}, u_{-3}, \ldots$ can then be generated.

Although the recursion formula together with the boundary values are sufficient to generate the successive values of the solution as just described, this approach does not give a general expression for the function u_x. For certain difference equations, a general expression for u_x can be obtained. Sections 9.3 through 9.6 give methods of finding such general expressions for certain of the simpler and more commonly encountered difference equations.

All of the methods developed in Sections 9.3 through 9.6 assume that the difference equation is *linear*. A difference equation is said to be linear if the functional values enter the recursion formula linearly. Thus, formula (9.2) for a linear difference equation can be expressed as

$$f(u_x, u_{x+1}, u_{x+2}, \ldots, u_{x+m}) = a_0(x)u_x + a_1(x)u_{x+1} + \cdots + a_m(x)u_{x+m} + b(x)$$
$$= 0 \qquad\qquad (9.3)$$

Formula (9.3) is general in the sense that the coefficients $a_0(x)$, $a_1(x)$, ..., $a_m(x)$ are expressed as functions of x. Some of the solution methods to be discussed in succeeding sections require that these coefficients be constant, that is, independent of x. When this is the case, the coefficients are written as $a_0, a_1, \ldots, a_m$.

The student should be careful to note that the assumption of linearity is made in succeeding sections only to simplify the development of a general solution for u_x. However, computation of the successive values of a particular solution can be done for nonlinear recursion formulas just as easily as for linear recursion formulas.

Example 9.1. *The following second-order difference equation is satisfied for all integral values of* x

$$[\Delta^2 + \Delta - 2]f(x) = 0$$

Given the two boundary values $f(0) = 0$ *and* $f(1) = 1$, *find* $f(6)$.

We first express the difference equation as a recursion formula

$$[\Delta^2 + \Delta - 2]f(x) = [(E - 1)^2 + (E - 1) - 2]f(x)$$
$$= [E^2 - E - 2]f(x)$$
$$= f(x + 2) - f(x + 1) - 2f(x) = 0$$

This can be rewritten as

$$f(x + 2) = 2f(x) + f(x + 1)$$

The above formula can be applied five times to successively generate $f(2)$, $f(3), f(4), f(5), f(6)$ from the two boundary values. The results are as follows:

x	$f(x)$
0	0
1	1
2	1
3	3
4	5
5	11
6	21

9.3. FIRST-ORDER EQUATIONS

First-order linear difference equations can readily be solved in terms of sums, just as the analogous first-order linear differential equations are solved in terms of integrals. The general first-order linear difference equation is given by

$$u_{x+1} = a_x u_x + b_x \tag{9.4}$$

Formula (9.4) can be obtained from formula (9.3) by setting $m = 1$, $a_0(x) = a_x$, $a_1(x) = -1$, and $b(x) = b_x$.

Assume that we wish to solve formula (9.4) using the boundary value $u_0 = A$. We can proceed recursively using formula (9.4) to obtain

$$u_1 = a_0 u_0 + b_0 = a_0 A + b_0$$
$$u_2 = a_1 u_1 + b_1 = a_0 a_1 A + a_1 b_0 + b_1$$
$$u_3 = a_2 u_2 + b_2 = a_0 a_1 a_2 A + a_1 a_2 b_0 + a_2 b_1 + b_2$$

and so forth, until in general

$$u_n = a_0 \cdots a_{n-1} A + a_1 \cdots a_{n-1} b_0 + \cdots + a_{n-1} b_{n-2} + b_{n-1} \qquad (9.5)$$

If we define p_n as

$$p_n = a_0 a_1 \cdots a_{n-1} a_n \qquad (9.6)$$

then formula (9.5) can alternatively be expressed as

$$u_n = p_{n-1}\left(A + \frac{b_0}{p_0} + \frac{b_1}{p_1} + \cdots + \frac{b_{n-1}}{p_{n-1}} \right)$$

$$= p_{n-1}\left(A + \sum_{i=0}^{n-1} \frac{b_i}{p_i} \right) \qquad (9.7)$$

Formula (9.7) assumes that the boundary value is a given value of u_0. If some other value of u_x is given as the boundary value, then an adjustment is required. One approach is to find u_0 using the recursion formula as many times as necessary. This approach works well unless the given value of x is not close to zero. A second approach is to modify the derivation of formula (9.7) to accommodate a boundary value other than u_0. For example, if the boundary value $u_3 = A$ is given, formula (9.7) becomes

$$u_n = p_{n-1}\left(A + \sum_{i=3}^{n-1} \frac{b_i}{p_i} \right)$$

where $p_n = a_3 a_4 \cdots a_n$. A third approach is to make a change in origin, so that the boundary value is transformed into u_0 after the change of variable is completed. This approach produces an answer for which the change of variable must be reversed to give the required answer.[2]

Formula (9.7) involves the summation of a number of terms of the form $\frac{b_i}{p_i}$. In certain problems the summation techniques developed in Chapter 6 are useful in summing these terms.

---

[2] The procedure for finding the general form of a function when a change in origin and scale has been made is illustrated in Example 3.4.

Formula (9.7) is the general solution of a first-order difference equation where a_x and b_x are both functions of x. An important special case of formula (9.7) is the general solution of a difference equation in which a_x and b_x are constant, that is, independent of x. Corresponding to formula (9.4), we have

$$u_{x+1} = au_x + b \qquad (9.8)$$

where $a_x = a$ and $b_x = b$ for all x. Formula (9.6) now becomes

$$p_n = a^{n+1} \qquad (9.9)$$

and formula (9.7) becomes

$$
\begin{aligned}
u_n &= a^n \left(A + \frac{b}{a} + \frac{b}{a^2} + \cdots + \frac{b}{a^n} \right) \\
&= a^n A + b(a^{n-1} + a^{n-2} + \cdots + 1) \\
&= a^n A + b \frac{a^n - 1}{a - 1} \qquad (9.10)
\end{aligned}
$$

Formula (9.10) is the general solution of a first-order difference equation in which a_x and b_x are constant.

First-order difference equations can also be solved by constructing a difference table instead of by using formula (9.7) or (9.10). In this approach the given recursion formula and the boundary value are used to construct a difference table. The general form of the function is then obtained from the difference table as discussed in Chapters 2 and 3.

When this latter approach is used, the student should be careful to remember the convention that differencing stops when a geometric progression appears in the difference table, since it is then assumed that the remaining function is an exponential at that point. Example 2.1 illustrates the manner in which the general form of a function can be determined under these conditions. The appearance of a geometric progression in the difference table is relatively common in the solution of first-order difference equations.

Naturally, if differences become constant at some point, a polynomial and not an exponential is indicated. Newton's advancing difference formula or any other finite difference interpolation formula can then be applied to find the general form of the function.

Example 9.2. **If $u_1 = 6$ and $\Delta u_x = 3u_x + 2$ for all positive integral values of x, find u_x. Use the general formula approach.**

We first write the difference equation as a recursion formula, obtaining

$$\Delta u_x = u_{x+1} - u_x = 3u_x + 2$$

or

$$u_{x+1} = 4u_x + 2$$

In this case a and b are constant, so we can use formula (9.10).

Since $u_1 = 6$ is the given boundary value and since formula (9.10) is based on a boundary value of u_0, an adjustment is required. In this case u_0 can be immediately obtained from the recursion formula based on $u_1 = 6$. It is evident that $u_0 = 1$.

Formula (9.10) is now applied with $A = u_0 = 1$, $a = 4$, $b = 2$ to give

$$u_n = 4^n + 2\frac{4^n - 1}{4 - 1}$$

$$= \frac{5 \cdot 4^n - 2}{3}$$

or replacing n with x

$$u_x = \frac{5 \cdot 4^x - 2}{3}$$

Example 9.3.   *Rework Example 9.2 using the difference table approach.*

The recursion formula

$$u_{x+1} = 4u_x + 2$$

and the boundary value

$$u_1 = 6$$

can be used to generate a few functional values: $u_1 = 6$, $u_2 = 26$, $u_3 = 106$, $u_4 = 426$, $u_5 = 1706$.

The following is a difference table based on these functional values:

x	u_x	Δu_x
1	6	
		20
2	26	
		80
3	106	
		320
4	426	
		1280
5	1706	

It is observed that the first difference column is a geometric progression with common ratio equal to four, so no further differences are computed. We now set

$$u_x = A + B \cdot 4^x$$

and

$$\Delta u_x = \qquad 3B \cdot 4^x$$

so that
$$u_1 = A + 4B = 6$$
and
$$\Delta u_1 = 12B = 20$$

The solution to this system of equations is $A = -\tfrac{2}{3}$, $B = \tfrac{5}{3}$. Thus, we have

$$u_x = \frac{5 \cdot 4^x - 2}{3}$$

which agrees with the answer obtained in Example 9.2.

9.4.   HOMOGENEOUS SECOND-ORDER EQUATIONS

The general second-order linear difference equation given by formula (9.3) is

$$a_0(x)u_x + a_1(x)u_{x+1} + a_2(x)u_{x+2} + b(x) = 0 \tag{9.11}$$

A *homogeneous* difference equation is one in which $b(x)$ vanishes for all x, that is, $b(x) \equiv 0$. The term "homogeneous" refers to the fact that the function $u_x \equiv 0$ is a solution.[3] Since we have a second-order equation, we know that $a_2(x) \neq 0$ and we can divide formula (9.11) by $a_2(x)$ in order to produce a coefficient on u_{x+2} equal to one. Thus, the general formula for a homogeneous linear second-order difference equation reduces to

$$u_{x+2} + a_1(x)u_{x+1} + a_0(x)u_x = 0 \tag{9.12}$$

The difference equation given by formula (9.12) will now be shown to have the family of solutions

$$u_x = c_1 s_x + c_2 t_x \tag{9.13}$$

where s_x and t_x are two different nontrivial particular solutions and where c_1 and c_2 are arbitrary constants. For a solution to exist the *Wronskian determinant*, w_x, defined by

$$w_x = \begin{vmatrix} s_{x+1} & t_{x+1} \\ s_x & t_x \end{vmatrix} \tag{9.14}$$

must be nonzero for all values of x.

Formula (9.13) is analogous to a well-known result for homogeneous linear second-order differential equations and is often called the *principle of superposition*. If two boundary values are given, then c_1 and c_2 can be determined to give a particular solution.

---

[3] The solution $u_x \equiv 0$ is called the *trivial solution*.

To show the above results, let s_x and t_x be two different nontrivial particular solutions. From formula (9.12) we have

$$s_{x+2} + a_1(x)s_{x+1} + a_0(x)s_x = 0$$

and

$$t_{x+2} + a_1(x)t_{x+1} + a_0(x)t_x = 0$$

From these two equations we obtain

$$c_1[s_{x+2} + a_1(x)s_{x+1} + a_0(x)s_x] + c_2[t_{x+2} + a_1(x)t_{x+1} + a_0(x)t_x]$$
$$= [c_1s_{x+2} + c_2 t_{x+2}] + a_1(x)[c_1s_{x+1} + c_2 t_{x+1}] + a_0(x)[c_1s_x + c_2 t_x]$$
$$= u_{x+2} + a_1(x)u_{x+1} + a_0(x)u_x = 0$$

which shows that u_x is also a solution.

It remains to show that the Wronskian determinant must be nonzero. Consider two successive values of u_x given by formula (9.13).

$$u_{x+1} = c_1s_{x+1} + c_2 t_{x+1}$$
$$u_x = c_1s_x + c_2 t_x$$

These two equations can be solved for the two constants c_1, c_2. Using *Cramer's rule*[4], we have

$$c_1 = \frac{\begin{vmatrix} u_{x+1} & t_{x+1} \\ u_x & t_x \end{vmatrix}}{\begin{vmatrix} s_{x+1} & t_{x+1} \\ s_x & t_x \end{vmatrix}} = \frac{\begin{vmatrix} u_{x+1} & t_{x+1} \\ u_x & t_x \end{vmatrix}}{w_x} \qquad (9.15a)$$

$$c_2 = \frac{\begin{vmatrix} s_{x+1} & u_{x+1} \\ s_x & u_x \end{vmatrix}}{\begin{vmatrix} s_{x+1} & t_{x+1} \\ s_x & t_x \end{vmatrix}} = \frac{\begin{vmatrix} s_{x+1} & u_{x+1} \\ s_x & u_x \end{vmatrix}}{w_x} \qquad (9.15b)$$

Thus, the Wronskian determinant w_x must be nonzero for all values of x. Formulas (9.15a) and (9.15b) can be applied for the two boundary values to uniquely determine c_1 and c_2.

The student should note that if either s_x or t_x is the trivial solution, that is, $s_x \equiv 0$ or $t_x \equiv 0$, then the value of the Wronskian determinant is immediately equal to zero.

Finding the functions s_x and t_x in order to solve the general difference equation given by formula (9.12) may prove to be difficult. However, an important special case of formula (9.12) is the general solution of a difference

---

[4]Cramer's rule is a well-known algebraic method of solving systems of linear equations. Students not familiar with Cramer's rule are referred to Section 13.2.

equation in which $a_0(x)$ and $a_1(x)$ are constant, that is, independent of x. Corresponding to formula (9.12), we have

$$u_{x+2} + a_1 u_{x+1} + a_0 u_x = 0 \tag{9.16}$$

where $a_0(x) = a_0$ and $a_1(x) = a_1$ for all x. It is assumed that $a_0 \neq 0$, so that formula (9.16) is second-order. Under these conditions s_x and t_x can readily be determined.

The first step in the solution of the difference equation given by formula (9.16) is to determine the roots of the *characteristic equation*

$$r^2 + a_1 r + a_0 = 0 \tag{9.17}$$

Denote the two roots of the characteristic equation by r_1 and r_2. There are three cases which must be considered in determining s_x and t_x.

Case 1.   Distinct real roots.

In Case 1 we have $a_1^2 > 4a_0$. The two roots r_1 and r_2 are real and distinct and are given by

$$r_1 = \frac{-a_1 + \sqrt{a_1^2 - 4a_0}}{2} \tag{9.18a}$$

$$r_2 = \frac{-a_1 - \sqrt{a_1^2 - 4a_0}}{2} \tag{9.18b}$$

The functions s_x and t_x are given by

$$s_x = r_1^x \tag{9.19a}$$

$$t_x = r_2^x \tag{9.19b}$$

To show that s_x is a solution of formula (9.16) we note that

$$s_{x+2} + a_1 s_{x+1} + a_0 s_x = r_1^{x+2} + a_1 r_1^{x+1} + a_0 r_1^x$$
$$= r_1^x(r_1^2 + a_1 r + a_0)$$
$$= 0$$

from formula (9.17). The proof for t_x is identical.

Case 2.   Equal real roots.

In Case 2 we have $a_1^2 = 4a_0$. The two roots r_1 and r_2 are real and equal. Let $r_1 = r_2 = r$ and the root is given by

$$r = -\tfrac{1}{2}a_1 \tag{9.20}$$

The functions s_x and t_x are given by

$$s_x = r^x \tag{9.21a}$$

$$t_x = xr^x \tag{9.21b}$$

The proof that s_x is a solution of formula (9.16) is identical to Case 1. To show that t_x is also a solution we note that

$$t_{x+2} + a_1 t_{x+1} + a_0 t_x = (x + 2)r^{x+2} + a_1(x + 1)r^{x+1} + a_0 xr^x$$
$$= r^x[x(r^2 + a_1 r + a_0) + (2r^2 + a_1 r)]$$
$$= 0$$

since $r^2 + a_1 r + a_0 = 0$ from formula (9.17) and $2r^2 + a_1 r = 0$ from formula (9.20).

Case 3.   Complex roots.

In Case 3 we have $a_1^2 < 4a_0$. The two roots r_1 and r_2 are a complex pair of conjugates and can be expressed in polar coordinates as

$$r_1 = R(\cos \theta + i \sin \theta) \tag{9.22a}$$

$$r_2 = R(\cos \theta - i \sin \theta) \tag{9.22b}$$

It is useful to express R, $\cos \theta$, and $\sin \theta$ in terms of a_0 and a_1. From formulas (9.17) and (9.22a) we have

$$r_1 = \frac{-a_1 + i\sqrt{4a_0 - a_1^2}}{2}$$

$$= R(\cos \theta + i \sin \theta)$$

Thus,

$$R \cos \theta = -\frac{a_1}{2}$$

and

$$R \sin \theta = \frac{\sqrt{4a_0 - a_1^2}}{2}$$

Squaring both expressions and adding gives

$$R^2(\cos^2 \theta + \sin^2 \theta) = \frac{a_1^2}{4} + \frac{4a_0 - a_1^2}{4}$$

which becomes

$$R^2 = a_0$$

Thus, we have

$$R = \sqrt{a_0} \tag{9.23a}$$

$$\cos \theta = -\frac{1}{2}\sqrt{\frac{a_1^2}{a_0}} \tag{9.23b}$$

$$\sin \theta = \frac{1}{2}\sqrt{\frac{4a_0 - a_1^2}{a_0}} \tag{9.23c}$$

The functions s_x and t_x are given by

$$s_x = R^x \sin x\theta \qquad (9.24a)$$

$$t_x = R^x \cos x\theta \qquad (9.24b)$$

To show that s_x is a solution of formula (9.16) we note that

$$s_{x+2} + a_1 s_{x+1} + a_0 s_x = R^x [R^2 \sin (x+2)\theta + a_1 R \sin (x+1)\theta + a_0 \sin x\theta]$$
$$= R^x \sin x\theta (R^2 \cos 2\theta + a_1 R \cos \theta + a_0)$$
$$+ R^x \cos x\theta (R^2 \sin 2\theta + a_1 R \sin \theta)$$

But we know from formulas (9.17) and (9.22a) that

$$r_1^2 + a_1 r_1 + a_0 = R^2(\cos 2\theta + i \sin 2\theta) + a_1 R(\cos \theta + i \sin \theta) + a_0$$
$$= (R^2 \cos 2\theta + a_1 R \cos \theta + a_0) + i(R^2 \sin 2\theta + a_1 R \sin \theta)$$
$$= 0$$

Thus,

$$R^2 \cos 2\theta + a_1 R \cos \theta + a_0 = 0$$

and

$$R^2 \sin 2\theta + a_1 R \sin \theta = 0$$

so that

$$s_{x+2} + a_1 s_{x+1} + a_0 s_x = 0$$

To show that t_x is a solution of formula (9.16) we note that

$$t_{x+2} + a_1 t_{x+1} + a_0 t_x = R^x [R^2 \cos (x+2)\theta + a_1 R \cos (x+1)\theta + a_0 \cos x\theta]$$
$$= R^x \cos x\theta (R^2 \cos 2\theta + a_1 R \cos \theta + a_0)$$
$$- R^x \sin x\theta (R^2 \sin 2\theta + a_1 R \sin \theta)$$

As above,

$$R^2 \cos 2\theta + a_1 R \cos \theta + a_0 = 0$$

and

$$R^2 \sin 2\theta + a_1 R \sin \theta = 0$$

so that

$$t_{x+2} + a_1 t_{x+1} + a_0 t_x = 0$$

Example 9.4.   *Find the general solution for each of the following difference equations:*
  (1) $u_{x+2} - 5u_{x+1} + 4u_x = 0$.
  (2) $u_{x+2} - 4u_{x+1} + 4u_x = 0$.
  (3) $u_{x+2} - 3u_{x+1} + 4u_x = 0$.

(1)   The characteristic equation is

$$r^2 - 5r + 4 = 0$$

which has roots $r_1 = 1$ and $r_2 = 4$. Thus, Case 1 applies and

$$u_x = c_1 + c_2 4^x$$

(2)   The characteristic equation is

$$r^2 - 4r + 4 = 0$$

which has roots $r_1 = r_2 = 2$. Thus, Case 2 applies and

$$\begin{aligned} u_x &= c_1 2^x + c_2 x 2^x \\ &= 2^x(c_1 + c_2 x) \end{aligned}$$

(3)   The characteristic equation is

$$r^2 - 3r + 4 = 0$$

which has roots $r_1 = \frac{1}{2}(3 + i\sqrt{7})$ and $r_2 = \frac{1}{2}(3 - i\sqrt{7})$. Thus, Case 3 applies with

$$R = \sqrt{4} = 2$$

$$\cos\theta = -\frac{1}{2}\sqrt{\frac{9}{4}} = -\frac{3}{4}$$

$$\sin\theta = \frac{1}{2}\sqrt{\frac{16-9}{4}} = \frac{\sqrt{7}}{4}$$

so that

$$\begin{aligned} u_x &= c_1 2^x \sin x\theta + c_2 2^x \cos x\theta \\ &= 2^x(c_1 \sin x\theta + c_2 \cos x\theta) \end{aligned}$$

Example 9.5.   *The following second-order difference equation is satisfied for all nonnegative integral values of* x

$$u_{x+2} = u_{x+1} + u_x$$

Given the two boundary values $u_0 = 0$ *and* $u_1 = 1$, *generate values of* u_x *through* $x = 10$ *and determine the general form of the function* u_x.

The recursion formula can be applied directly by noting that each functional value is the sum of the two immediately preceding functional values. The values of u_x through $x = 10$ are given by the following:[5]

---

[5]These functional values comprise the first 10 terms of the well-known *Fibonacci series* in mathematics.

x	u_x
0	0
1	1
2	1
3	2
4	3
5	5
6	8
7	13
8	21
9	34
10	55

To find the general form of the function u_x we find the roots of the characteristic equation

$$r^2 - r - 1 = 0$$

which are $r_1 = \frac{1}{2}(1 + \sqrt{5})$ and $r_2 = \frac{1}{2}(1 - \sqrt{5})$. Thus Case 1 applies and

$$u_x = c_1 \left(\frac{1 + \sqrt{5}}{2}\right)^x + c_2 \left(\frac{1 - \sqrt{5}}{2}\right)^x$$

It remains to find c_1 and c_2. Using the two boundary values $u_0 = 0$ and $u_1 = 1$, we have

$$u_0 = \qquad c_1 + \qquad c_2 = 0$$

$$u_1 = \left(\frac{1 + \sqrt{5}}{2}\right) c_1 + \left(\frac{1 - \sqrt{5}}{2}\right) c_2 = 1$$

Thus, $c_1 = -c_2 = \dfrac{1}{\sqrt{5}}$, so that the solution is

$$u_x = \frac{1}{\sqrt{5}} \left[\left(\frac{1 + \sqrt{5}}{2}\right)^x - \left(\frac{1 - \sqrt{5}}{2}\right)^x \right]$$

The student should note that c_1 and c_2 could also have been found using formulas (9.15a) and (9.15b). However, in this example the two equations in two unknowns can be solved by inspection, so that Cramer's rule is not required.

9.5. NONHOMOGENEOUS SECOND-ORDER EQUATIONS

It is possible to extend the solution for homogeneous second-order difference equations developed in Section 9.4 to nonhomogeneous equations. A *nonhomogeneous* difference equation is one in which $b(x)$ in formula (9.11) does not vanish. Thus, the general formula for a nonhomogeneous second-order difference equation corresponding to formula (9.16) for homogeneous equations is given by

$$u_{x+2} + a_1 u_{x+1} + a_0 u_x + b(x) = 0 \qquad (9.25)$$

The difference equation given by formula (9.25) can be shown to have the family of solutions

$$u_x = c_1 s_x + c_2 t_x + v_x \qquad (9.26)$$

where c_1, c_2, s_x, t_x are the same as for homogeneous equations in Section 9.4 and where v_x is one particular solution of the nonhomogeneous equation.

To prove formula (9.26) let u_x be the general solution of formula (9.25), let v_x be one particular solution, and define d_x by

$$d_x = u_x - v_x \qquad (9.27)$$

We then have

$$u_{x+2} + a_1 u_{x+1} + a_0 u_x + b(x) = 0$$
$$v_{x+2} + a_1 v_{x+1} + a_0 v_x + b(x) = 0$$

and subtracting

$$d_{x+2} + a_1 d_{x+1} + a_0 d_x = 0$$

Now d_x is seen to be a solution of the homogeneous equation from formula (9.16) and can be written as

$$d_x = c_1 s_x + c_2 t_x$$

which gives

$$u_x = d_x + v_x = c_1 s_x + c_2 t_x + v_x$$

Thus, the solution of nonhomogeneous second-order equations is complete upon finding a particular solution v_x. Although it is difficult to give general methods for finding a solution v_x, simple results can readily be obtained if $b(x)$ is a polynomial or an exponential, which is often the case in practice.

If $b(x)$ is a polynomial of degree n, then v_x is set equal to a polynomial of degree n with unknown coefficients. Since v_x must be a solution of the difference equation, formula (9.25) can be solved by the *method of undetermined coefficients* to find the unknown coefficients of v_x. This procedure is illustrated in Example 9.6.

If $b(x)$ is an exponential, then v_x is set equal to an exponential with the same common ratio as $b(x)$ and with an unknown coefficient. The unknown coefficient can be directly determined by applying formula (9.25) and solving for the unknown coefficient. This procedure is illustrated in Example 9.7.

Example 9.6. **Find the general solution of the difference equation $u_{x+2} - u_{x+1} - u_x = 2x^2 + 1$.**

The solution of the corresponding homogeneous equation is known to be

$$u_x = c_1 \left(\frac{1 + \sqrt{5}}{2} \right)^x + c_2 \left(\frac{1 - \sqrt{5}}{2} \right)^x$$

from Example 9.5.

Since $b(x) = -2x^2 - 1$ is a quadratic, we set v_x equal to the quadratic

$$v_x = Ax^2 + Bx + C$$

Now v_x is a solution of the original difference equation, so that

$$
\begin{aligned}
v_{x+2} - v_{x+1} - v_x = \ & A(x+2)^2 + B(x+2) + C \\
& - A(x+1)^2 - B(x+1) - C \\
& - Ax^2 \qquad\quad - Bx \qquad\quad - C \\
= \ & -Ax^2 + (2A - B)x + (3A + B - C) \\
= \ & 2x^2 + 1
\end{aligned}
$$

Applying the method of undetermined coefficients we have

$$
\begin{aligned}
-A \qquad\quad &= 2 \\
2A - B \quad &= 0 \\
3A + B - C &= 1
\end{aligned}
$$

which gives $A = -2$, $B = -4$, $C = -11$. Thus, $v_x = -(2x^2 + 4x + 11)$ and the general solution of the difference equation is

$$u_x = c_1 \left(\frac{1 + \sqrt{5}}{2} \right)^x + c_2 \left(\frac{1 - \sqrt{5}}{2} \right)^x - (2x^2 + 4x + 11)$$

Example 9.7. **Find the general solution of the difference equation $u_{x+2} - u_{x+1} - u_x = 3^x$.**

The solution of the corresponding homogeneous equation is the same as in Example 9.6.

Since $b(x) = -3^x$ is an exponential, we set v_x equal to the exponential

$$v_x = A3^x$$

Now v_x is a solution of the original difference equation, so that

$$
\begin{aligned}
v_{x+2} - v_{x+1} - v_x &= A\left(3^{x+2} - 3^{x+1} - 3^x\right) \\
&= A3^x\left(3^2 - 3 - 1\right) \\
&= 5A3^x \\
&= 3^x
\end{aligned}
$$

Thus, $A = \frac{1}{5}$ and the general solution of the difference equation is

$$
u_x = c_1\left(\frac{1 + \sqrt{5}}{2}\right)^x + c_2\left(\frac{1 - \sqrt{5}}{2}\right)^x + \frac{1}{5}\cdot 3^x
$$

The approach just shown will work as long as the common ratio in $b(x)$ and v_x, that is, 3 in this example, is different from the common ratios in s_x and t_x, that is, $\dfrac{1 + \sqrt{5}}{2}$ and $\dfrac{1 - \sqrt{5}}{2}$ in this example. If we have Case 1 and the common ratio in $b(x)$ and v_x is equal to the common ratio in either s_x or t_x, then v_x should be set equal to Axr^x instead of just Ar^x where r is this common ratio. Furthermore, if we have Case 2, so that $s_x = r^x$ and $t_x = xr^x$, then v_x should be set equal to Ax^2r^x.

9.6. HIGHER-ORDER EQUATIONS

Linear difference equations of order higher than two can be handled in a similar fashion to second-order equations.

The first step is to solve the corresponding homogeneous equation. This requires the determination of the roots of the characteristic equation. As with second-order equations, distinct real roots enter the solution in the form

$$
r^x
$$

with a constant coefficient. Real roots of multiplicity $n(n \geq 2)$ enter the solution n times in the forms

$$
r^x, xr^x, x^2r^x, \ldots, x^{n-1}r^x
$$

each with a constant coefficient. Complex roots always occur in pairs and enter the solution in the forms

$$
R^x \sin x\theta \qquad R^x \cos x\theta
$$

where the two complex roots are given by

$$
R(\cos\theta + i\sin\theta) \quad \text{and} \quad R(\cos\theta - i\sin\theta)
$$

The second step which is required for nonhomogeneous equations is to determine v_x. The approach described in Section 9.5 for second-order equations can also be applied for higher-order equations.

Example 9.8.   Find the general solution of the difference equation u_{x+4}
$- 2u_{x+3} + 2u_{x+2} - 2u_{x+1} + u_x = x$.

The characteristic equation is

$$r^4 - 2r^3 + 2r^2 - 2r + 1 = 0$$

which has roots $r_1 = 1$, $r_2 = 1$, $r_3 = i$, $r_4 = -i$. Thus, the general solution of
the associated homogeneous equation is

$$u_x = c_1 + c_2 x + c_3 \sin \frac{\pi x}{2} + c_4 \cos \frac{\pi x}{2}$$

To find the general solution of the nonhomogeneous equation we note
that $b(x) = -x$ is a first-degree polynomial, so that normally we would let v_x be
a first-degree polynomial. However, as indicated at the end of Example 9.7,
since a first-degree polynomial already appears in the general solution of the
homogeneous equation, we must let v_x be a higher degree polynomial.

We first let v_x be a first-degree polynomial multiplied by x, that is,

$$v_x = Ax^2 + Bx$$

and since v_x is a solution of the original difference equation

$$v_{x+4} - 2v_{x+3} + 2v_{x+2} - 2v_{x+1} + v_x = 4A$$
$$= x$$

upon simplification. Since all coefficients of x and x^2 drop out, the result
cannot equal x. Thus, this approach fails to produce a solution.

We next let v_x be a first-degree polynomial multiplied by x^2, that is,

$$v_x = Ax^3 + Bx^2$$

and since v_x is a solution of the original difference equation

$$v_{x+4} - 2v_{x+3} + 2v_{x+2} - 2v_{x+1} + v_x = 12Ax + 24A + 4B$$
$$= x$$

upon simplification. Applying the method of undetermined coefficients we
have

$$12A \qquad = 1$$
$$24A + 4B = 0$$

which gives $A = \frac{1}{12}$, $B = -\frac{1}{2}$.

Thus,

$$v_x = \frac{x^3 - 6x^2}{12}$$

and the general solution of the difference equation is

$$u_x = c_1 + c_2 x + c_3 \sin \frac{\pi x}{2} + c_4 \cos \frac{\pi x}{2} + \frac{x^3 - 6x^2}{12}$$

EXERCISES

9.1. Introduction; 9.2. Recursion formulas

1. Find the order of the following difference equations:
   (a) $(\Delta^3 + 3\Delta^2 + 5\Delta + 3)f(x) = 0$.
   (b) $(2\Delta^2 + 3\Delta + 1)f(x) = 0$.

2. If $f(0) = 2$ and $f(x) = 1 + \nabla f(x+1)$, find $f(7)$.

3. If $u_0 = 0$, $u_1 = 0$, and $4\Delta^2 u_x + 12\Delta u_x + 9u_x = x^2$, find u_4.

4. If $u_0 = 0$, $u_3 = 5$, and $\Delta^2 u_x = 2u_x$, find u_6.

9.3. First-order equations

5. Find the special case of formula (9.10) if $A = b = 1$.

6. Formula (9.10) is not defined if $a = 1$. Find the formula to be used instead of formula (9.10) if $a = 1$.

7. Find the general form for $f(x)$ in Exercise 2.

8. If $u_1 = 2$ and $x\Delta u_x = 3u_x$, find u_x.

9. Generalize Exercise 8 to show that if $x\Delta u_x = mu_x$, then u_x is a polynomial of degree m.

10. Find a difference equation with a boundary value specified for $f(0)$ which will be satisfied by $f(x) = x!$.

11. If $u_2 = 0$ and $\Delta u_x = (4x - 2)3^x$, find u_x.

12. If $u_1 = 2$ and $\Delta u_x = x \cdot x!$, show that $u_x = x! + 1$.

13. It is known that a solution of $(x + 1)u_{x+1} + xu_x = 2x - 3$ is of the form $u_x = \dfrac{x - \lambda}{x}$. Find λ.

14. If $u_1 = 1$ and $\Delta u_x = \dfrac{1 - x}{x} u_x$, find u_x.

9.4. Homogeneous second-order equations

15. Consider the general homogeneous second-order difference equation with constant coefficients

$$u_{x+2} + a_1 u_{x+1} + a_0 u_x = 0$$

Find the value of the Wronskian determinant, w_x, and show that it is nonzero for:
(a) Case 1.
(b) Case 2.
(c) Case 3.

16. Consider the general homogeneous first-order difference equation with constant coefficients

$$u_{x+1} = au_x$$

Assume that the boundary value $u_0 = A$ is given.
    (a) Find u_x using formula (9.10).
    (b) What is the characteristic equation?
    (c) Modify Case 1 to solve for u_x based on the root of the characteristic equation.
    (d) Are the two approaches given in (a) and (c) consistent?

17. Find the general solution for $u_{x+2} - 2u_{x+1} + 2u_x = 0$.

18. (a) Find the general solution for $u_{x+2} - 2u_{x+1} + u_x = 0$.
    (b) Is the answer to (a) consistent with the observation that the given difference equation can be written $\Delta^2 u_x = 0$? Explain your answer.

19. Find the general form for u_x in Exercise 4.

20. Show that if $\lim_{x \to \infty} u_x = 0$, then all the roots of the characteristic equation (real or complex) are less than one in absolute value.

21. Find the general solution for $\delta^2 f(x) = 2f(x)$.

22. If $u_0 = 2$, $u_1 = 12$, and $u_{x+2} - 8u_{x+1} + 16u_x = 0$, find u_x.

23. If $u_0 = 0$, $u_5 = 1$, and $u_{x+2} - u_x = 0$, find u_x.

24. If $u_0 = 2$, $u_1 = 5$, and $u_{x+2} - 5u_{x+1} + 6u_x = 0$, find $\sum_{x=0}^{n-1} u_x$.

9.5. Nonhomogeneous second-order equations

25. Use the result of Exercise 17 to find the general solution for $u_{x+2} - 2u_{x+1} + 2u_x = x^2 - 2$.

26. Find the general solution for $u_{x+2} - 4u_{x+1} + 4u_x = 3 \cdot 2^x$.

27. Find the general solution for $u_{x+2} - 4u_{x+1} + 3u_x = x2^x$.

28. Find the particular solution for Exercise 26 which satisfies the boundary values $u_0 = 2$ and $u_1 = 6$.

29. Find the particular solution for Exercise 27 which satisfies the boundary values $u_0 = 5$ and $u_1 = 7$.

30. Find the general form for u_x in Exercise 3.

9.6. Higher-order equations

31. Find the general solution for $u_{x+3} - 6u_{x+2} + 11u_{x+1} - 6u_x = 0$.

32. If $u_0 = 1$, $u_1 = 6$, $u_2 = 10$ and $u_{x+3} - 3u_{x+2} - 4u_{x+1} + 12u_x = 0$, find u_x.

Miscellaneous problems

33. If $u_n = n(1 + k)u_{n-1} - n(n-1)ku_{n-2}$ and $u_2 = 2u_1 k$, express u_n in terms of u_1.

34. If $u_0 = 1$ and $u_{x+1} = \dfrac{x+1}{k} u_x + 1$, find $\lim_{n \to \infty} \dfrac{u_n}{p_{n-1}}$ where p_n is defined by formula (9.6).

35. Find the solution of the nonlinear difference equation

$$F(x+1) = \frac{F(x)}{1+F(x)}$$

with boundary value $F(0) = A$ by making the change of variable $f(x) = 1/F(x)$.

36. The *nested method* of evaluating polynomials is defined in Exercise 33(b) in Chapter 1. Show that this technique can be expressed as a difference equation in which we find b_0 if $b_{i-1} = (b_i + a_{i-1}k)$ given the boundary value $b_n = a_n k$. What modification in the difference equation is required in the last step when b_0 is computed?

37. A loan of amount $L = P_0$ is being amortized by regular payments of R at the end of each year at an annual rate of interest i. The outstanding loan balance at time t is denoted by P_t and it is known that $P_{t+1} = P_t(1+i) - R$.

   (a) Show that

$$P_t = L(1+i)^t - R\frac{(1+i)^t - 1}{i}$$

   (b) Show that to reduce the outstanding loan balance to exactly zero after n payments, that is, $P_n = 0$, the payment R is given by

$$R = \frac{Li}{1 - (1+i)^{-n}}$$

10

Differential equations

10.1. INTRODUCTION

As THE STUDENT recalls from previous courses in calculus, a *differential equation* is an equation to be solved involving derivatives. Standard mathematics courses in differential equations attempt to develop closed-form *analytical* solutions for as many types of differential equations as possible. However, a large number of differential equations encountered in practice do not possess analytical solutions, and *numerical* solutions based on the techniques of numerical analysis are required.

A general differential equation involving some function of x denoted by $y = y(x)$ and derivatives of y through order m can be expressed as

$$F(x, y, y^{(1)}, \ldots, y^{(m)}) = 0 \qquad (10.1)$$

where $y^{(i)}$ is the ith derivative of y. The *order* of a differential equation is the highest-order derivative involved in the equation. Thus, the differential equation given in formula (10.1) is of order m.

We shall first focus our attention on the general first-order differential equation written as

$$y' = \frac{dy}{dx} = f(x, y) \qquad (10.2)$$

The differential equation given by formula (10.2) may be either linear or non-linear.

A number of algorithms have been developed for the numerical solution of the general first-order differential equation. Sections 10.2 through 10.5 respectively consider four of these methods, or family of methods:

1. Picard's method.
2. Methods based on Taylor series.
3. Runge-Kutta methods.
4. Predictor-corrector methods.

The description of these four methods is followed by an analysis, extension, and comparison of them in Section 10.6. Finally, Section 10.7 considers the solution of higher-order equations. It will be seen that the above four methods can readily be applied to higher-order equations as well.

It is assumed in this chapter that y is a function of one variable. Although it is possible to develop numerical techniques for the solution of *partial differential equations* for functions of more than one variable, such a development is beyond the scope of this book.[1]

Although it is helpful if the student has previously taken course work in differential equations, it is not essential for a mastery of this chapter. Since we are concerned with the numerical solution of differential equations, the wide variety of analytical methods developed for certain special types of equations are not considered here. However, on occasion the analytical solution of an equation being used as an example is given to serve as a standard of comparison for the numerical solution of the same equation.

As with difference equations, the *general solution* of a differential equation of order m involves m constants. If m *boundary values* are specified, then a unique *particular solution* is determined. Since this chapter is concerned with the development of numerical solutions, it is assumed that the boundary values are given. A numerical solution is, of necessity, a particular solution and not a general solution.

The form of the numerical solution of a first-order differential equation is generally expressed as a series of points based on the given boundary value which satisfy the equation. This series of solution points is said to lie on an *integral curve* of the differential equation.

The series of solution points is generally tabulated from the boundary value at equal intervals of differencing, h. It will be seen that the appropriate

---

[1] The interested reader is referred to Bruce W. Arden and Kenneth N. Astill, *Numerical Algorithms: Origins and Applications* (Boston: Addison-Wesley Publishing Co., Inc., 1970), pp. 246–92; and to Carl-Erik Fröberg, *Introduction to Numerical Analysis* (Boston: Addison-Wesley Publishing Co., Inc., 1969), pp. 294–319, for a discussion of partial differential equations.

choice of h is important in the efficient and accurate solution of a differential equation. Certain of the methods to be discussed do not require equal intervals and are thus more flexible in this respect.

The solution points are generally found recursively in the methods to be discussed. Thus, the numerical solution of a differential equation often assumes the form of a difference equation as developed in Chapter 9. In fact, the numerical solution of differential equations is one of the primary motivations for the study of difference equations.

Differential equations appear widely in applied disciplines, particularly in scientific areas. The numerical solution of differential equations which cannot be solved analytically is one of the more important contributions of numerical analysis to applied mathematics.

10.2. PICARD'S METHOD

Picard's method of solving a first-order differential equation is a method of successive approximation. A series of functions which converge to the true solution is generated.

Assume we have a differential equation as given in formula (10.2) and a boundary value (x_0, y_0). Substituting t for x in formula (10.2) and integrating from x_0 to x gives

$$\int_{x_0}^{x} dy = \int_{x_0}^{x} f(t, y)\, dt$$

or

$$y(x) - y_0 = \int_{x_0}^{x} f(t, y)\, dt \qquad (10.3)^2$$

If $f(t, y)$ in formula (10.3) is strictly a function of t, then $y(x)$ is immediately obtained as a result of the integration

$$y(x) = y_0 + \int_{x_0}^{x} f(t)\, dt \qquad (10.4)$$

More typically, $f(t, y)$ is a function of y as well as t and the integration cannot be directly performed. In this case we adopt a recursive procedure as follows:

$$_1y(x) = y_0 + \int_{x_0}^{x} f(t, {}_0y)\, dt$$

$$_2y(x) = y_0 + \int_{x_0}^{x} f(t, {}_1y)\, dt$$

$$\vdots$$

$$_ny(x) = y_0 + \int_{x_0}^{x} f(t, {}_{n-1}y)\, dt \qquad (10.5)$$

---

[2] The student should note that $y(x_0) = y_0$, since (x_0, y_0) is the boundary value.

The series of functions $_1y, _2y, \ldots, _ny$ constitute a set of successive approxima-tions to the true solution y. The student should note that $_0y, _1y, \ldots, _{n-1}y$ appearing under the integral signs above are functions of t.

In the above approach the constant function $_0y(x) \equiv y_0$ is the starting function on which the series of successive approximations is based. Although this is the conventional starting function, alternative starting functions can be used. If information is available which indicates a function more nearly equal to the true solution y, then it should be used instead of a constant y. This will generally accelerate the rate of convergence.

Formula (10.5) is then applied for the values of x at which numerical solutions of the differential equation are required. For example, if solutions at the points

$$x_0 = x_0$$
$$x_1 = x_0 + h$$
$$x_2 = x_0 + 2h$$
$$x_3 = x_0 + 3h$$

are required, formula (10.5) is applied n times to give[3]

$$(x_0, {}_ny_0) = (x_0, y_0)$$
$$(x_1, {}_ny_1)$$
$$(x_2, {}_ny_2)$$
$$(x_3, {}_ny_3)$$

as solution points.

In applying formula (10.5) the value of n can be selected depending on the number of successive approximations deemed appropriate. Typically, values of $_ny(x)$ for $n = 1, 2, 3, \ldots$ are generated until no changes are noted to a certain number of decimal places for successive values of n. The greater the difference between the value of x being used and x_0, the larger n will have to be to produce a given level of accuracy.

The above approach can readily be applied when $f(t, y)$ can be directly integrated. However, in practice $f(t, y)$ often cannot be integrated analytically, and approximate integration is required. A wide variety of approximate integration methods developed in Chapter 8 can be adapted to this situation.

As an example, again assume that we require solutions for x_0, x_1, x_2, x_3, as defined above. We can derive approximate integration formulas based on Newton's advancing difference formula. Since four values of x are involved, we carry Newton's formula as far as third differences.

---

[3] Throughout the rest of the chapter $y(x_i)$ will sometimes be written as y_i for convenience.

From formula (10.3) we have

$$y(x_i) = y_0 + \int_{x_0}^{x_i} f(t, y)\, dt = y_0 + \int_{x_0}^{x_i} dy\,(t)$$

and making a change of variable $r = \dfrac{t - x_0}{h}$ with $dr = \dfrac{dt}{h}$, we have

$$y(x_i) = y_0 + h \int_0^i \left[y'(x_0) + r\underset{h}{\Delta y'(x_0)} + \frac{r^2 - r}{2} \underset{h}{\Delta^2 y'(x_0)} \right.$$
$$\left. + \frac{r^3 - 3r^2 + 2r}{6} \underset{h}{\Delta^3 y'(x_0)} \right] dr \quad (10.6)$$

Formula (10.6) evaluated for $i = 1, 2, 3$, respectively, gives

$$y(x_1) = y_0 + h \left[y'(x_0) + \tfrac{1}{2}\underset{h}{\Delta y'(x_0)} - \tfrac{1}{12}\underset{h}{\Delta^2 y'(x_0)} + \tfrac{1}{24}\underset{h}{\Delta^3 y'(x_0)} \right] \quad (10.7a)$$

$$y(x_2) = y_0 + h \left[2y'(x_0) + 2\underset{h}{\Delta y'(x_0)} + \tfrac{1}{3}\underset{h}{\Delta^2 y'(x_0)} \right] \quad (10.7b)$$

$$y(x_3) = y_0 + h \left[3y'(x_0) + \tfrac{9}{2}\underset{h}{\Delta y'(x_0)} + \tfrac{9}{4}\underset{h}{\Delta^2 y'(x_0)} + \tfrac{3}{8}\underset{h}{\Delta^3 y'(x_0)} \right] \quad (10.7c)$$

These equations can be expressed in terms of values of the derivatives to give

$$y(x_1) = y_0 + \frac{h}{24} \left[9y'(x_0) + 19y'(x_1) - 5y'(x_2) + y'(x_3) \right] \quad (10.8a)$$

$$y(x_2) = y_0 + \frac{h}{3} \left[y'(x_0) + 4y'(x_1) + y'(x_2) \right] \quad (10.8b)$$

$$y(x_3) = y_0 + \frac{3h}{8} \left[y'(x_0) + 3y'(x_1) + 3y'(x_2) + y'(x_3) \right] \quad (10.8c)$$

These equations can be applied recursively in a similar manner to formula (10.5). Thus, the terms $y(x_j)$ for $j = 1, 2, 3$ on the left-hand side of the equations become $_n y(x_j)$, while the terms $y'(x_j)$ for $j = 0, 1, 2, 3$ on the right-hand side become $_{n-1} y'(x_j)$. The derivatives on the right-hand side can be directly evaluated using formula (10.2). The choice of n, that is, the number of successive approximations, is based on similar considerations to the direct integration case discussed above.

Example 10.1. **Use Picard's method to solve the differential equation** $y' = x - y$ **given the boundary value** $(0, 1)$. **Use formula** *(10.5)* **as far as three successive approximations. Find numerical solutions for** $x = 0, .1, .2, .3$. **Carry four decimal places in the answers.**

The differential equation can be solved analytically to give

$$y = 2e^{-x} + x - 1$$

This exact result can be used as a standard of comparison for the approximate results to be developed.

We now apply formula (10.5) three times to obtain

$$_1y(x) = 1 + \int_0^x (t - 1)\, dt = 1 - x + \frac{x^2}{2}$$

$$_2y(x) = 1 + \int_0^x \left[t - \left(1 - t + \frac{t^2}{2}\right) \right] dt = 1 - x + x^2 - \frac{x^3}{6}$$

$$_3y(x) = 1 + \int_0^x \left[t - \left(1 - t + t^2 - \frac{t^3}{6}\right) \right] dt = 1 - x + x^2 - \frac{x^3}{3} + \frac{x^4}{24}$$

If we compute the series expansion of $2e^{-x} + x - 1$, we find that the series for $_3y(x)$ agrees as far as the third-degree term.

Table 10.1 is a comparison of values of $_1y$, $_2y$, $_3y$ together with the true values of y for $x = 0, .1, .2, .3$.

TABLE 10.1

x	$_1y(x)$	$_2y(x)$	$_3y(x)$	$y(x)$
0	1.0000	1.0000	1.0000	1.0000
.1	.9050	.9098	.9097	.9097
.2	.8200	.8387	.8374	.8375
.3	.7450	.7855	.7813	.7816

Example 10.2. Rework Example 10.1 using the approximate integration approach given in formulas (10.8a) to (10.8c).

On the first successive approximation we assume that $y \equiv 1$ is constant, and from the differential equation $y' = x - y$ we have

$$_0y'(.1) = -.9 \qquad _0y'(.2) = -.8 \qquad _0y'(.3) = -.7$$

The first application of the formulas with $h = .1$ gives

$$_1y(.1) = 1 + \frac{.1}{24}[(9)(-1) + (19)(-.9) - (5)(-.8) - .7] = .9050$$

$$_1y(.2) = 1 + \frac{.1}{3}[-1 + (4)(-.9) - .8] = .8200$$

$$_1y(.3) = 1 + \frac{.3}{8}[-1 + (3)(-.9) + (3)(-.8) - .7] = .7450$$

On the second successive approximation, from the differential equation we have

$$_1y'(.1) = -.8050 \qquad _1y'(.2) = -.6200 \qquad _1y'(.3) = -.4450$$

The second application of the formulas gives

$$_2y(.1) = 1 + \frac{.1}{24}[(9)(-1) + (19)(-.8050) - (5)(-.6200) - .4450] = .9098$$

$$_2y(.2) = 1 + \frac{.1}{3}[-1 + (4)(-.8050) - .6200] = .8387$$

$$_2y(.3) = 1 + \frac{.3}{8}[-1 + (3)(-.8050) + (3)(-.6200) - .4450] = .7855$$

On the third successive approximation, from the differential equation we have

$$_2y'(.1) = -.8098 \qquad _2y'(.2) = -.6387 \qquad _2y'(.3) = -.4855$$

The third application of the formulas gives

$$_3y(.1) = 1 + \frac{.1}{24}[(9)(-1) + (19)(-.8098) - (5)(-.6387) - .4855] = .9097$$

$$_3y(.2) = 1 + \frac{.1}{3}[-1 + (4)(-.8098) - .6387] = .8374$$

$$_3y(.3) = 1 + \frac{.3}{8}[-1 + (3)(-.8098) + (3)(-.6387) - .4855] = .7813$$

In all cases the results agree with those obtained in Example 10.1 to four decimal places.

10.3. METHODS BASED ON TAYLOR SERIES

It is possible to adapt the *Taylor series* to the numerical solution of differential equations. The Taylor series given by formula (5.20) as far as mth derivatives can be expressed as

$$y(x_{i+1}) = y(x_i) + hy^{(1)}(x_i) + \frac{h^2}{2!}y^{(2)}(x_i) + \cdots + \frac{h^m}{m!}y^{(m)}(x_i) \qquad (10.9)$$

where $h = x_{i+1} - x_i$.

Formula (10.9) is applied as a recursion formula to generate a series of solution points to the differential equation $(x_1, y_1), (x_2, y_2), \ldots$, based on the given boundary value (x_0, y_0). The first derivative term in formula (10.9) is obtained from the differential equation, while the higher-order derivatives are obtained by differentiating the differential equation.

In applying formula (10.9) both m and h can be selected. Generally, a small value of m requires a small value of h to produce a given level of accuracy, that is, more applications of the method are necessary to cover an interval of a certain length. Conversely, a large value of m permits a larger value of h to be chosen and still produce a given level of accuracy.

An important special case of formula (10.9) carries the Taylor series only as far as first derivatives to give

$$y(x_{i+1}) = y(x_i) + hy'(x_i) = y(x_i) + hf(x_i, y_i) \qquad (10.10)$$

Formula (10.10) is called *Euler's method*. Although Euler's method requires a smaller value of h than would be required if more terms are carried, the method does avoid the computation of higher-order derivatives.

The truncation error involved in formula (10.9) is given by the remainder term of the Taylor series

$$\text{truncation error} = \frac{h^{m+1}}{(m+1)!} y^{(m+1)}(\xi) \qquad (10.11)$$

where ξ lies between x_i and x_{i+1}. As a special case, the truncation error of Euler's method is

$$\text{truncation error} = \frac{h^2}{2} y^{(2)}(\xi) \qquad (10.12)$$

The truncation errors given by formulas (10.11) and (10.12) are sometimes called *local truncation errors*, since they arise on each application of the method. Thus, in generating the solution point (x_k, y_k) the truncation error appears k times, once for each application of the method.

Example 10.3. **Use Euler's method to solve the differential equation** $y' = x - y$ **given the boundary value** $(0, 1)$. **Find numerical solutions for** $x = 0, .1, .2, .3$.

Formula (10.10) becomes

$$y(x_{i+1}) = y(x_i) + .1(x_i - y_i)$$

Now applying this formula with the boundary value $(0, 1)$, we obtain the following:

$$y(.1) = \ 1 + .1(0 - \ 1) = .9$$
$$y(.2) = \ .9 + .1(.1 - \ .9) = .82$$
$$y(.3) = .82 + .1(.2 - .82) = .758$$

This example involves the same differential equation as Examples 10.1 and 10.2. Clearly, the results using Euler's method are inferior to those obtained in these earlier examples. However, the accuracy of Euler's method could be considerably improved by using a smaller value of h than .1.

Example 10.4. **Rework Example 10.3 carrying the Taylor series as far as third derivatives.**

The first three orders of derivatives are

$$\begin{align}
y' &= x - y \\
y'' &= 1 - y' = 1 - x + y \\
y''' &= -1 + y' = -1 + x - y
\end{align}$$

Formula (10.9) now becomes

$$y(x_{i+1}) = y(x_i) + .1(x_i - y_i) + .005(1 - x_i + y_i) + \tfrac{1}{6}(.001)(-1 + x_i - y_i)$$

This formula applied three times gives the following results:

$$\begin{align}
y(.1) &= .9097 \\
y(.2) &= .8375 \\
y(.3) &= .7817
\end{align}$$

If these results are compared with the true solutions given in Table 10.1 it is seen that they are entirely correct to four decimal places except for an error of only .0001 in the answer for $y(.3)$. The extra accuracy obtained in carrying the Taylor series to third derivatives instead of only to first derivatives, as in Example 10.3, is apparent.

10.4. RUNGE-KUTTA METHODS

The *Runge-Kutta methods* are a family of methods derived from the methods based on Taylor series. The Runge-Kutta methods replace the second and higher-order derivatives in the Taylor series expansion with expressions involving $f(x, y)$ which are equivalent to a certain order of derivative.

The most common and widely used Runge-Kutta formula is given by

$$y(x_{i+1}) = y(x_i) + \tfrac{1}{6}(k_1 + 2k_2 + 2k_3 + k_4) \tag{10.13}$$

where k_1, k_2, k_3, k_4 are defined by

$$k_1 = hf(x_i, y_i) \tag{10.14a}$$

$$k_2 = hf(x_i + \tfrac{1}{2}h, y_i + \tfrac{1}{2}k_1) \tag{10.14b}$$

$$k_3 = hf(x_i + \tfrac{1}{2}h, y_i + \tfrac{1}{2}k_2) \tag{10.14c}$$

$$k_4 = hf(x_i + h, y_i + k_3) \tag{10.14d}$$

Formula (10.13) is applied as a recursion formula to generate a series of solution points to the differential equation $(x_1, y_1), (x_2, y_2), \ldots$, based on the given boundary value (x_0, y_0).

Formula (10.13) is equivalent to the Taylor series method carried as far as fourth derivatives. The derivation of formula (10.13) is extremely laborious and is not given here.[4] The truncation error of formula (10.13) is the same as for a fourth-degree Taylor series method and thus involves the fifth derivative of y.

Another Runge-Kutta formula, which reproduces the Taylor series as far as third derivatives, is given by

$$y(x_{i+1}) = y(x_i) + \tfrac{1}{6}(k_1 + 4k_2 + k_3) \tag{10.15}$$

where k_1, k_2, k_3 are defined by

$$k_1 = hf(x_i, y_i) \tag{10.16a}$$

$$k_2 = hf(x_i + \tfrac{1}{2}h, y_i + \tfrac{1}{2}k_1) \tag{10.16b}$$

$$k_3 = hf(x_i + h, y_i + 2k_2 - k_1) \tag{10.16c}$$

As expected, the truncation error of formula (10.15) is the same as for a third-degree Taylor series method and thus involves the fourth derivative of y.

A third Runge-Kutta formula, which reproduces the Taylor series only as far as second derivatives, is given by

$$y(x_{i+1}) = y(x_i) + \tfrac{1}{2}(k_1 + k_2) \tag{10.17}$$

where k_1, k_2 are defined by

$$k_1 = hf(x_i, y_i) \tag{10.18a}$$

$$k_2 = hf(x_i + h, y_i + k_1) \tag{10.18b}$$

Again, the truncation error of formula (10.17) is the same as for a second-degree Taylor series method and thus involves the third derivative of y.

The student should note that if $f(x, y)$ is strictly a function of x, then both formulas (10.13) and (10.15) reduce to Simpson's rule. Similarly, under this condition formula (10.17) reduces to the trapezoidal rule.

Formulas (10.13), (10.15), and (10.17) respectively give fourth-, third-, and second-degree Runge-Kutta formulas. Other Runge-Kutta formulas can also be derived. A first-degree Runge-Kutta formula is seen to be equivalent to Euler's method. Runge-Kutta formulas of degree higher than four also exist but are difficult to derive. Finally, variations of the standard formulas given

---

[4]For a derivation of formula (10.13) see Francis Scheid, *Theory and Problems of Numerical Analysis* (New York: McGraw-Hill Book Co., 1968), pp. 202–3.

above exist. Two of these, *Gill's method* and *Heun's method*, are defined in Exercises 14 and 15, respectively.

Example 10.5. *Use the standard Runge-Kutta method given in formula (10.13) to solve the differential equation $y' = x - y$ given the boundary value $(0, 1)$. Find numerical solutions for $x = 0, .1, .2, .3$.*

The first application of (10.14) gives

$$k_1 = .1(0 - 1) = -.1$$
$$k_2 = .1(.05 - .95) = -.09$$
$$k_3 = .1(.05 - .955) = -.0905$$
$$k_4 = .1(.1 - .9095) = -.08095$$

Formula (10.13) then yields

$$y(.1) = 1 + \tfrac{1}{6}[-.1 - 2(.09) - 2(.0905) - .08095] = .9097$$

A second application of (10.14) gives

$$k_1 = .1(.1 - .9097) = -.08097$$
$$k_2 = .1(.15 - .8692) = -.07192$$
$$k_3 = .1(.15 - .8737) = -.07237$$
$$k_4 = .1(.2 - .8373) = -.06373$$

Formula (10.13) then yields

$$y(.2) = .9097 + \tfrac{1}{6}[-.08097 - 2(.07192) - 2(.07237) - .06373] = .8375$$

A final application of (10.14) gives

$$k_1 = .1(.2 - .8375) = -.06375$$
$$k_2 = .1(.25 - .8056) = -.05556$$
$$k_3 = .1(.25 - .8097) = -.05597$$
$$k_4 = .1(.3 - .7815) = -.04815$$

Formula (10.13) then yields

$$y(.3) = .8375 + \tfrac{1}{6}[-.06375 - 2(.05556) - 2(.05597) - .04815] = .7817$$

If these results are compared with the true solutions given in Table 10.1, it is seen that they are entirely correct to four decimal places except for an error of only .0001 in the answer for $y(.3)$.

10.5. PREDICTOR-CORRECTOR METHODS

The *predictor-corrector methods* are a family of methods involving a pair of formulas: (1) the predictor, and (2) the corrector. The following are the steps involved in applying a predictor-corrector method:

1. The predictor formula is applied to predict a value for $y(x_{i+1})$. This formula for $y(x_{i+1})$ is based on values of $y(x_j)$ and $y'(x_j)$ for $j \le i$.
2. The differential equation is used to find the value of $y'(x_{i+1})$ corresponding to the value of $y(x_{i+1})$ just obtained.
3. The corrector formula is then applied to improve the estimate for $y(x_{i+1})$. This formula for $y(x_{i+1})$ is based on values of $y(x_j)$ and $y'(x_j)$ for $j \le i$ and on the value of $y'(x_{i+1})$ determined in step 2.
4. Steps 2–3 are repeated as many times as necessary to produce the required level of accuracy in $y(x_{i+1})$, that is, until further applications do not change the answer obtained to a certain number of decimal places. The student should note that although the corrector may be applied repeatedly, the predictor is only applied once at the beginning of the process.
5. Steps 1–4 are then repeated to generate additional solution points past (x_{i+1}, y_{i+1}), that is, the points (x_{i+2}, y_{i+2}), (x_{i+3}, y_{i+3}), . . . , as required.

There are a wide variety of predictor-corrector methods which have been developed. This section examines the three most common methods used in practice.

The first method is the *modified Euler method*. As the student would anticipate, this method is an extension of Euler's method given in Section 10.3. The predictor and corrector formulas are defined as follows:

$$\text{Predictor: } y(x_{i+1}) = y(x_i) + hy'(x_i) \tag{10.19}$$

$$\text{Corrector: } y(x_{i+1}) = y(x_i) + \tfrac{1}{2}h[y'(x_i) + y'(x_{i+1})] \tag{10.20}$$

The predictor formula (10.19) is the same as formula (10.10) which is the formula for Euler's method. The corrector formula (10.20) is immediately obtained from formula (10.3) by an application of the trapezoidal rule.

It is instructive to compare the local truncation errors of the predictor and corrector formulas.

$$\text{truncation error of predictor} = \frac{h^2}{2} y^{(2)}(\xi) \tag{10.21}$$

$$\text{truncation error of corrector} = -\frac{h^3}{12} y^{(3)}(\xi) \tag{10.22}$$

Formula (10.21) is the same as formula (10.12). Formula (10.22) is obtained from formula (8.15) applied to the function y'. It is apparent that the truncation error of the corrector is of the opposite sign and is substantially smaller in absolute value than the truncation error of the predictor. This is a pattern which is generally true for predictor-corrector methods.

The second method is the *Milne method*. The predictor and corrector formulas are defined as follows:

Predictor: $y(x_{i+1}) = y(x_{i-3}) + \frac{4}{3}h[2y'(x_i) - y'(x_{i-1}) + 2y'(x_{i-2})]$ (10.23)

Corrector: $y(x_{i+1}) = y(x_{i-1}) + \frac{1}{3}h[y'(x_{i+1}) + 4y'(x_i) + y'(x_{i-1})]$ (10.24)

The formulas for the Milne method differ significantly in one respect from those for the modified Euler method. The formulas for $y(x_{i+1})$ in the modified Euler method involve only values of y and y' evaluated at x_i and x_{i+1}. However, the formulas for $y(x_{i+1})$ in the Milne method involve values of y and y' evaluated at x_{i-3}, x_{i-2}, x_{i-1}, x_i, x_{i+1}. Thus, starting values at x_{i-3}, x_{i-2}, x_{i-1}, x_i are required to generate values at x_{i+1}. These starting values must be obtained from one of the other methods discussed in this chapter and the given boundary value.

The predictor formula (10.23) is obtained from formula (10.3) by applying the approximate integration formula

$$\int_{-3h}^{h} f(x)\, dx = \frac{4}{3}h[2f(0) - f(-h) + 2f(-2h)]$$ (10.25)

Formula (10.25) can be derived by the method of undetermined coefficients as discussed in Section 8.4. The derivation is left as Exercise 16(a).

The corrector formula (10.24) is immediately obtained from formula (10.3) by an application of Simpson's rule.

The local truncation errors for the Milne method are given by

$$\text{truncation error of predictor} = \frac{14}{45}h^5 y^{(5)}(\xi)$$ (10.26)

$$\text{truncation error of corrector} = -\frac{1}{90}h^5 y^{(5)}(\xi)$$ (10.27)

Formula (10.26) can be derived by a method analogous to the derivations of the truncation errors for the first three Newton-Cotes formulas given in Section 8.3 applied to formula (10.25). We note that formula (10.25) involves three terms and that it is symmetrical. Since it involves an odd number of terms and is symmetrical, it gains one additional degree of accuracy and thus is accurate for third-degree polynomials. Choosing four points of collocation $-2h$, $-h$, 0, h, we obtain the truncation error of formula (10.25) as

$$\frac{f^{(4)}(\xi)}{4!} \int_{-3h}^{h} \pi(x)\, dx = \frac{f^{(4)}(\xi)}{24} \int_{-3h}^{h} (x + 2h)(x + h)x(x - h)\, dx$$

$$= \frac{f^{(4)}(\xi)}{24} \int_{-3h}^{h} (x^4 + 2hx^3 - h^2x^2 - 2h^3x)\, dx$$

$$= \frac{14}{45}h^5 f^{(4)}(\xi)$$ (10.28)

Formula (10.26) is obtained from formula (10.28) applied to the function y'.

Similarly, formula (10.27) is obtained from formula (8.16) applied to the function y'.

Again it is noted that the truncation error of the corrector is of the opposite sign and is smaller in absolute value than the truncation error of the predictor. In this case the ratio of the truncation errors is approximately 1 to -28.[5]

The third method is the *Adams-Bashforth method*. The predictor and corrector formulas are defined as follows:

Predictor: $y(x_{i+1}) = y(x_i) + \frac{1}{24}h[55y'(x_i) - 59y'(x_{i-1})$
$$+ 37y'(x_{i-2}) - 9y'(x_{i-3})] \quad (10.29)$$

Corrector: $y(x_{i+1}) = y(x_i) + \frac{1}{24}h[9y'(x_{i+1}) + 19y'(x_i)$
$$- 5y'(x_{i-1}) + y'(x_{i-2})] \quad (10.30)$$

The Adams-Bashforth method requires starting values at x_{i-3}, x_{i-2}, x_{i-1}, x_i similarly to the Milne method.

The predictor formula (10.29) is obtained from formula (10.3) by applying the approximate integration formula

$$\int_0^h f(x)\,dx = \frac{1}{24}h[55f(0) - 59f(-h) + 37f(-2h) - 9f(-3h)] \quad (10.31)$$

Formula (10.31) can be derived by the method of undetermined coefficients. The derivation is left as Exercise 16(b).

The corrector formula (10.30) is obtained from formula (10.3) by applying the approximate integration formula

$$\int_0^h f(x)\,dx = \frac{1}{24}h[9f(h) + 19f(0) - 5f(-h) + f(-2h)] \quad (10.32)$$

Formula (10.32) can also be derived by the method of undetermined coefficients. The derivation is left as Exercise 16(c).

The local truncation errors for the Adams-Bashforth method are given by

$$\text{truncation error of predictor} = \frac{251}{720}h^5 y^{(5)}(\xi) \quad (10.33)$$

$$\text{truncation error of corrector} = -\frac{19}{720}h^5 y^{(5)}(\xi) \quad (10.34)$$

Formulas (10.33) and (10.34) can be derived similarly to formula (10.28).

---

[5] The relationship is only approximate since the values of ξ in formulas (10.26) and (10.27) need not be equal.

To derive formula (10.33) we use the points of collocation $-3h$, $-2h$, $-h$, 0 to give

$$\frac{f^{(4)}(\xi)}{4!} \int_0^h \pi(x)\, dx = \frac{f^{(4)}(\xi)}{24} \int_0^h (x + 3h)(x + 2h)(x + h)x\, dx$$

$$= \frac{f^{(4)}(\xi)}{24} \int_0^h (x^4 + 6hx^3 + 11h^2 x^2 + 6h^3 x)\, dx$$

$$= \tfrac{251}{720} h^5 f^{(4)}(\xi) \tag{10.35}$$

Formula (10.33) is obtained from formula (10.35) applied to the function y'.

Similarly, to derive formula (10.34) we use the points of collocation $-2h$, $-h$, 0, h to give

$$\frac{f^{(4)}(\xi)}{4!} \int_0^h \pi(x)\, dx = \frac{f^{(4)}(\xi)}{24} \int_0^h (x + 2h)(x + h)x(x - h)\, dx$$

$$= \frac{f^{(4)}(\xi)}{24} \int_0^h (x^4 + 2hx^3 - h^2 x^2 - 2h^3 x)\, dx$$

$$= -\tfrac{19}{720} h^5 f^{(4)}(\xi) \tag{10.36}$$

Formula (10.34) is obtained from formula (10.36) applied to the function y'.

As expected, the truncation error of the corrector is of the opposite sign and is smaller in absolute value than the truncation error of the predictor. In this case the ratio of the truncation errors is approximately 19 to -251.

Example 10.6.   *Use the modified Euler method to solve the differential equation $y' = x - y$ given the boundary value $(0, 1)$. Find numerical solutions for $x = 0, .1, .2, .3$.*

The given differential equation and formulas (10.19) and (10.20) are required for the working process. The values of y to four decimal places are given in Table 10.2.

TABLE 10.2

Trial	$x = .1$ Predictor	$x = .1$ Corrector	$x = .2$ Predictor	$x = .2$ Corrector	$x = .3$ Predictor	$x = .3$ Corrector
1	.9000	.9100	.8286	.8376	.7735	.7817
2		.9095		.8371		.7813
3		.9095		.8372		.7813
4				.8372		

The derivative at the boundary value $(0, 1)$ is equal to $0 - 1 = -1$ from the differential equation. The predictor formula (10.19) applied for $x = .1$ gives

$$1 + .1(-1) = .9$$

The derivative at $(.1, .9)$ is equal to $-.8$, and the corrector formula (10.20) then gives

$$1 + \tfrac{1}{2}(.1)(-1 - .8) = .91$$

A second application of the corrector gives

$$1 + \tfrac{1}{2}(.1)(-1 - .81) = .9095$$

A third application of the corrector also gives .9095 which stops the process. The values for $x = .2$ are similarly computed based on the solution point $(.1, .9095)$. Finally, the values for $x = .3$ are based on the solution point $(.2, .8372)$.

The results in Table 10.2 can be compared with the exact results given in Table 10.1. It is apparent that the corrector does indeed substantially improve the results obtained from the predictor. The error in $y(.1)$ is equal to .0002, while the errors in $y(.2)$ and $y(.3)$ are both equal to .0003.

Example 10.7.   Use the Milne method to find the solution of the differential equation considered in Example 10.6. for $x = .4$. Use as starting values the true solutions for $x = 0, .1, .2, .3$ given in Table 10.1.

The values of x, y, y' for $x = 0, .1, .2, .3$ from Table 10.1 are as follows:

x	y	$y' = x - y$
0	1.0000	-1.0000
.1	.9097	$-.8097$
.2	.8375	$-.6375$
.3	.7816	$-.4816$

The predictor formula (10.23) gives

$$y(.4) = 1.0000 + \tfrac{4}{3}(.1)[-2(.4816) - .6375 - 2(.8097)] = .6786$$

The corrector formula (10.24) applied twice gives

$$y(.4) = .8375 + \tfrac{1}{3}(.1)[-.2786 - 4(.4816) - .6375] = .7428$$
$$y(.4) = .8375 + \tfrac{1}{3}(.1)[-.3428 - 4(.4816) - .6375] = .7406$$

This result agrees with the true solution $2e^{-.4} + .4 - 1 = .7406$.

Example 10.8.   Rework Example 10.7 using the Adams-Bashforth method.

Using the same starting values as in Example 10.7, the predictor formula (10.29) gives

$$y(.4) = .7816 + \tfrac{1}{24}(.1)[55(-.4816) - 59(-.6375) + 37(-.8097) - 9(-1)]$$
$$= .7406$$

The corrector formula (10.30) applied once gives

$$y(.4) = .7816 + \tfrac{1}{24}(.1)[9(-.3406) + 19(-.4816) - 5(-.6375) - .8097]$$
$$= .7406$$

The power of the Adams-Bashforth method is evident in this example. The predictor obtains the true solution to four decimal places, and the corrector verifies the answer. This performance is definitely superior to the performance of the Milne method in Example 10.7.

10.6. ANALYSIS, EXTENSION, AND COMPARISON OF METHODS

Sections 10.2 through 10.5 have presented descriptions of the manner in which the four methods of solving first-order differential equations under consideration are applied. However, these previous sections have not discussed the advantages and disadvantages inherent in these methods.

Picard's method is of considerable theoretical interest in the solution of differential equations. It is also an efficient practical method when $f(x, y)$ can be analytically integrated. However, in most practical problems $f(x, y)$ is not readily integrable. Although approximate integration can be used when $f(x, y)$ is not readily integrable, as illustrated in Section 10.2, alternative methods are generally preferable. Nevertheless, Picard's method is sometimes used to generate the starting values which are required for the Milne method or the Adams-Bashforth method. Both of these methods require four starting values which can be obtained from the boundary value together with formulas (10.8a)–(10.8c).

Methods based on Taylor series are excellent when the higher-order derivatives can be found. The main disadvantage of these methods is that the higher-order derivatives may be quite complex and difficult to find. Since these methods are based on derivatives rather than differences, as with predictor-corrector methods, roundoff error tends to be less of a problem with methods based on Taylor series than with certain of the other methods.

Euler's method does not require the computation of the higher-order derivatives which can be a problem with the higher-order Taylor methods and is thus quite simple to apply. However, as illustrated in Example 10.3, the interval h will generally have to be quite small for Euler's method to produce satisfactory results. This requires a great many applications of the method and can also lead to roundoff error accumulation. Generally, Euler's method is not as popular in practical work as many of the other methods.

Runge-Kutta methods possess many of the advantages of the methods based on Taylor series but avoid the problems inherent in computing higher-order derivatives. The Runge-Kutta methods have proven to be very popular in practice. As with the methods based on Taylor series, the Runge-Kutta methods do not require starting values, that is, they are "self-starting." Also, the interval h can readily be changed during the solution process if appropriate. However, the method does not indicate when a change in h is advisable.

The Runge-Kutta methods are quite efficient in terms of computer storage requirements. In particular, Gill's method is designed to minimize storage requirements. The Runge-Kutta methods do require that several values of $f(x, y)$ must be computed at intermediate points for each step. This may be time-consuming if $f(x, y)$ is quite complicated. Although the Runge-Kutta methods do not require the evaluation of higher-order derivatives as with the methods based on Taylor series, these latter methods do provide an easier control over error accumulation, since the results can continually be checked back on results computed earlier.

The Runge-Kutta methods have the advantage of using only information contained in the interval under consideration. By comparison, the predictor-corrector methods require information outside the interval under consideration which may not be as relevant to the solution.

Predictor-corrector methods are the most important competitor to the Runge-Kutta methods in practical applications. Predictor-corrector methods have the advantage that a tight control over error is possible at each step by repeated applications of the corrector. This feature is not present with any of the other methods. The predictor-corrector methods also tend to have shorter computation times than the Runge-Kutta methods, since they avoid the computation of values of $f(x, y)$ at intermediate points.

The predictor-corrector methods have the disadvantage that starting values are required. Thus, one of the other methods must be used to generate starting values. All of the other methods discussed are self-starting. Actually the modified Euler method is also self-starting. Nevertheless, the results produced by the modified Euler method are inferior to the results produced by more powerful predictor-corrector methods.

The other major disadvantage of the predictor-corrector methods is that changing the interval h during the working process is difficult. Changing the interval h is relatively simple for any of the other methods.

Although one of the advantages of the predictor-corrector methods is the ability to improve accuracy by applying the corrector repeatedly, it is desirable to have the predictor be as accurate as possible to minimize the number of applications of the corrector. Repeated applications of the corrector consume computer time and increase the chances for an accumulation of roundoff error. The Adams-Bashforth method is the most widely used predictor-corrector method in practice. Example 10.8 illustrates the power of the predictor in this method.

All four of the algorithms presented in this chapter are subject to both truncation error and roundoff error. Local truncation errors for one application of each of the methods considered are given in the preceding sections. The total truncation error is an accumulation of the local truncation errors. Attempts to analyze the total truncation error accumulation have generally not been successful in developing practical results which are readily applied. It is not safe to say that errors are additive, since an error arising early in a solution may grow, decay, or remain relatively constant throughout the remainder of the solution depending on conditions.

One of the advantages of the predictor-corrector methods which has not previously been considered is the possibility of developing acceleration methods for the convergence of the successive applications of the corrector. It is possible to use the expressions for the truncation errors of the predictor and the corrector to develop estimates of the true solution. The concept involved is quite similar to the development of Romberg's method of approximate integration developed in Section 8.6 and is called *extrapolation to the limit*.[6]

Let P be the value given by the predictor, C be the value obtained on the last application of the corrector, T be the true solution, E_p be the truncation error of the predictor, and E_c be the truncation error of the corrector. Then we have

$$E_p = T - P \tag{10.37}$$

$$E_c = T - C \tag{10.38}$$

We first consider extrapolation to the limit for the Milne method. From formulas (10.26) and (10.27) we have

$$\frac{E_p}{E_c} \doteqdot -28 \tag{10.39}$$

Now applying formula (10.39) we have

$$P - C = (T - E_p) - (T - E_c)$$
$$= E_c - E_p$$
$$\doteqdot 29E_c$$

---

[6]Extrapolation to the limit is sometimes given the picturesque name *mop-up*.

so that

$$T = C + E_c$$
$$\doteqdot C + \tfrac{1}{29}(P - C) \tag{10.40}$$

Formula (10.40) is the formula for extrapolation to the limit for the Milne method.

Similarly, for the Adams-Bashforth method we have

$$T \doteqdot C + \tfrac{19}{270}(P - C) \tag{10.41}$$

The derivation of formula (10.41) is similar to formula (10.40) and is left as Exercise 20.

All four of the algorithms considered are subject to roundoff error as well as truncation error. Unfortunately, roundoff error in these algorithms has proven to be elusive to analyze and, like total truncation error, readily applied practical results are not available.

However, it should be noted that both truncation and roundoff error are highly dependent upon the interval h chosen. From all the truncation error formulas considered in the previous sections, it is apparent that decreasing h also decreases truncation error, and conversely. However, decreasing h increases roundoff error because of the larger volume of computations required, and conversely.

These results suggest the possibility that the interval h can be chosen either too large or too small, and that some optimum intermediate value of h for which total error is minimized may exist. Conceptually the situation is somewhat analogous to subdivision of intervals for approximate differentiation considered in Section 7.5. Although examples can be constructed in which total error first decreases, reaches a minimum, and then increases as h decreases, analytical results which could approximate the optimum choice of h in advance are not available because of the difficulty in analyzing total truncation and roundoff error.

In analyzing error in the solution of differential equations the primary emphasis should be on relative error rather than actual error. For example, a growth in actual error which parallels the growth in the true solution such that relative error remains constant or even decreases is quite acceptable. On the other hand, the same actual error growth arising in a situation in which the true solution is decreasing is usually disastrous.

A method is called *relatively stable* if an error introduced into the solution process behaves similarly to the true solution, that is, increases or decreases in proportion to the true solution so that relative error remains constant or even decreases. *Instability* is a situation in which the relative error increases significantly throughout the solution process. The stability of a solution method is dependent on three factors:

1. The method used.
2. The differential equation being solved.
3. The interval h chosen.

Several methods are available in analyzing stability. The first is to perform the solution algorithm twice carrying more decimal places the second time than the first, that is, in computer terminology to do the solution both in single precision and double precision. This is one way of controlling the accumulation of the roundoff error component of total error.

The second is to choose several values of h and obtain a set of solution points for each h. As seen above, reducing h generally improves accuracy, although a point may be reached where accuracy is worsened from a large accumulation of roundoff error. This is one way of controlling the accumulation of the truncation error component of total error.

The third is to use a differential equation, for which an analytical solution is known, as a test case. For example, the differential equation $y' = Ay$ has the solution $y = e^{Ax}$ and can be used to illustrate what may happen for other differential equations. If A is chosen to be positive, the solution is increasing; whereas if A is chosen to be negative, the solution is decreasing. The differential equation chosen as a test case can be analyzed either algebraically or computationally to determine the nature of error behavior for the various methods under consideration.[7]

10.7. HIGHER-ORDER EQUATIONS

The previous sections of this chapter have considered only the solution of first-order differential equations. It remains to consider higher-order equations. This section demonstrates that all of the methods discussed previously can easily be extended to solve higher-order equations.

As a prelude to the solution of higher-order equations we consider the solution of systems of first-order equations. It will be shown that the solution of higher-order equations can be reduced to the solution of systems of first-order equations.

We first consider a system of two first-order differential equations. An extension of formula (10.2) gives the system

$$y' = \frac{dy}{dx} = f(x, y, z) \tag{10.42a}$$

$$z' = \frac{dz}{dx} = g(x, y, z) \tag{10.42b}$$

---

[7]For an interesting algebraic analysis of error behavior using the differential equation $y' = Ay$ and several of the solution methods discussed in this chapter see Scheid, *Theory and Problems of Numerical Analysis*, pp. 216–19.

where $y = y(x)$ and $z = z(x)$ are two functions of x. It is assumed that the two boundary values $y(x_0) = y_0$ and $z(x_0) = z_0$ are given. These two boundary values can be expressed as the single point (x_0, y_0, z_0).

For Picard's method formula (10.5) is replaced by the system

$$_n y(x) = y_0 + \int_{x_0}^x f(t, {}_{n-1}y, {}_{n-1}z) \, dt \tag{10.43a}$$

$$_n z(x) = z_0 + \int_{x_0}^x g(t, {}_{n-1}y, {}_{n-1}z) \, dt \tag{10.43b}$$

The working process is then quite similar to that described in Section 10.2.

For the methods based on Taylor series, formula (10.9) is replaced by the system

$$y(x_{i+1}) = y(x_i) + hy^{(1)}(x_i) + \frac{h^2}{2!} y^{(2)}(x_i) + \cdots + \frac{h^m}{m!} y^{(m)}(x_i) \tag{10.44a}$$

$$z(x_{i+1}) = z(x_i) + hz^{(1)}(x_i) + \frac{h^2}{2!} z^{(2)}(x_i) + \cdots + \frac{h^m}{m!} z^{(m)}(x_i) \tag{10.44b}$$

The higher-order derivatives for y are obtained by differentiating formula (10.42a), while the higher-order derivatives for z are obtained by differentiating formula (10.42b).

For the Runge-Kutta methods formula (10.13) is replaced by the system

$$y(x_{i+1}) = y(x_i) + \tfrac{1}{6}(k_1 + 2k_2 + 2k_3 + k_4) \tag{10.45a}$$

$$z(x_{i+1}) = z(x_i) + \tfrac{1}{6}(l_1 + 2l_2 + 2l_3 + l_4) \tag{10.45b}$$

where $k_1, l_1, k_2, l_2, k_3, l_3, k_4, l_4$ are defined by

$$k_1 = hf(x_i, y_i, z_i) \tag{10.46a}$$

$$l_1 = hg(x_i, y_i, z_i) \tag{10.46b}$$

$$k_2 = hf(x_i + \tfrac{1}{2}h, y_i + \tfrac{1}{2}k_1, z_i + \tfrac{1}{2}l_1) \tag{10.46c}$$

$$l_2 = hg(x_i + \tfrac{1}{2}h, y_i + \tfrac{1}{2}k_1, z_i + \tfrac{1}{2}l_1) \tag{10.46d}$$

$$k_3 = hf(x_i + \tfrac{1}{2}h, y_i + \tfrac{1}{2}k_2, z_i + \tfrac{1}{2}l_2) \tag{10.46e}$$

$$l_3 = hg(x_i + \tfrac{1}{2}h, y_i + \tfrac{1}{2}k_2, z_i + \tfrac{1}{2}l_2) \tag{10.46f}$$

$$k_4 = hf(x_i + h, y_i + k_3, z_i + l_3) \tag{10.46g}$$

$$l_4 = hg(x_i + h, y_i + k_3, z_i + l_3) \tag{10.46h}$$

The other Runge-Kutta formulas are similarly extended.

For the Adams-Bashforth method formula (10.29) is replaced by the system

Predictor y: $y(x_{i+1}) = y(x_i) + \frac{1}{24}h[55y'(x_i) - 59y'(x_{i-1})$

$$+ 37y'(x_{i-2}) - 9y'(x_{i-3})] \quad (10.47a)$$

Predictor z: $z(x_{i+1}) = z(x_i) + \frac{1}{24}h[55z'(x_i) - 59z'(x_{i-1})$

$$+ 37z'(x_{i-2}) - 9z'(x_{i-3})] \quad (10.47b)$$

while formula (10.30) is replaced by the system

Corrector y: $y(x_{i+1}) = y(x_i) + \frac{1}{24}h[9y'(x_{i+1}) + 19y'(x_i)$

$$- 5y'(x_{i-1}) + y'(x_{i-2})] \quad (10.48a)$$

Corrector z: $z(x_{i+1}) = z(x_i) + \frac{1}{24}h[9z'(x_{i+1}) + 19z'(x_i)$

$$- 5z'(x_{i-1}) + z'(x_{i-2})] \quad (10.48b)$$

The other predictor-corrector formulas are similarly extended.

The above equations have extended formulas for the four methods under consideration to systems of two first-order equations. The generalizations to systems of m first-order equations are obvious.

The solution of higher-order differential equations can now be obtained by replacing the higher-order equation with a system of first-order equations.

An extension of formula (10.2) for a second-order differential equation is

$$y'' = g(x, y, y') \quad (10.49)$$

Formula (10.49) can be written as two first-order equations by letting $y' = z$ and applying formulas (10.42a) and (10.42b) to give

$$y' = z \quad (10.50a)$$

$$z' = g(x, y, z) \quad (10.50b)$$

It is apparent that formula (10.50b) is identical to formula (10.42b) and that formula (10.50a) is identical to formula (10.42a) in which $f(x, y, z) = z$. The solution of the above system requires two boundary values $y(x_0) = y_0$ and $y'(x_0) = z(x_0) = z_0$.

In general, the solution of a differential equation of order m can be obtained by expressing the differential equation as a system of m first-order equations. A total of m boundary values is then required for a solution. As above, the boundary values usually are expressed as functional values and successive derivatives at one point.

However, in some cases the boundary values are expressed as functional values at more than one point. Many problems of this type can be handled by replacing the differential equation with a difference equation equivalent to a certain order of differences. The problem then reduces to solving a system of linear equations. This type of problem is illustrated in Exercise 30.

The student should note that although the solution of systems of first-order equations appears to be presented only for its application to the solution of higher-order equations, such systems of first-order equations often arise in applications directly. Thus, the solution of systems of first-order equations is of considerably broader utility than just leading to the solution of higher-order equations.

Example 10.9.   Express the differential equation $y'' - 3y' + 2y = 0$ as a system of first-order differential equations.

Formula (10.49) becomes

$$y'' = g(x, y, y') = -2y + 3y'$$

and letting $y' = z$ we have the system

$$y' = \qquad z$$
$$z' = -2y + 3z$$

The general solution of this differential equation is

$$y = c_1 e^x + c_2 e^{2x}$$

The numerical solution of this differential equation for two given boundary values is contained in Exercises 25–27.

EXERCISES

10.1.   Introduction;   10.2.   Picard's method

1.   Use Picard's method to solve the differential equation $y' = x^2 - 1$ given the boundary value (2, 3).

2.   Use Picard's method to solve the differential equation $y' = 2xy$ given the boundary value (0, 1). Express the answer as a series expansion in x as far as x^6.

3.   Use the series derived in Exercise 2 to find numerical solutions corresponding to $x = 0, .1, .2, .3$, respectively. Carry four decimal places in the answers.

4.   Rework Exercise 3 using the approximate integration method given by formulas (10.8a) to (10.8c).

5.   Derive formulas corresponding to formulas (10.8a) to (10.8c) if only values of x_0, x_1, x_2 are required instead of x_0, x_1, x_2, x_3.

10.3.   Methods based on Taylor series

6.   Rework Exercise 3 using Euler's method.

7.   (a)   Find expressions for y', y'', y''' strictly in terms of x and y for the differential equation given in Exercise 2.
      (b)   Rework Exercise 3 using the Taylor series as far as third derivatives.

8. (a) Use formula (10.12) to find an expression for the local truncation error involved in one application of Euler's method to Exercise 3.
    (b) Find a numerical upper bound for (a) for the interval $0 \leq x \leq .1$, assuming the answers obtained in Exercise 3 are true solutions.
    (c) What is the actual truncation error for $y(.1)$?
    (d) Are the answers to (b) and (c) consistent?
    (e) Find a numerical upper bound for (a) for the interval $.1 \leq x \leq .2$, again assuming the answers obtained in Exercise 3 are true solutions.
    (f) What is the actual truncation error for $y(.2)$?
    (g) How do you explain answer (f) exceeding answer (e)?

9. Show that if $y' = f(x, y) = f$, then—

    (a) $y'' = \dfrac{\partial f}{\partial x} + f \dfrac{\partial f}{\partial y}.$

    (b) $y''' = \dfrac{\partial^2 f}{\partial x^2} + \dfrac{\partial f}{\partial x}\dfrac{\partial f}{\partial y} + 2f \dfrac{\partial^2 f}{\partial x\,\partial y} + f\left(\dfrac{\partial f}{\partial y}\right)^2 + f^2\dfrac{\partial^2 f}{\partial y^2}.$

10.4. Runge-Kutta methods

10. Consider the differential equation $y' = f(x, y) = x - y$.
     (a) Find an expression for $y(x_{i+1})$ in terms of x_i and $y(x_i)$ using formula (10.9), that is, the Taylor series, as far as second derivatives.
     (b) Verify that the Runge-Kutta method given in formula (10.17) agrees with the answer in (a).

11. Rework Exercise 3 using the standard Runge-Kutta method given in formula (10.13).

12. Consider the differential equation given in Example 10.5. Find $y(.1)$ using the Runge-Kutta method given in formula (10.15).

13. Rework Exercise 12 using the Runge-Kutta method given in formula (10.17).

14. *Gill's method* is defined by

$$y(x_{i+1}) = y(x_i) + \frac{1}{6}\left[k_1 + 2\left(1 - \frac{1}{\sqrt{2}}\right)k_2 + 2\left(1 + \frac{1}{\sqrt{2}}\right)k_3 + k_4\right]$$

where

$$k_1 = hf(x_i, y_i)$$
$$k_2 = hf(x_i + \tfrac{1}{2}h, y_i + \tfrac{1}{2}k_1)$$
$$k_3 = hf\left[x_i + \tfrac{1}{2}h, y_i + \left(1 - \frac{1}{\sqrt{2}}\right)k_2 - \left(\frac{1}{2} - \frac{1}{\sqrt{2}}\right)k_1\right]$$
$$k_4 = hf\left[x_i + h, y_i + \left(1 + \frac{1}{\sqrt{2}}\right)k_3 - \frac{1}{\sqrt{2}}k_2\right]$$

Rework Exercise 12 using Gill's method.

15. *Heun's method* is defined by

$$y(x_{i+1}) = y(x_i) + \tfrac{1}{4}(k_1 + 3k_3)$$

where

$$k_1 = hf(x_i, y_i)$$
$$k_2 = hf(x_i + \tfrac{1}{3}h, y_i + \tfrac{1}{3}k_1)$$
$$k_3 = hf(x_i + \tfrac{2}{3}h, y_i + \tfrac{2}{3}k_2)$$

Rework Exercise 12 using Heun's method.

10.5. Predictor-corrector methods

16. (a) Derive formula (10.25).
    (b) Derive formula (10.31).
    (c) Derive formula (10.32).

17. Find the value of $y(.4)$ for the differential equation given in Exercise 2 based on the four starting values given in Exercise 3—
    (a) Using the predictor of the Milne method.
    (b) Using the corrector of the Milne method applied once.

18. Rework Exercise 17 using the Adams-Bashforth method.

19. For the differential equation $y' = 2y$ it is known that the local truncation error of the corrector for the modified Euler method is equal to $-\tfrac{1}{10}$ times the local truncation error of the predictor for that method. Find h, the interval at which solution points are tabulated. Assume that the values of ξ in the two truncation error terms are equal.

10.6. Analysis, extension, and comparison of methods

20. Derive formula (10.41).

21. Apply formula (10.40) to the results obtained in Example 10.7 from the predictor and the first application of the corrector.

22. Using the results in Example 10.7 and Exercise 21 compare the absolute errors to four decimal places in —
    (a) The predictor.
    (b) The first application of the corrector.
    (c) Extrapolation to the limit.

23. In predictor-corrector methods the predictor is generally more subject to roundoff error than the corrector. Assume that values of y' are subject to a maximum roundoff error of E. For the Milne method:
    (a) Find the maximum roundoff error of the terms involving y' in the predictor.
    (b) Find the maximum roundoff error of the terms involving y' in the corrector.

24. Rework Exercise 23 for the Adams-Bashforth method.

10.7. Higher-order equations

25. (a) Use Picard's method to solve the differential equation $y'' - 3y' + 2y = 0$ given the boundary values $y(0) = 2$ and $y'(0) = 3$. Express your answer as a series expansion in x as far as x^3.

    (b) Verify that the answer to (a) agrees as far as x^3 with the series expansion of $e^x + e^{2x}$, which is the true solution of the differential equation.

26. (a) For the differential equation given in Exercise 25 give expressions for the following strictly as functions of y and z:

      (1) y'.
      (2) z'.
      (3) y''.
      (4) z''.

    (b) Use the Taylor series as far as second derivatives to find numerical solutions corresponding to $x = 0, .1, .2$.

27. Use the Runge-Kutta method defined in Section 10.7 to solve the differential equation in Exercise 25 for a value of y corresponding to $x = .1$.

28. Express the differential equation

$$y''' + 2y'' = 3x - y$$

as a system of first-order differential equations.

Miscellaneous Problems

29. (a) Show that the predictor of the Adams-Bashforth method given by formula (10.29) can be expressed in operators as

$$y(x_{i+1}) = y(x_i) - \frac{\dfrac{h\nabla}{h}}{\left(1 - \nabla\limits_h\right) \log_e \left(1 - \nabla\limits_h\right)} y'(x_i)$$

    (b) Show that the corrector of the Adams-Bashforth method given by formula (10.30) can be expressed in operators as

$$y(x_{i+1}) = y(x_i) - \frac{\dfrac{h\nabla}{h}}{\log_e \left(1 - \nabla\limits_h\right)} y'(x_{i+1})$$

30. The approach for handling higher-order equations developed in Section 10.7 assumes that all boundary values are expressed as functional values and successive derivatives at one point. Alternative approaches are required if the boundary values are expressed as functional values at more than one point.

    The differential equation $y'' = xy + 2$ is to be solved given the boundary values $y(0) = 1$ and $y(6) = 7$. The solution is to consist of a set of points (x_i, y_i) for $x_i = 0, 1, 2, 3, 4, 5, 6$.

    (a) Using formula (7.5a) as far as third differences show that the differential equation can be expressed as the following difference equation

$$y(x_{i-1}) - (2 + x_i)y(x_i) + y(x_{i+1}) - 2 = 0$$

(b) By successively applying the difference equation derived in (a) show that values of $y(x_i)$ for $x_i = 1, 2, 3, 4, 5$ can be obtained by solving the following system of linear equations:

$$
\begin{aligned}
-3y(1) + y(2) &= 1 \\
y(1) - 4y(2) + y(3) &= 2 \\
y(2) - 5y(3) + y(4) &= 2 \\
y(3) - 6y(4) + y(5) &= 2 \\
y(4) - 7y(5) &= -5
\end{aligned}
$$

31. Another formula which is sometimes used to numerically solve differential equations is the *midpoint formula* defined by

$$y(x_{i+1}) = y(x_{i-1}) + 2hf(x_i, y_i)$$

Derive the midpoint formula.[8]

---

[8] The student should be careful not to confuse the *midpoint formula* with the *midpoint rule* discussed in Section 8.1.

11

Iteration

11.1. INTRODUCTION

ITERATION is a technique in which roots of equations are found by *successive approximation*. For many equations direct analytical solutions may be difficult or even impossible to implement, and iterative methods provide a workable alternative. For example, analytical solutions to find the roots of third- and fourth-degree polynomials exist, but are inconvenient to apply. For fifth- and higher-degree polynomials, analytical solutions do not even exist, and iteration is the only practical approach. Iteration can conveniently be used not only for polynomials but also for nonpolynomials, such as exponential, logarithmic, and trigonometric functions.

Iterative techniques can be considered as "trial and error" methods, in which the trials are selected to maximize the efficiency of the method, and the errors determine the subsequent steps of the process. Several different iterative methods have been developed in practice, and some of the more common of these are discussed in this chapter.

Iteration is readily applied on modern high-speed digital computers. Most iterative methods are characterized by a rather simple working formula which is applied recursively. In computer applications it is convenient to write an iteration subroutine and then loop through it as many times as necessary to produce the required level of accuracy. This procedure is often more efficient than devising complicated direct analytical solutions.

A general formula for an equation to be solved using iteration is given by

$$f(x) = 0 \qquad (11.1)$$

The *roots* of formula (11.1), which are the required answers, are those values of x for which $f(x) = 0$.

In many iterative methods it is convenient to express the formula for the equation to be solved in the following form

$$x = F(x) \qquad (11.2)$$

rather than in the form given by formula (11.1). The student should be careful to use lower-case letters for functions in the context of formula (11.1) and capital letters in the context of formula (11.2).

A typical iterative method requires a *starting value*, labeled x_0. Using this starting value, the first iterated value x_1 is generated. Then using x_1, the second iterated value x_2 is generated. This process is continued until the required level of accuracy is obtained. The iteration formula is generally expressed as a recursion formula based on formula (11.2)

$$x_{n+1} = F(x_n) \qquad (11.3)$$

The above procedure computes each iterated value from the immediately preceding iterated value. However, some iterative methods are based on more than one preceding iterated value. For example, if two values are required, then formula (11.3) becomes

$$x_{n+2} = F(x_n, x_{n+1}) \qquad (11.4)$$

where F is a function of two variables. Such an iterative method requires two starting values, x_0 and x_1. Generalizations to more than two values are obvious.

One practical problem in using iteration is determining when to stop the iterative process. An iterative method may converge or diverge, and if divergence is occurring, the iteration should be stopped. The question of convergence or divergence is considered in Section 11.2.

Even if the iteration is converging, it is usually not possible to find a value of x for which formulas (11.1) and (11.2) are exactly satisfied because of the presence of roundoff error. One possible criterion is to perform a fixed number of iterations and then stop the process. Although this approach is well defined and consumes a fixed amount of human or computer time, it does not guarantee any given level of accuracy.

A second criterion is to stop the iteration when the difference between successive iterated values is less than some predetermined quantity, that is, until

$$|x_{n+1} - x_n| < E \qquad (11.5)$$

An alternative, which is often preferable, is to base the criterion on the *relative difference*, that is,

$$\left| \frac{x_{n+1} - x_n}{x_{n+1}} \right| < E \qquad x_{n+1} \neq 0 \qquad (11.6)$$

rather than the *actual difference*.[1]

One significant characteristic of iteration is that roundoff error affects only the last iterated value, since the process is continually self-correcting until it is ended. Thus, iteration involves no accumulation of roundoff error from step to step. In fact, if an iteration is being performed by hand and a mistake is made in one iteration, the mistake will ultimately be corrected in later iterations as long as the mistake is not severe enough to cause divergence. However, in the event of a mistake of this type, a much larger number of iterations may be required.

The following assumptions are made in this chapter:

1. The equation to be solved has at least one root, that is, at least one value of x exists such that formulas (11.1) and (11.2) are satisfied.
2. The roots to be found are real, that is, iteration for complex roots is not considered.
3. The function for which roots are being determined is continuous. Thus, if values x_i and x_j are found such that $f(x_i)f(x_j) < 0$, that is, the two functional values are of opposite signs, then it is certain that a root x exists such that x lies between x_i and x_j.
4. Only functions of one variable are considered, that is, the roots to be determined arise from one equation in one unknown. Systems of linear equations are considered in Chapter 13, while systems of nonlinear equations are beyond the scope of this book.

Rapid convergence of iterative methods generally require the presence of good starting values. Obtaining good starting values is a problem for which no universal approaches exist. Any available information which might give an indication of the location of a root should obviously be used.

If no information concerning the roots of an equation is available, then a systematic search to locate the general vicinity of the roots is necessary. One

---

[1] The student should note that formulas (11.5) and (11.6) are closely related to formulas (1.2) and (1.3). However, x_{n+1} is not exactly the "true answer," but is merely an approximation to it.

possible approach is to compute values of $f(x)$ at fixed intervals and inspect them to determine the general location of roots. As mentioned above, if two successive values of x can be found such that the sign of $f(x)$ changes, then a root lies somewhere in between.

If the above search method is adopted, some judgment is required concerning the interval of x at which values of $f(x)$ should be tabulated. If values are tabulated at small intervals, excessive amounts of computer time may be involved. If values are tabulated at large intervals, some roots may be missed if more than one root lies in one of the intervals. The possibility of more than one root lying in a short interval is one of the more frustrating problems encountered in iteration. For example, the Legendre polynomial $O_{10}^L(x) = 0$ has 10 roots packed in the interval $-1 \leq x \leq 1$.

Another problem in isolating all the roots is the uncertainty of how many roots exist. It is known that a polynomial of degree n has n roots. However, this count includes complex roots and roots of multiplicity m where $m > 1$. Thus, the number of distinct, real roots of a polynomial of degree n can be any integer from 0 through n, inclusive. *Descartes' rule of signs* is useful in determining the nature of the roots of a polynomial and is discussed in Exercise 34. For nonpolynomial functions, the situation is even more complex, and the number of roots is often difficult to determine. Frequently elementary graphing of the function is useful in finding the general location of roots.

One approach that is sometimes used to isolate the roots of polynomials is *deflation*. Under this approach when one of the roots is determined, the factor theorem[2] is applied to produce a new polynomial of one lower degree. Deflation is popular since in general the lower the degree of the polynomial the easier it is to find the roots. Deflation is subject to substantial roundoff error after several applications. However, this roundoff problem is not usually very serious, since iteration can be used to improve the accuracy of any of the roots to whatever extent is required.

Example 11.1.   Isolate the real roots of $f(x) = x^3 + x - 5 = 0$ ***between successive integers.***

By computation, we have the following:

x	$f(x)$
0	-5
1	-3
2	5
3	25

---

[2] The factor theorem is discussed in Section 1.6.

Thus, a root lies between $x = 1$ and $x = 2$. This is the only real root, since $f'(x) = 3x^2 + 1 > 0$ for all x, showing that f is monotonic increasing.

11.2. RATE OF CONVERGENCE

Iteration algorithms may either converge or diverge. Furthermore, the rate of convergence may differ significantly among various convergent algorithms when applied to a specific problem.

The convergence or divergence of an iterative method can be analyzed in terms of the value of the first derivative of F in the neighborhood of the root. Let x_0 be the starting value of an iteration, $x_1, x_2, \ldots, x_n, x_{n+1}$ be the successive iterated values of x, and r be the true root. From formulas (11.2) and (11.3) we know that

$$F(r) = r$$

and

$$F(x_n) = x_{n+1}$$

We can now apply the Taylor series to give

$$F(x_n) = F(r) + (x_n - r)F'(\xi)$$

or

$$F(x_n) - F(r) = (x_n - r)F'(\xi)$$

or

$$|x_{n+1} - r| = |x_n - r||F'(\xi)|$$

where ξ lies between x_n and r.

We now define L to be

$$L = |F'(r)| \qquad (11.7)$$

and note that L can serve as an approximation to $|F'(\xi)|$. If $L < 1$, then convergence results, since the error in x_{n+1} is less than the error in x_n for all n. Similarly, if $L \geq 1$, then divergence results, since the error magnifies instead of diminishes for increasing n.[3]

If the error in x_n is denoted by E_n, we have

$$E_n = r - x_n \qquad (11.8)$$

Then the above result can be written as

$$|E_{n+1}| \doteq |E_n|L \qquad (11.9)$$

---

[3]This approach assumes that $|F'(r)|$ cannot be less than one if $|F'(\xi)|$ is greater than one, and conversely. This is almost always true in practice, particularly if x_n and r are "close."

Thus, if we have convergence, the error is reduced by approximately a factor of L each time an iteration is performed, so that the convergence is sometimes called "linear."

From the above analysis it is apparent that the rate of convergence is highly dependent on L. If L is close to one, convergence is quite slow; whereas if L is close to zero, convergence is more rapid. For example, if $L = .9$, the error in x_0 will be reduced by a factor of only $(.9)^{10} = .349$ after 10 iterations. This may not be enough to improve the accuracy of the solution by even one decimal place. However, if $L = .1$, the error in x_0 will be reduced by a factor of $(.1)^{10}$ after 10 iterations, which improves the accuracy of the solution by 10 decimal places.

Especially rapid convergence occurs if $F'(r) = 0$. In this case the Taylor series gives

$$F(x_n) = F(r) + (x_n - r)F'(r) + \tfrac{1}{2}(x_n - r)^2 F''(\xi)$$

or

$$F(x_n) - F(r) = \tfrac{1}{2}(x_n - r)^2 F''(\xi)$$

or

$$\left| x_{n+1} - r \right| = \tfrac{1}{2}\left| x_n - r \right|^2 \left| F''(\xi) \right|$$

This shows that

$$\left| E_{n+1} \right| \doteq \left| E_n \right|^2 K \tag{11.10}$$

where K is an approximation for $\left| \tfrac{1}{2}F''(\xi) \right|$. Formula (11.10) can be interpreted by saying that the error in each iterated value is proportional to the square of the error in the previous iterated value. For this reason convergence in this case is sometimes called "quadratic."

The "linear" convergence which occurs when $0 < \left| F'(r) \right| < 1$ is often called *first-order convergence*, and the "quadratic" convergence which occurs when $F'(r) = 0$ is often called *second-order convergence*. With second-order convergence the number of correct decimal places is doubled at each step of the iteration. Thus, if an iteration is first-order and p applications increase the accuracy from 3 decimal places to 6 decimal places, another p applications will increase the accuracy to 9 decimal places. However, if the iteration is second-order, the accuracy will be increased to 12 decimal places.

Actually, from the derivation of formula (11.10) second-order convergence requires that $F''(r) \neq 0$. If $F'(r) = F''(r) = 0$, then the convergence is of higher order than second. In general if $F^{(1)}(r) = F^{(2)}(r) = \cdots = F^{(m-1)}(r) = 0$ and $F^{(m)}(r) \neq 0$, then the convergence is said to be *mth-order*.

We shall show in Section 11.6 that it is often possible to develop easily applied second-order methods. The rapid convergence of these methods makes them very popular to use in practice. Theoretically, higher-order convergence methods would produce still more rapid convergence. However, these higher-order methods are difficult to develop and are generally complicated. The convergence of second-order methods is so rapid in most situations that attempting to develop yet more powerful methods is usually not worth the trouble. A third-order method is developed in Exercise 36.

Example 11.2.   Test the following iteration formulas to be used in finding the roots of the equation given in Example 11.1 for convergence or divergence:

(1) $x_{n+1} = 5 - x_n^3$.

(2) $x_{n+1} = (5 - x_n)^{1/3}$.

The two iteration formulas are obtained by simple algebraic manipulation of $f(x)$ as given in Example 11.1.

(1)   In this case $F(x) = 5 - x^3$, so that $|F'(x)| = |-3x^2|$. Since the root r satisfies $1 < r < 2$ from Example 11.1, it is clear that $|F'(r)| > 1$. Thus, the iteration diverges.

(2)   In this case $F(x) = (5 - x)^{1/3}$, so that $|F'(x)| = |-\frac{1}{3}(5 - x)^{-2/3}|$. Over the interval $1 \le x \le 2$ we have $|F'(1)| = .132$ and $|F'(2)| = .160$. Since $|F'(x)|$ is monotonic over this interval, we have $|F'(x)| < 1$ for all x, $1 \le x \le 2$. Thus, the iteration converges. Since $|F'(x)| > 0$, the convergence is first-order.

11.3.   METHOD OF SUCCESSIVE SUBSTITUTIONS

The *method of successive substitutions* is one of the most elementary and basic methods of iteration. The iteration formula is developed on an ad hoc basis by algebraic manipulation of the given function, $f(x)$. The two iteration formulas tested for convergence in Example 11.2 were developed by this method.

When developing iteration formulas in this fashion it is important to test them for convergence as in Example 11.2. It is quite easy to develop iteration formulas which diverge. If convergence occurs, it is generally first-order. One would be unusually fortunate to develop a second-order method by simple algebraic manipulations of this type.

The convergence in which $|F'(x)| < 1$ can be illustrated graphically. Figure 11.1 illustrates the situation in which $0 < F'(x) < 1$ and Figure 11.2 illustrates the situation in which $-1 < F'(x) < 0$. In both Figures 11.1 and 11.2 the starting value x_0 is to the right of the root r. The demonstration that convergence to r also occurs in both cases if starting values to the left of r are used is left as Exercise 9(a). It is interesting to note in these examples that with positive slopes of $F(x)$ the convergence is from one side, while with negative slopes successive values alternate sides.

FIGURE 11.1

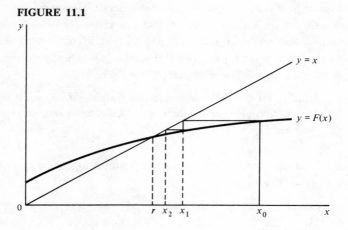

FIGURE 11.2

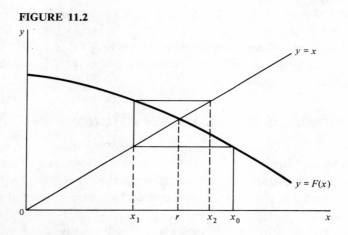

Figure 11.3 shows the divergence if $F'(x) > 1$, and Figure 11.4 shows the divergence if $F'(x) < -1$. The demonstration that divergence in both Figures 11.3 and 11.4 occurs if starting values to the left of r are used is left as Exercise 9(b).

Example 11.3. Use the method of successive substitutions to find the real root of the equation given in Example 11.1 using the second (convergent) iteration formula developed in Example 11.2. Develop an answer accurate to three decimal places and use as a starting value $x_0 = 2$.

The iteration formula is given by

$$x_{n+1} = (5 - x_n)^{1/3}$$

FIGURE 11.3

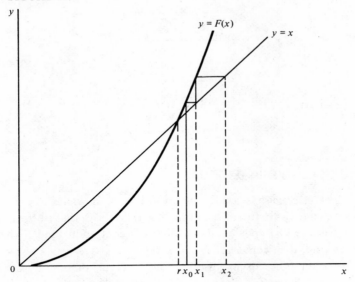

FIGURE 11.4

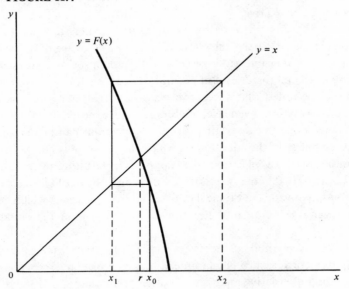

The successive iterated values are given in Table 11.1. The values of x obtained on the fourth and fifth iterations do not change, so that further iterations will not develop any more accuracy unless more decimal places are carried. The root to three decimal places is 1.516.

TABLE 11.1

n	x_n
0	2.000
1	1.442
2	1.527
3	1.514
4	1.516
5	1.516

11.4. SUCCESSIVE BISECTION

Successive bisection[4] is a straightforward method which can be applied whenever x_0 and x_1 are found such that $f(x_0)f(x_1) < 0$, that is, the two functional values are of opposite signs. If $f(x)$ is continuous and if the above conditions hold, then convergence is certain.

The iteration proceeds by bisecting the interval, that is, by computing $\bar{x} = \frac{1}{2}(x_0 + x_1)$. Then $f(\bar{x})$ is computed. In rare cases $f(\bar{x})$ is exactly zero and the iteration is finished, since $\bar{x}$ is the required root. More typically, $f(\bar{x})$ is either positive or negative.

The next iteration proceeds by choosing as starting values $\bar{x}$ and either x_0 or x_1, such that $f(\bar{x})$ and the choice of $f(x_0)$ or $f(x_1)$ are of opposite sign. Then the new smaller interval is bisected. The same process continues in this manner until the required level of accuracy is obtained.

It should be noted that successive bisection involves a fixed number of applications to attain a given level of accuracy, since the length of the resulting interval remaining after a certain number of bisections can be determined in advance regardless of the equation on which the iteration is being performed. For example, after n applications the length of the resulting interval in which the answer lies is $\left(\frac{1}{2}\right)^n$, if the two starting values differ by one. The fixed number of iterations involved is not typical of other iterative methods which generally involve a variable number of iterations depending upon the equation involved.

Although the method of successive bisection is quite easy to implement and although convergence is guaranteed, the rate of convergence is usually quite slow and alternative methods are often preferable.

Example 11.4. Use successive bisection to solve $x^2 = e^x$ correct to five decimal places. Use as starting values $x_0 = -1$ and $x_1 = 0$.

---

[4]Successive bisection is sometimes called the *Bolzano algorithm.*

The equation for which a root is required is

$$f(x) = x^2 - e^x = 0$$

It is apparent that a root exists between -1 and 0, since $f(-1) = .63212$ and $f(0) = -1.00000$.

The iteration is given in Table 11.2. Thus, the root lies in the interval $-.70347 < r < -.70346$. Without performing further bisections, the answer to five decimal places is chosen as $-.70347$, since $|f(-.70347)| < |f(-.70346)|$.

TABLE 11.2

Iteration	Bisection point, $\bar{x}$	$f(\bar{x})$	Interval	
			x_0	x_1
0			-1.00000	.00000
1	$-.50000$	$-.35653$	-1.00000	$-.50000$
2	$-.75000$	.09013	$-.75000$	$-.50000$
3	$-.62500$	$-.14464$	$-.75000$	$-.62500$
4	$-.68750$	$-.03018$	$-.75000$	$-.68750$
5	$-.71875$	.02924	$-.71875$	$-.68750$
6	$-.70312$	$-.00066$	$-.71875$	$-.70312$
7	$-.71094$	.01425	$-.71094$	$-.70312$
8	$-.70703$	.00678	$-.70703$	$-.70312$
9	$-.70508$	.00307	$-.70508$	$-.70312$
10	$-.70410$	.00120	$-.70410$	$-.70312$
11	$-.70361$	.00027	$-.70361$	$-.70312$
12	$-.70336$	$-.00020$	$-.70361$	$-.70336$
13	$-.70348$	.00002	$-.70348$	$-.70336$
14	$-.70342$	$-.00009$	$-.70348$	$-.70342$
15	$-.70345$	$-.00003$	$-.70348$	$-.70345$
16	$-.70346$	$-.000014$	$-.70348$	$-.70346$
17	$-.70347$	.000005	$-.70347$	$-.70346$

11.5. SUCCESSIVE INVERSE INTERPOLATION

Successive inverse interpolation is another method which can be applied whenever x_0 and x_1 are found such that $f(x_0)f(x_1) < 0$, that is, the two functional values are of opposite signs. Generally, the rate of convergence of successive inverse interpolation is superior to successive bisection. Successive inverse interpolation is often called *regula falsi* (the method of false position).

The iteration formula can be derived from geometric considerations based on Figure 11.5. Equating slopes of the line through $(x_0, f(x_0))$, $(x_1, f(x_1))$ and $(x_2, 0)$, we have

$$\frac{f(x_1) - 0}{x_1 - x_2} = \frac{f(x_1) - f(x_0)}{x_1 - x_0}$$

FIGURE 11.5

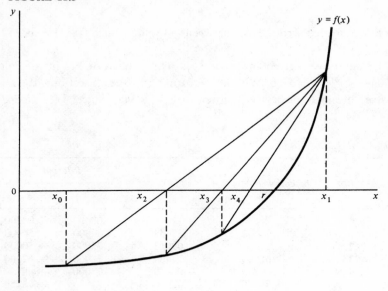

We now take reciprocals and obtain

$$\frac{x_1 - x_2}{f(x_1)} = \frac{x_1 - x_0}{f(x_1) - f(x_0)}$$

or

$$x_2 = x_1 - \frac{(x_1 - x_0)f(x_1)}{f(x_1) - f(x_0)} \qquad (11.11a)$$

The above formula can also be expressed in the more symmetrical form

$$x_2 = \frac{x_0 f(x_1) - x_1 f(x_0)}{f(x_1) - f(x_0)} \qquad (11.11b)$$

The general iteration formula is then given by

$$x_{n+2} = x_{n+1} - \frac{(x_{n+1} - x_n)f(x_{n+1})}{f(x_{n+1}) - f(x_n)} \qquad (11.12a)$$

or

$$x_{n+2} = \frac{x_n f(x_{n+1}) - x_{n+1} f(x_n)}{f(x_{n+1}) - f(x_n)} \qquad (11.12b)$$

The student should note that the above formulas are of the type given by formula (11.4), that is, each iterated value of x is based on two previous iterated values rather than just one.

In applying successive inverse interpolation after the first iteration, the two most recently computed values of x for which the values of $f(x)$ are of opposite sign are used. For example, in Figure 11.5, x_4 is based on x_1 and x_3 rather than on x_2 and x_3, since $f(x_2)$ and $f(x_3)$ have the same sign. Thus, there is no problem of division by zero in the denominator of formula (11.12b) since $f(x_{n+1}) - f(x_n)$ is really the addition of two numbers of the same sign. The convergence using this approach is often strictly from one side after the first iteration, as illustrated in Figure 11.5.

It is possible to develop a more precise set of conditions for convergence of successive inverse interpolation. The following three conditions, called *Fourier conditions*, are sufficient, but not necessary, to have convergence[5]:

$$(1) \quad f(x_0)f(x_1) < 0$$
$$(2) \quad f(x_0)f''(x_0) > 0$$
$$(3) \quad f''(x) \neq 0 \qquad \text{for } x_0 < x < x_1$$

The rate of convergence of successive inverse interpolation cannot be directly ascertained from the results developed in Section 11.2, since F is a function of two variables. However, if the iteration proceeds from one side only as illustrated in Figure 11.5, then in formula (11.12b) either $(x_n, f(x_n))$ or $(x_{n+1}, f(x_{n+1}))$ is fixed and equal to either $(x_0, f(x_0))$ or $(x_1, f(x_1))$ throughout the entire iteration. Thus, if $x_0 = A$ and $f(x_0) = B$ are fixed values throughout the entire iteration for x_n and $f(x_n)$, formula (11.12b) becomes

$$x_{n+2} = \frac{Af(x_{n+1}) - Bx_{n+1}}{f(x_{n+1}) - B}$$

This expression can be written as

$$F(x) = \frac{Af(x) - Bx}{f(x) - B}$$

and can be analyzed as in Section 11.2. In general it is seen that $F'(x) \neq 0$, so that the convergence is first-order. The details are left as Exercise 18.

In practice several variations of successive inverse interpolation have been developed to accelerate convergence. Two of these variations are described here.

The first variation is the *secant method*. In this method, formula (11.12b) is always applied to the last two iterated values of x regardless of sign. Thus, convergence is accelerated by avoiding the one-sided approach so often encountered in successive inverse interpolation. The last two iterated values of x

---

[5]For a proof of the Fourier conditions see W. Jennings, *First Course in Numerical Methods* (New York: The Macmillan Co., 1964), pp. 24–25.

generally are closest to the root. If the secant method converges, then convergence is usually more rapid than for unmodified successive inverse interpolation. However, there is a greater chance for divergence. Since the two values of x may be on the same side of the root, there is no guarantee of convergence as with unmodified successive inverse interpolation.

The second variation is *Wheeler's modification*. This modification is designed to accelerate convergence when the iteration proceeds from one side only, as is typically the case. In this variation if a functional value is repeatedly being used $\left(e.g., f(x_1) \text{ in Figure 11.5}\right)$, then it is replaced by $\frac{1}{2}f(x_1)$ which may make the convergence two-sided or may at least accelerate the one-sided convergence. This modification can be applied once, at scattered intervals, or on every iteration, as conditions warrant. Figure 11.6 illustrates the effect of this modification for the iteration illustrated in Figure 11.5.

Example 11.5.　*Rework Example 11.4 using successive inverse interpolation.*

The working formulas are

$$f(x) = x^2 - e^x$$

and

$$x_{n+2} = \frac{x_n f(x_{n+1}) - x_{n+1} f(x_n)}{f(x_{n+1}) - f(x_n)}$$

FIGURE 11.6

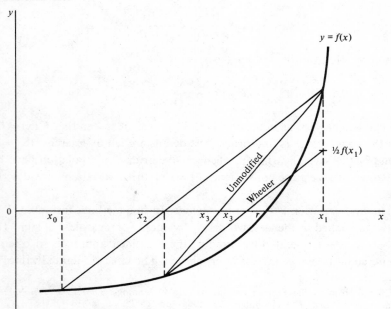

The iteration is given in Table 11.3. Thus the correct answer to five decimal places is $-.70347$.

The student should note the superior convergence of successive inverse interpolation in comparison with successive bisection.

TABLE 11.3

Step n	Values of x used		Functional values used		Iterated values	
	x_n	x_{n+1}	$f(x_n)$	$f(x_{n+1})$	x_{n+2}	$f(x_{n+2})$
0	-1.00000	.00000	.63212	-1.00000	$-.61270$	$-.16649$
1	-1.00000	$-.61270$	.63212	$-.16649$	$-.69344$	$-.01899$
2	-1.00000	$-.69344$	.63212	$-.01899$	$-.70238$	$-.00207$
3	-1.00000	$-.70238$	.63212	$-.00207$	$-.70335$	$-.00022$
4	-1.00000	$-.70335$	.63212	$-.00022$	$-.70345$	$-.00003$
5	-1.00000	$-.70345$	.63212	$-.00003$	$-.70346$	$-.000014$
6	-1.00000	$-.70346$	.63212	$-.000014$	$-.70347$	.000005
7	-1.00000	$-.70347$	.63212	.000005	$-.70347$	.000005

11.6. NEWTON-RAPHSON METHOD

The *Newton-Raphson method* is an iterative method which produces second-order convergence. Since it is a rather simple method producing rapid convergence, it may be the most widely used of all iterative methods.

The iteration formula for the Newton-Raphson method is obtained by using only the first two terms in the Taylor series to give

$$f(x_{n+1}) = f(x_n) + (x_{n+1} - x_n)f'(x_n)$$

If $f(x_{n+1})$ is set equal to zero, then

$$x_{n+1} = x_n - \frac{f(x_n)}{f'(x_n)} \tag{11.13}$$

Formula (11.13) is the standard form of the iteration formula for the Newton-Raphson method.

Geometrically, the Newton-Raphson method involves a succession of tangent lines as illustrated in Figure 11.7. Generally the convergence of the Newton-Raphson method is from one side similar to successive inverse interpolation. This is the case in the iteration illustrated in Figure 11.7.

The second-order convergence of the Newton-Raphson method can be

FIGURE 11.7

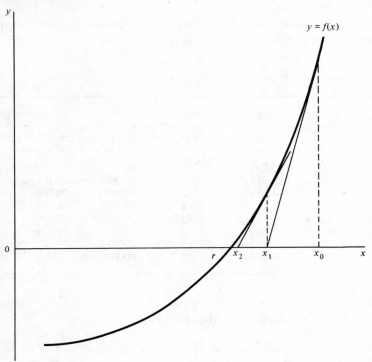

demonstrated using the results developed in Section 11.2. From formula (11.13), we have

$$F(x) = x - \frac{f(x)}{f'(x)}$$

so that

$$F'(x) = 1 - \frac{[f'(x)]^2 - f(x)f''(x)}{[f'(x)]^2}$$

$$= \frac{f(x)f''(x)}{[f'(x)]^2} \tag{11.14}$$

Thus, $F'(r) = 0$ since $f(r) = 0$, as long as first and second derivatives of f exist and as long as $f'(r) \neq 0$.

One disadvantage of the Newton-Raphson method is that it requires the derivative of $f(x)$ to be readily obtainable. In some cases the derivative may be difficult or even impossible to find. One possibility is to use the same value of the derivative for several iterations in succession rather than recomputing it each time. A second possibility is to use approximate differentiation. These alternatives are not particularly effective, however, and an iterative method other than Newton-Raphson is generally employed if the derivative of $f(x)$ poses a problem.

It is possible for the Newton-Raphson method to fail in certain instances:

1. The method may converge to some other root than the one sought, particularly if roots are tightly packed together. Thus, convergence to the correct root may require excellent starting values which have to be obtained by one of the other iterative methods.
2. If $f'(x) = 0$ then the method fails, since division by zero is impossible. This can be interpreted geometrically in Figure 11.7 as a horizontal tangent line which never intersects the x-axis. This problem can sometimes be avoided by choosing a different starting value (but not always, for example, see number 4 below).
3. The method can fail if $f(x)$ is not defined for all values of x. Figure 11.8 illustrates this possibility for a situation in which $f(x)$ is defined for only positive values of x.
4. The method can fail in the vicinity of a maximum or minimum of $f(x)$. This is illustrated in Figure 11.9.
5. The method can fail in the vicinity of a point of inflection of $f(x)$. This is illustrated in Figure 11.10.

FIGURE 11.8

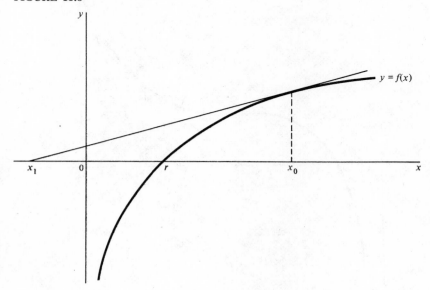

FIGURE 11.9

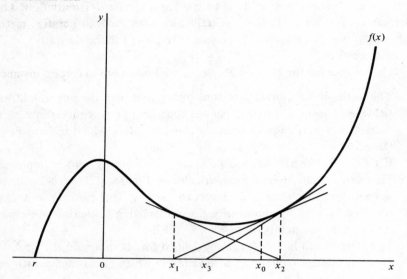

FIGURE 11.10

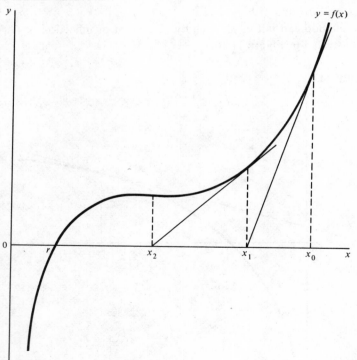

Example 11.6.   Rework Example 11.4 using the Newton-Raphson method.
Use $x_0 = -1$ as the starting value.

The formulas for $f(x)$ and $f'(x)$ are

$$f(x) = x^2 - e^x$$
$$f'(x) = 2x - e^x$$

so that

$$x_{n+1} = x_n - \frac{f(x_n)}{f'(x_n)}$$

$$= \frac{x_n(2x_n - e^{x_n}) - (x_n^2 - e^{x_n})}{2x_n - e^{x_n}}$$

$$= \frac{x_n^2 + (1 - x_n)e^{x_n}}{2x_n - e^{x_n}}$$

The iteration is given in Table 11.4. Thus, the correct answer to five decimal places is $-.70347$. The one-sided approach and rapid convergence of the Newton-Raphson method are evident.

TABLE 11.4

n	x_n
0	-1.00000
1	$-\ .73304$
2	$-\ .70381$
3	$-\ .70347$
4	$-\ .70347$

11.7.   AITKEN'S Δ^2 PROCESS

Aitken's Δ^2 process, sometimes called *extrapolation to the limit,*[6] is a method of accelerating the rate of convergence in an iteration. It is a general method and can be used for any iteration in which three consecutive iterated values of x are available. However, it is primarily used as an acceleration method for slow first-order convergence methods, such as might arise from the method of successive substitutions.

From the derivation of formula (11.9) we have

$$r - x_{n+1} \doteqdot F'(r)(r - x_n)$$

and

$$r - x_{n+2} \doteqdot F'(r)(r - x_{n+1})$$

---

[6]The student should note the use of the term "extrapolation to the limit" in Sections 8.6 and 10.6. Thus, the term is used in several different contexts in numerical analysis.

from which we obtain

$$\frac{r - x_{n+1}}{r - x_{n+2}} \doteq \frac{r - x_n}{r - x_{n+1}}$$

or

$$(r - x_{n+1})^2 \doteq (r - x_n)(r - x_{n+2})$$

or

$$r^2 - 2rx_{n+1} + x_{n+1}^2 \doteq r^2 - r(x_n + x_{n+2}) + x_n x_{n+2}$$

If we solve for r, we obtain

$$r \doteq \frac{x_n x_{n+2} - x_{n+1}^2}{x_n - 2x_{n+1} + x_{n+2}}$$

or

$$r \doteq x_{n+2} - \frac{(\Delta x_{n+1})^2}{\Delta^2 x_n} \tag{11.15}$$

Formula (11.15) is the standard version of Aitken's Δ^2 process applied to three consecutive iterated values x_n, x_{n+1}, and x_{n+2}.

Example 11.7.   *Apply Aitken's Δ^2 process to the iteration given in Example 11.5 based on x_2, x_3, and x_4.*

From Example 11.5 we have the following:

n	x_n	Δx_n	$\Delta^2 x_n$
2	$-.61270$		
		$-.08074$	
3	$-.69344$		$.07180$
		$-.00894$	
4	$-.70238$		

Aitken's Δ^2 process gives

$$r \doteq x_4 - \frac{(\Delta x_3)^2}{\Delta^2 x_2}$$

$$= -.70238 - \frac{(-.00894)^2}{.07180}$$

$$= -.70349$$

The correct answer, as we have seen before, is $-.70347$. Thus, Aitken's Δ^2 process produces an answer in error by only $.00002$. This can be compared with the error in the next iterated value of x in Example 11.5, that is, x_5, of $-.00012$.

The original iteration can now resume and convergence will occur more rapidly. Aitken's Δ^2 process can be applied at any point in an iteration to accelerate convergence in this fashion.

11.8. INVERSE INTERPOLATION

In Section 5.8 *inverse interpolation* is defined as interpolation in which the roles of x and y are reversed, that is, a value of x is sought corresponding to a given value of y.

One of the approaches discussed in Section 5.8 is to assume that y is a polynomial in x. This assumption produces a polynomial in which the roots must be determined to obtain the values of x.

The iterative methods developed in this chapter offer a variety of approaches to find these roots. In Example 5.8 an ad hoc method is developed based on Bessel's formula. Alternative approaches for this example using methods developed in this chapter are generally preferable to the method developed in Example 5.8. Exercises 32 and 33 apply the Newton-Raphson method to this example.

EXERCISES

11.1. Introduction

1. (a) Isolate the real roots of $f(x) = x^3 - 4x + 2$ between successive integers.
    (b) Isolate the real roots of $f(x) = 12x^3 - 7x^2 - 6x + 2$ between successive integers.
2. The roots of $f(x) = x^3 + x^2 - 5x - 2$ are to be found.
    (a) Find one root which is an integer.
    (b) If $f(x)$ is deflated based on the root found in (a), find the resulting quadratic.
    (c) Find the remaining two roots to three decimal places.
3. How many real roots do the following equations have?
    (a) $xe^x = 1$.
    (b) $e^x = 4x + 1$.
    (c) $e^x = \log_e x + 1$.
    (d) $\tan x = x$.
    (e) $\sin x + \frac{1}{4}x = 1$.

11.2. Rate of convergence

4. Test the following iteration formulas to be used in finding a root of $xe^x = 1$ between 0 and 1 for divergence or convergence. If the iteration converges, find the order of convergence.
    (a) $x_{n+1} = e^{-x_n}$.
    (b) $x_{n+1} = -\log_e x_n$.
    (c) $x_{n+1} = \dfrac{x_n^2 + e^{-x_n}}{x_n + 1}$.

5. The following iteration formula is given for finding $a^{-1/2}$, $a \neq 0$

$$x_{n+1} = \tfrac{1}{2}(3x_n - ax_n^3)$$

Show that second-order convergence results.

6. It is known that $E_{n+1} = \lambda E_n$ for all positive integers n, where $-1 < \lambda < 1$, $\lambda \neq 0$, and $E_n = r - x_n$. If $x_2 = 1$, $x_3 = 2$, $x_4 = 2.5$, find r.

7. The roots of the equation $x = a - bx^2$ for $a > 0$, $b > 0$, are to be found by using the following iteration formula

$$x_{n+1} = a - bx_n^2$$

Show that a necessary and sufficient condition for convergence is $ab < \tfrac{3}{4}$.

11.3.   Method of successive substitutions

8. It is known that $x^3 = x^2 + x + 1$ has a root, r, between 1 and 2.
    (a) Find two iterative formulas by ad hoc methods which will converge to this root.

    (b) Use both of these two formulas to find this root to two decimal places. Use as a starting value $x_0 = 2$.

    (c) Compare the rates of convergence of the two methods. Is this borne out by an examination of $|F'(r)|$ for both methods?

9. (a) Demonstrate geometrically that convergence occurs in Figures 11.1 and 11.2 if $x_0 < r$, that is, if the starting value is to the left of the root.
    (b) Demonstrate geometrically that divergence occurs in Figures 11.3 and 11.4 if $x_0 < r$, that is, if the starting value is to the left of the root.

10. (a) Use the iteration formula given in Exercise 5 with a starting value $x_0 = .5$ to find $1/\sqrt{3}$ to five decimal places.
    (b) The correct answer is first obtained on which iteration?

11. (a) Find the real nonzero root obtained by the following iteration formula

$$x_{n+1} = \sqrt{x_n}(3 - \sqrt{x_n})$$

(b) What is the order of convergence?

11.4.   Successive bisection

12. Rework Example 11.3 using successive bisection with starting values 1 and 2.

13. How many applications of successive bisection must be made in order that the length of the resulting interval is less than .0005 in general? Assume that the two starting values are successive integers.

14. The root of $x = e^{-x}$ which lies between 0 and 1 is to be found. If two applications of successive bisection are made, what is the resulting interval?

11.5. Successive inverse interpolation

15. Rework Example 11.3 using successive inverse interpolation with starting values 1 and 2.

16. The root of $x^3 - x - 1 = 0$ which lies between 1 and 2 is to be found by successive inverse interpolation. On one of the iterations the iterated value is 1.24038. If one of the two values of x on which this iteration is based is 1, find the other value of x (to one decimal place).

17. The following equations have a root at $x = 1$. If starting values 0 and 2 are chosen and successive inverse interpolation is applied, which of the Fourier conditions is (are) *not* satisfied for each?
    (a) $x^2 - 2x + 1 \doteq 0$.
    (b) $x^3 - 3x^2 + 2 = 0$.

18. If an iteration by successive inverse interpolation proceeds from one side, then we can write

$$F(x) = \frac{Af(x) - Bx}{f(x) - B}$$

    (a) Find an expression for $F'(r)$ and thereby show that generally convergence is first-order.
    (b) Under what conditions on $f'(r)$ will the convergence be second-order?

19. In Example 11.5 it is decided to use the secant method for the third iteration, that is, x_4 is to be based on x_2 and x_3, instead of on x_0 and x_3.
    (a) Find x_4 to five decimal places.
    (b) Find the error in x_4, $|E_4|$, using the secant method.
    (c) Find the error in x_4, $|E_4|$, using unmodified successive inverse interpolation.

20. The root of $x^3 - x - 1 = 0$ which lies between 1 and 2 is to be found.
    (a) Use successive inverse interpolation to find the first iterated value, x_2, based on starting values $x_0 = 1$ and $x_1 = 2$, to three decimal places.
    (b) Use unmodified successive inverse interpolation to find the second iterated value, x_3, to three decimal places.
    (c) Rework the second iteration in (b) using Wheeler's modification. In this case the functional value $f(x)$ which is used in (a) and is again used in (b) is halved in (b).
    (d) The correct answer to three decimal places is 1.325. Find the error in x_3, $|E_3|$, using unmodified successive inverse interpolation in (b).
    (e) Find the error in x_3, $|E_3|$, using Wheeler's modification in (c).

11.6. Newton-Raphson method

21. Rework Example 11.3 using the Newton-Raphson method with a starting value $x_0 = 2$.

22. Derive Newton-Raphson iteration formulas for the four equations given in Exercise 3 which have roots.

23. Given the following Newton-Raphson iteration formulas, find $f(x)$:

   (a) $x_{n+1} = \dfrac{3x_n^4 + 10}{4x_n^3 - 1}$.

   (b) $x_{n+1} = -\frac{1}{2}x_n + 15x_n^{1/3}$.

   (c) $x_{n+1} = \dfrac{x_n + 1}{e^{x_n} + 1}$.

24. Values of $a^{-1/2}$, $a \neq 0$, are to be found by using a Newton-Raphson iteration.
   (a) Derive an iteration formula, using $f(x) = x^2 - a^{-1} = 0$.
   (b) Derive an iteration formula, using $f(x) = x^{-2} - a = 0$. Show that the answer agrees with the iteration formula given in Exercise 5. Thus, Newton-Raphson iteration formulas are not unique and can be derived in a variety of forms.

25. (a) Derive the following Newton-Raphson iteration formula for finding $a^{1/m}$

   $$x_{n+1} = \frac{(m-1)x_n^m + a}{mx_n^{m-1}}$$

   (b) Derive the following Newton-Raphson iteration formula for finding $\sqrt{a}$

   $$x_{n+1} = \frac{1}{2}\left(x_n + \frac{a}{x_n}\right)$$

   This formula is widely used for generating square roots.

   (c) Derive the following alternative Newton-Raphson iteration formula for finding $\sqrt{a}$

   $$x_{n+1} = \frac{3ax_n - x_n^3}{2a}$$

26. Derive a Newton-Raphson iteration formula for finding reciprocals, that is, $1/a$, which involves no divisions.

27. The root of $x^2 - x - 1 = 0$ which lies between 1 and 2 is to be found by using Newton-Raphson iteration. If the third iterated value $x_3 = 34/21$, find the second iterated value x_2.

28. For the Newton-Raphson iteration given in Exercise 27, express x_{n+2} as a function of x_n.

29. Generalize the derivation of the Newton-Raphson method by carrying the Taylor series to three terms to find x_{n+1}, if it is known that $x_n = 2$, $f(x_n) = 5$, $f'(x_n) = 13$, and $f''(x_n) = 12$.

11.7. Aitken's Δ^2 process

30. (a) Apply Aitken's Δ^2 process to the first three iterated values in Exercise 15, which are 1.375, 1.481, 1.508.
   (b) What is the absolute error in the answer to (a)?

31. Aitken's Δ^2 process, is applied to three consecutive iterated values x_0, x_1, x_2 to generate x_3, an approximation to r. If $x_1 = 2$, $x_2 = 2.5$, and $x_3 = 3$, find x_0.

11.8. Inverse interpolation

32. Find the polynomial which passes through the following points given in Example 5.8:

x	y	Δy	$\Delta^2 y$	$\Delta^3 y$
1	1.000			
		.414		
2	1.414		$-.096$	
		.318		.046
3	1.732		$-.050$	
		.268		
4	2.000			

33. Use the result from Exercise 32 and the Newton-Raphson method to find the value of x which corresponds to $y = 1.5$. Use as a starting value $x_0 = 2$.

Miscellaneous problems

34. *Descartes' rule of signs* is a method of helping to determine the nature of the roots of a polynomial. The polynomial to be tested, $p(x)$, is written in standard notation given by formula (1.10). Then Descartes' rule of signs states:

   (1) The maximum number of positive roots is the number of sign changes of the a_i's in $p(x)$.
   (2) The maximum number of negative roots is the number of sign changes of the a_i's in $p(-x)$.

   The actual number of positive or negative roots may be less than the numbers given by rules 1 and 2, respectively, but cannot be greater. Use Descartes' rule of signs and deflation to isolate the roots of the following polynomials and determine how many are—

   (1) Positive.
   (2) Negative.
   (3) Complex.

   It should be noted that complex roots always occur in pairs.
   (a) $x^4 + 1$.
   (b) $x^3 + 2x + 5$.
   (c) $x^4 - x - 10$.
   (d) $x^4 - 3x^3 + 2x^2 - 10x + 12$.

35. It is possible to generalize successive inverse interpolation to pass a quadratic through three iterated values instead of a straight line through two iterated values. Show that if x_3 is to be based on x_0, x_1, x_2, then

$$x_3 = x_2 - \frac{2C}{B \pm \sqrt{B^2 - 4AC}}$$

where

$$A = \frac{(x_1 - x_0)f(x_2) + (x_0 - x_2)f(x_1) + (x_2 - x_1)f(x_0)}{(x_2 - x_1)(x_1 - x_0)^2}$$

$$B = \frac{(x_1 - x_0)(2x_2 - x_1 - x_0)f(x_2) - (x_2 - x_0)^2 f(x_1) + (x_2 - x_1)^2 f(x_0)}{(x_2 - x_1)(x_1 - x_0)^2}$$

$$C = \frac{x_2 - x_0}{x_1 - x_0} f(x_2)$$

36. *Bailey's method* of iteration has third-order convergence and is given by

$$x_{n+1} = x_n - \left[\frac{f(x_n)}{f'(x_n) - \dfrac{f(x_n)f''(x_n)}{2f'(x_n)}} \right]$$

Derive the above formula by carrying three terms of the Taylor series in the derivation of the Newton-Raphson formula instead of only two terms. *Hint*: The Newton-Raphson formula must be used in the derivation.

37. *Steffensen's method* of iteration is defined as follows:
   (1) Find values of x_0, x_1, x_2 by any method of iteration.
   (2) Find x_3 by Aitken's Δ^2 process based on x_0, x_1, x_2.
   (3) Find values of x_4, x_5 by the iterative method in step 1.
   (4) Find x_6 by Aitken's Δ^2 process based on x_3, x_4, x_5.
   (5) Continue the above process for as many cycles as necessary to obtain the required level of accuracy. Every third iterated value, x_9, x_{12}, ... is found by Aitken's Δ^2 process.

Steffensen's method is to be used to find the root of $x^3 = x^2 + x + 1$ which lies between 1 and 2. The iteration formula

$$x_{n+1} = 1 + \frac{1}{x_n} + \frac{1}{x_n^2}$$

is used. Four decimal places are carried in all computations and a starting value $x_0 = 2$ is used.
   (a) Find x_3.
   (b) Find x_6.

12

Matrices and determinants

12.1. INTRODUCTION

CHAPTER 12 contains a brief review of some of the more important properties of matrices and determinants. This material serves as important background information for Chapters 13 and 14.

Students who have had course work in linear algebra will find that much of this chapter is familiar to them. The theoretical development of matrices and determinants is not considered here and can be found in textbooks in linear algebra. Thus, in this chapter the key properties of matrices and determinants are stated without proof.

12.2. MATRIX DEFINITIONS

A *matrix* is a rectangular array of numbers. If the matrix A has m rows and n columns, then it is said to be of *order* $m \times n$. Square brackets are placed around the rectangular array of numbers so that it is clear that the array is to be interpreted as a matrix.

Each of the mn numbers in the matrix A is said to be an *element* of A. The elements of a matrix are denoted by lower-case letters with two subscripts. The first subscript denotes the row of the element and the second subscript denotes the column. Thus, a_{ij} is the element in the ith row and jth column. The general form of an $m \times n$ matrix is given by

$$A = \begin{bmatrix} a_{11} & a_{12} & \cdots & a_{1n} \\ a_{21} & a_{22} & \cdots & a_{2n} \\ \vdots & \vdots & & \vdots \\ a_{m1} & a_{m2} & \cdots & a_{mn} \end{bmatrix} \tag{12.1}$$

For ease in writing, the matrix given by formula (12.1) is often written as A or as $[a_{ij}]$, if there can be no ambiguity about its order.

Two matrices A and B are said to be *equal*, if the corresponding elements are all equal, that is, if $a_{ij} = b_{ij}$ for all i and j. Note that for two matrices to be equal they must be of the same order.

The *transpose* of a matrix A is denoted by A^T and is the matrix obtained by interchanging the rows and columns of A. Thus A^T is given by

$$A^T = \begin{bmatrix} a_{11} & a_{21} & \cdots & a_{m1} \\ a_{12} & a_{22} & \cdots & a_{m2} \\ \vdots & \vdots & & \vdots \\ a_{1n} & a_{2n} & \cdots & a_{mn} \end{bmatrix} \tag{12.2}$$

If the order of A is $m \times n$, then the order of A^T is $n \times m$.

Sometimes a matrix consists of only a single row or column. The former is often called a *row matrix* and the latter a *column matrix*. These special matrices are *vectors*, which appear in many areas of mathematics.

A *square matrix* is a matrix in which the number of rows is equal to the number of columns, that is, $m = n$. The elements lying on the path from the upper left-hand corner to the lower right-hand corner are called the *diagonal* of the matrix. A square matrix is said to be *symmetric*, if $A^T = A$, that is, if $a_{ij} = a_{ji}$ for all i and j.

A square matrix is said to be *upper triangular* if all the elements below the diagonal are equal to zero. Thus,

$$A = \begin{bmatrix} a_{11} & a_{12} & a_{13} & \cdots & a_{1n} \\ 0 & a_{22} & a_{23} & \cdots & a_{2n} \\ 0 & 0 & a_{33} & \cdots & a_{3n} \\ \vdots & \vdots & \vdots & & \vdots \\ 0 & 0 & 0 & \cdots & a_{nn} \end{bmatrix} \tag{12.3}$$

is upper triangular. Similarly,

$$A = \begin{bmatrix} a_{11} & 0 & 0 & \cdots & 0 \\ a_{21} & a_{22} & 0 & \cdots & 0 \\ a_{31} & a_{32} & a_{33} & \cdots & 0 \\ \vdots & \vdots & \vdots & & \vdots \\ a_{n1} & a_{n2} & a_{n3} & \cdots & a_{nn} \end{bmatrix} \tag{12.4}$$

is said to be *lower triangular*. A matrix which is both upper and lower triangular has elements only on the diagonal and is called a *diagonal matrix*.

A matrix whose elements are all equal to zero is called the *zero matrix* or the *null matrix*. The zero matrix is usually denoted by 0, although this notation does not distinguish it from the real number zero.

A diagonal matrix whose diagonal elements are all equal to one is called the *identity matrix*. Note that the identity matrix must be square and that all its nondiagonal elements are equal to zero. The identity matrix is usually denoted by I.

12.3. ELEMENTARY MATRIX OPERATIONS

The addition of two matrices A and B is defined if both matrices have the same order. The sum is obtained by adding corresponding elements. Thus,

$$\begin{bmatrix} a_{11} & \cdots & a_{1n} \\ \vdots & & \vdots \\ a_{m1} & \cdots & a_{mn} \end{bmatrix} + \begin{bmatrix} b_{11} & \cdots & b_{1n} \\ \vdots & & \vdots \\ b_{m1} & \cdots & b_{mn} \end{bmatrix} = \begin{bmatrix} a_{11} + b_{11} & \cdots & a_{1n} + b_{1n} \\ \vdots & & \vdots \\ a_{m1} + b_{m1} & \cdots & a_{mn} + b_{mn} \end{bmatrix} \quad (12.5)$$

Subtraction is similarly defined.

It is easy to show that the following algebraic properties are satisfied by matrix addition when the operation is defined:

1. Commutative law of addition: $A + B = B + A$.
2. Associative law of addition: $A + (B + C) = (A + B) + C$.

In considering multiplication two cases arise: (a) multiplication of a matrix by a scalar, and (b) multiplication of two matrices.

To multiply a matrix A by a scalar k each element of A is multiplied by k. Thus,

$$k \begin{bmatrix} a_{11} & \cdots & a_{1n} \\ \vdots & & \vdots \\ a_{m1} & \cdots & a_{mn} \end{bmatrix} = \begin{bmatrix} ka_{11} & \cdots & ka_{1n} \\ \vdots & & \vdots \\ ka_{m1} & \cdots & ka_{mn} \end{bmatrix} \quad (12.6)$$

The multiplication of two matrices is more complex. The product AB is defined only if the number of columns in A is equal to the number of rows in B. The number of rows in the answer is equal to the number of rows in A, and the number of columns in the answer is equal to the number of columns in B. Thus, if A is of order $m \times n$ and B is of order $n \times p$, the product AB is defined and is of order $m \times p$.

The multiplication is performed by taking the sum of the products of elements in rows of A and columns of B. More specifically, if $AB = C$, then each element of the answer C is obtained by

$$c_{ij} = a_{i1}b_{1j} + a_{i2}b_{2j} + \cdots + a_{in}b_{nj} \quad (12.7)$$

Since the number of columns in A and the number of rows in B are both equal to n, there is an exact correspondence of elements each time the sum given by formula (12.7) is computed. An illustration of matrix multiplication is given in Example 12.1.

It can be shown that the following algebraic properties are satisfied by matrix multiplication when the operation is defined:

1. Associative law of multiplication: $A(BC) = (AB)C$.
2. Distributive law: $A(B + C) = AB + AC$ or $(B + C)A = BA + CA$.

However, the commutative law of multiplication does not hold, that is, AB does not necessarily equal BA. In fact, BA is not even defined unless B happens to be of order $n \times m$. If $AB = BA$, then A and B are said to be *commutative*.

The role of the zero matrix in matrix addition is analogous to the number zero in the addition of real numbers. Thus,

$$A + 0 = 0 + A = A \tag{12.8}$$

Similarly, the role of the identity matrix in matrix multiplication is analogous to the number one in the multiplication of real numbers. Thus,

$$AI = IA = A \tag{12.9}$$

The student should note that the two identity matrices in formula (12.9) are of different orders if $m \neq n$.

It is interesting to note that if each element of the identity matrix is multiplied by some constant k, then

$$A(kI) = (kI)A = kA$$

This produces the same answer as scalar multiplication given by formula (12.6). The matrix kI is often called a *scalar matrix* for this reason.

It is instructive to consider the transpose of a sum and product. The following results can easily be obtained:

$$(A + B)^T = A^T + B^T \tag{12.10}$$

$$(AB)^T = B^T A^T \tag{12.11}$$

The reverse order of formula (12.11) should be carefully noted.

One other matrix operation which is quite important is *matrix inversion*. However, a thorough treatment of matrix inversion requires the use of determinants which are considered in Sections 12.4 and 12.5. Matrix inversion is then considered in Section 12.6.

Example 12.1. *Compute $AB - 3C$ if*

$$A = \begin{bmatrix} 2 & -1 & 3 \\ -4 & 1 & 0 \end{bmatrix} \qquad B = \begin{bmatrix} 1 & -1 & 2 \\ 3 & 0 & 5 \\ -2 & -2 & 4 \end{bmatrix} \qquad C = \begin{bmatrix} 2 & -4 & 3 \\ 1 & -1 & 5 \end{bmatrix}$$

We have

$$AB = \begin{bmatrix} 2 & -1 & 3 \\ -4 & 1 & 0 \end{bmatrix} \begin{bmatrix} 1 & -1 & 2 \\ 3 & 0 & 5 \\ -2 & -2 & 4 \end{bmatrix}$$

$$= \begin{bmatrix} (2)(1) + (-1)(3) + (3)(-2) \\ (-4)(1) + (1)(3) + (0)(-2) \end{bmatrix}$$

$$\begin{bmatrix} (2)(-1) + (-1)(0) + (3)(-2) & (2)(2) + (-1)(5) + (3)(4) \\ (-4)(-1) + (1)(0) + (0)(-2) & (-4)(2) + (1)(5) + (0)(4) \end{bmatrix}$$

$$= \begin{bmatrix} -7 & -8 & 11 \\ -1 & 4 & -3 \end{bmatrix}$$

and

$$3C = 3 \begin{bmatrix} 2 & -4 & 3 \\ 1 & -1 & 5 \end{bmatrix} = \begin{bmatrix} 6 & -12 & 9 \\ 3 & -3 & 15 \end{bmatrix}$$

so that

$$AB - 3C = \begin{bmatrix} -7 & -8 & 11 \\ -1 & 4 & -3 \end{bmatrix} - \begin{bmatrix} 6 & -12 & 9 \\ 3 & -3 & 15 \end{bmatrix} = \begin{bmatrix} -13 & 4 & 2 \\ -4 & 7 & -18 \end{bmatrix}$$

12.4. DETERMINANTS

A *determinant* is a specific functional value given to a square array of numbers. If the determinant has n rows and columns, it is called a determinant of *order n*. Vertical lines are placed on each side of the array so that it will be interpreted as a determinant. The elements in a determinant are given labels in the same manner as the elements in a matrix. Thus, the general form of a determinant is given by

$$|A| = \begin{vmatrix} a_{11} & a_{12} & \cdots & a_{1n} \\ a_{21} & a_{22} & \cdots & a_{2n} \\ \vdots & \vdots & & \vdots \\ a_{n1} & a_{n2} & \cdots & a_{nn} \end{vmatrix} \qquad (12.12)$$

For ease in writing, the determinant given by formula (12.12) is often written as $|A|$ or $|a_{ij}|$, if there can be no ambiguity about its order.

It is important to distinguish determinants from matrices. A matrix is an array of numbers and nothing more. However, a determinant is one number, that is, a functional value assigned to the numbers in a square array. It is true that a square matrix has a determinant value which is important in certain applications, but there is still a basic conceptual difference between the two. In order to avoid any ambiguity, matrices are denoted by square brackets and determinants by vertical lines.

The basic definition of the determinant value is rather complex. It is found by taking the sum of all possible products of n of the elements of the determinant, such that one and only one element comes from each row and each column, and then attaching the proper sign. Thus, a general formula for the determinant value is given by

$$|A| = \sum \epsilon_{j_1 j_2 \cdots j_n} a_{1 j_1} a_{2 j_2} \cdots a_{n j_n} \tag{12.13}$$

In formula (12.13) the factor $\epsilon_{j_1 j_2 \cdots j_n}$ is either $+1$ or -1. One element is chosen from each row successively and then a column is chosen such that no column is used twice. From elementary permutation theory there are $n!$ ways of selecting the values $j_1, j_2, \ldots, j_n$, and thus the summation contains $n!$ terms.

The sign of each term, $\epsilon_{j_1 j_2 \cdots j_n}$, is determined by counting the number of interchanges of two labels which are necessary to arrange $j_1, j_2, \ldots, j_n$ in ascending order $1, 2, \ldots, n$. If an even number of interchanges is required, then $\epsilon = +1$; and if an odd number of interchanges is required, then $\epsilon = -1$. It can be shown that the number of interchanges required in any given problem is either even or odd, regardless of the order in which they are performed.

Evaluating a determinant of order two from the definition is straightforward. We have $2! = 2$ terms to evaluate which are given by

$$\begin{vmatrix} a_{11} & a_{12} \\ a_{21} & a_{22} \end{vmatrix} = a_{11} a_{22} - a_{12} a_{21} \tag{12.14}$$

Evaluating a determinant of order three from the definition is more complex but still feasible. We have $3! = 6$ terms to evaluate which are given by

$$\begin{vmatrix} a_{11} & a_{12} & a_{13} \\ a_{21} & a_{22} & a_{23} \\ a_{31} & a_{32} & a_{33} \end{vmatrix} = \begin{matrix} a_{11} a_{22} a_{33} + a_{12} a_{23} a_{31} + a_{13} a_{21} a_{32} \\ - a_{13} a_{22} a_{31} - a_{12} a_{21} a_{33} - a_{11} a_{23} a_{32} \end{matrix} \tag{12.15}$$

Evaluating determinants of order higher than three from the definition is quite involved and is usually avoided in practice. More efficient algorithms for evaluating these higher-order determinants are developed in Sections 12.5 and 13.3. In fact, even third-order determinants are often evaluated by these methods rather than by resorting to formula (12.15).

The following is a list of properties which are useful in working with determinants:

1. The determinant value is unchanged if the transpose is taken, that is, $|A^T| = |A|$. Thus, any statement made about rows applies to columns, and conversely.
2. If every element in any one row is equal to zero, then $|A| = 0$.
3. If every element in any one row is multiplied by some scalar k, then the value of the determinant is $k|A|$.
4. The interchange of any two rows of a determinant changes the sign of the determinant value.
5. If any two rows are equal to each other, then $|A| = 0$.
6. Any multiple of one row added to another row leaves the value of the determinant unchanged.
7. If A and B are two square matrices of the same order, then the determinant of the product (in either order) is equal to the product of the determinants, that is, $|AB| = |BA| = |A||B|$.
8. If the elements of $|A|$ are functions of x, then the derivative of $|A|$ with respect to x is obtained as the sum of n determinants in which all the row elements in the first row are differentiated in the first determinant, second row elements are differentiated in the second determinant, and so forth. For example, if $n = 3$ we have

$$\frac{d}{dx}\begin{vmatrix} a_{11} & a_{12} & a_{13} \\ a_{21} & a_{22} & a_{23} \\ a_{31} & a_{32} & a_{33} \end{vmatrix} = \begin{vmatrix} a'_{11} & a'_{12} & a'_{13} \\ a_{21} & a_{22} & a_{23} \\ a_{31} & a_{32} & a_{33} \end{vmatrix}$$

$$+ \begin{vmatrix} a_{11} & a_{12} & a_{13} \\ a'_{21} & a'_{22} & a'_{23} \\ a_{31} & a_{32} & a_{33} \end{vmatrix}$$

$$+ \begin{vmatrix} a_{11} & a_{12} & a_{13} \\ a_{21} & a_{22} & a_{23} \\ a'_{31} & a'_{32} & a'_{33} \end{vmatrix}$$

Example 12.2. ***Evaluate***

$$\begin{vmatrix} 13 & -14 & 26 \\ -26 & 27 & -50 \\ 39 & -42 & 76 \end{vmatrix}$$

The determinant could be evaluated directly using formula (12.15) to give

$$(13)(27)(76) + (-14)(-50)(39) + (26)(-26)(-42)$$
$$- (26)(27)(39) - (-14)(-26)(76) - (13)(-50)(-42) = 26$$

However, the volume of computation involved in this approach can be substantially reduced.

This is accomplished by using property 6 before applying formula (12.15). We have[1]

$$\begin{vmatrix} 13 & -14 & 26 \\ -26 & 27 & -50 \\ 39 & -42 & 76 \end{vmatrix} = \begin{vmatrix} 13 & -14 & 26 \\ 0 & -1 & 2 \\ 39 & -42 & 76 \end{vmatrix} \quad (R2 + 2R1)$$

$$= \begin{vmatrix} 13 & -14 & 26 \\ 0 & -1 & 2 \\ 0 & 0 & -2 \end{vmatrix} \quad (R3 - 3R1)$$

$$= \begin{vmatrix} 13 & -1 & 26 \\ 0 & -1 & 2 \\ 0 & 0 & -2 \end{vmatrix} \quad (C2 + C1)$$

$$= (13)(-1)(-2) = 26 \text{ (formula (12.15))}$$

Example 12.3. **Evaluate**

$$\begin{vmatrix} 14 & -7 & 5 \\ -8 & 4 & 7 \\ 6 & -3 & -3 \end{vmatrix}$$

From property 3 the value of the determinant is multiplied by -2, if the second column is multiplied by -2. But then the first and second columns are equal, so the determinant is equal to zero from property 5.

12.5. EVALUATION OF DETERMINANTS BY MINORS

One common method of evaluating determinants is by *minors*. This method replaces a determinant of order n with n determinants of order $n - 1$.

A *minor* is associated with each element a_{ij} of a determinant of order n and is the determinant of order $n - 1$ which is obtained by deleting the ith row and jth column while retaining the rest of the determinant. This minor is usually denoted by M_{ij}.

The *cofactor* associated with the element a_{ij} is denoted by A_{ij} and is a signed minor. Its definition is

$$A_{ij} = (-1)^{i+j} M_{ij} \tag{12.16}$$

The determinant $|A|$ can be expressed in terms of elements and cofactors as

$$|A| = \sum_{\substack{i=1,\,...,\,n \\ \text{or } j=1,\,...,\,n}} a_{ij} A_{ij} \tag{12.17}$$

where the summation is understood to be along any row or column.

---

[1] We shall use shorthand notations in the right margin to describe the row and column operations. " R2 + 2R1 " means add 2 times row 1 to row 2. Similarly, " C1 − ½C3 " would mean subtract ½ of column 3 from column 1.

For example, the third-order determinant

$$|A| = \begin{vmatrix} a_{11} & a_{12} & a_{13} \\ a_{21} & a_{22} & a_{23} \\ a_{31} & a_{32} & a_{33} \end{vmatrix}$$

can be expanded by minors along the first row to give

$$|A| = a_{11}A_{11} + a_{12}A_{12} + a_{13}A_{13}$$

$$= a_{11}\begin{vmatrix} a_{22} & a_{23} \\ a_{32} & a_{33} \end{vmatrix} - a_{12}\begin{vmatrix} a_{21} & a_{23} \\ a_{31} & a_{33} \end{vmatrix} + a_{13}\begin{vmatrix} a_{21} & a_{22} \\ a_{31} & a_{32} \end{vmatrix}$$

As one of several alternatives, it can also be expanded down the second column to give

$$|A| = a_{12}A_{12} + a_{22}A_{22} + a_{32}A_{32}$$

$$= -a_{12}\begin{vmatrix} a_{21} & a_{23} \\ a_{31} & a_{33} \end{vmatrix} + a_{22}\begin{vmatrix} a_{11} & a_{13} \\ a_{31} & a_{33} \end{vmatrix} - a_{32}\begin{vmatrix} a_{11} & a_{13} \\ a_{21} & a_{23} \end{vmatrix}$$

It may not be apparent that expansion by minors is really an efficient procedure. For example, an expansion of a fifth-order determinant will lead to five fourth-order determinants. If each of these is expanded once again, 20 third-order determinants are produced, hardly a pleasant prospect.

However, in practice, expansion by minors is usually combined with the properties discussed in the preceding section. For example, property 6 can be used to transform the determinant so that one row or column has only one nonzero element. Then expansion by minors becomes much more efficient.

Upper triangular, lower triangular, or diagonal determinants, are particularly easy to evaluate. As Exercise 28, the student will show that the determinant value for any of these particular forms is merely the product of the elements on the diagonal.

An alternative approach to the evaluation of large-order determinants is based on Gaussian elimination and is discussed in Chapter 13.

Example 12.4.   Evaluate

$$\begin{vmatrix} 1 & 1 & 1 & 1 \\ 1 & 2 & 3 & 4 \\ 1 & 3 & 6 & 10 \\ 1 & 4 & 10 & 20 \end{vmatrix}$$

We have

$$
\begin{vmatrix}
1 & 1 & 1 & 1 \\
1 & 2 & 3 & 4 \\
1 & 3 & 6 & 10 \\
1 & 4 & 10 & 20
\end{vmatrix}
=
\begin{vmatrix}
1 & 1 & 1 & 1 \\
0 & 1 & 2 & 3 \\
0 & 2 & 5 & 9 \\
0 & 3 & 9 & 19
\end{vmatrix}
\quad
\begin{matrix}
\\ (R2 - R1) \\ (R3 - R1) \\ (R4 - R1)
\end{matrix}
$$

$$
=
\begin{vmatrix}
1 & 2 & 3 \\
2 & 5 & 9 \\
3 & 9 & 19
\end{vmatrix}
\qquad \text{(expand C1 by minors)}
$$

$$
=
\begin{vmatrix}
1 & 2 & 3 \\
0 & 1 & 3 \\
0 & 3 & 10
\end{vmatrix}
\qquad
\begin{matrix}
(R2 - 2R1) \\ (R3 - 3R1)
\end{matrix}
$$

$$
=
\begin{vmatrix}
1 & 3 \\
3 & 10
\end{vmatrix}
\qquad \text{(expand C1 by minors)}
$$

$$
= 10 - 9 = 1
$$

Example 12.5. **Expand the determinant given in Example 12.3 by minors in the second row and verify that the determinant value is zero.**

We have

$$
\begin{vmatrix}
14 & -7 & 5 \\
-8 & 4 & 7 \\
6 & -3 & -3
\end{vmatrix}
= -(-8)\begin{vmatrix} -7 & 5 \\ -3 & -3 \end{vmatrix} + 4\begin{vmatrix} 14 & 5 \\ 6 & -3 \end{vmatrix} - 7\begin{vmatrix} 14 & -7 \\ 6 & -3 \end{vmatrix}
$$

$$
= (8)(36) + (4)(-72) - (7)(0)
$$

$$
= 0
$$

the anticipated answer.

12.6. THE INVERSE OF A MATRIX

If a matrix B exists such that $AB = BA = I$, then B is called the *inverse* of A and is denoted by $B = A^{-1}$. In a certain sense, the process of matrix inversion is analogous to division for real numbers.

Not every matrix has an inverse. For an inverse to exist, two conditions must hold:

1. The matrix must be square.
2. The determinant value of the matrix must be nonzero.

Matrices which possess inverses are called *nonsingular*. All other matrices are called *singular*. It can be shown that if an inverse exists, then it is unique.

It is instructive to consider the inverse of the product of two matrices. The following result can easily be derived by the student

$$(AB)^{-1} = B^{-1}A^{-1} \qquad (12.18)$$

The reverse order of formula (12.18) should be carefully noted.

There are several different approaches which can be used to find the inverse of a matrix. Four of the more common ones are discussed in this section.

The first approach for matrix inversion is the *elimination method*. In this method the matrix to be inverted, A, and the identity matrix of the same order, I, are written side by side. A series of row operations[2] is performed upon A to change it into I. If the same operations are performed upon I, then it is changed into A^{-1}. Legitimate row operations are:

1. Any row can be multiplied by a nonzero scalar.
2. Any multiple of one row can be added to any other row.

In theory, the order in which row operations are performed makes no difference in the answer obtained. However, in practice, if roundoff error is involved, then the order in which computations are performed may have a significant impact. This point is discussed further in Section 13.4.

The second approach for matrix inversion is the *adjoint method*, which is based on determinants. Arrange the cofactors in matrix form

$$\begin{bmatrix} A_{11} & A_{12} & \cdots & A_{1n} \\ A_{21} & A_{22} & \cdots & A_{2n} \\ \vdots & \vdots & & \vdots \\ A_{n1} & A_{n2} & \cdots & A_{nn} \end{bmatrix}$$

The *adjoint*, denoted by adj A, is the transpose of the above matrix, that is,

$$\text{adj } A = \begin{bmatrix} A_{11} & A_{21} & \cdots & A_{n1} \\ A_{12} & A_{22} & \cdots & A_{n2} \\ \vdots & \vdots & & \vdots \\ A_{1n} & A_{2n} & \cdots & A_{nn} \end{bmatrix} \qquad (12.19)$$

The inverse of A is then given by

$$A^{-1} = \frac{\text{adj } A}{|A|} \qquad (12.20)$$

The third approach for matrix inversion is the *exchange method*. In this method the following operations are performed n times, once for each element

---

[2]Column operations could also be used. However, it is not necessary to do so, and nothing is gained thereby. Standard practice is to use all row operations.

on the diagonal. Each of these diagonal elements in turn, when it is used, is called the *pivot*. The n diagonal elements may be used as pivots in any order. The rules of the exchange method are as follows:

1. Replace the pivot a_{ii} by its reciprocal.
2. Each other element in the same column is divided by a_{ii}.
3. Each other element in the same row is divided by a_{ii} and the sign is changed.
4. All other elements, a_{kl}, are replaced by

$$a_{kl} - \frac{a_{ki}a_{il}}{a_{ii}} \tag{12.21}$$

For example, if we have the matrix

$$\begin{bmatrix} a_{11} & a_{12} & a_{13} \\ a_{21} & a_{22} & a_{23} \\ a_{31} & a_{32} & a_{33} \end{bmatrix}$$

and perform the operations above for row 2, column 2, we have

$$\begin{bmatrix} a_{11} - \dfrac{a_{12}a_{21}}{a_{22}} & \dfrac{a_{12}}{a_{22}} & a_{13} - \dfrac{a_{12}a_{23}}{a_{22}} \\[3ex] -\dfrac{a_{21}}{a_{22}} & \dfrac{1}{a_{22}} & -\dfrac{a_{23}}{a_{22}} \\[3ex] a_{31} - \dfrac{a_{32}a_{21}}{a_{22}} & \dfrac{a_{32}}{a_{22}} & a_{33} - \dfrac{a_{32}a_{23}}{a_{22}} \end{bmatrix}$$

Formula (12.21) is often called the *rectangular rule*.[3] This name refers to the fact that the two elements in the numerator are on opposite corners of a rectangle in which the other two corners are a_{kl} and a_{ii}, that is, the element being replaced and the pivot. This property makes the formula particularly easy to apply in practice.

When the above operations have been performed n times, once for each pivot on the diagonal, the resulting matrix is A^{-1}. It should be noted that the operations cannot be performed for a pivot of zero, since division by zero is undefined. However, if the matrix is nonsingular, this problem can be avoided by using the pivots in a different order, if a zero appears on the diagonal at some stage in the exchange method when it would normally become the next pivot.

---

[3]The student should not confuse this rectangular rule with the rectangular rule for approximate integration introduced in Section 8.1.

The fourth approach for matrix inversion is the *iterative method*. This method requires that a good approximation, B, to the true inverse, A^{-1}, be available. Then the inverse is given by

$$A^{-1} = (I + R + R^2 + \cdots)B \tag{12.22}$$

where

$$R = I - BA \tag{12.23}[4]$$

If B is sufficiently close to A^{-1}, then the infinite series given by formula (12.22) will converge.

Formulas (12.22) and (12.23) can be derived by noting that if the matrix series is convergent, then

$$(I - R)(I + R + R^2 + \cdots) = I$$

so that

$$I + R + R^2 + \cdots = (I - R)^{-1}$$

Then we have

$$
\begin{aligned}
(I + R + R^2 + \cdots)B &= (I - R)^{-1}B \\
&= (BA)^{-1}B \\
&= A^{-1}B^{-1}B \\
&= A^{-1}
\end{aligned}
$$

All four of the above approaches are used in practice. The adjoint method is not commonly used for inverting high-order matrices because of the large number of computations involved in evaluating the determinants, although it is a useful method for inverting small matrices by hand. Either the elimination method or the exchange method generally involves fewer computations than the adjoint method for high-order matrices.

In finding the inverses of high-order matrices, roundoff error can be a serious problem. Methods of minimizing roundoff error for the elimination method and for the exchange method are discussed in Section 13.4.

The iterative method is frequently used after one of the other approaches has already been used. It is particularly effective in developing more accurate answers when roundoff error has appeared in one of the other approaches. No roundoff error appears in the iterative method until the last iteration.

---

[4] R stands for "residual," a topic which will be discussed further in Chapter 13.

Example 12.6.   Find the inverse of

$$A = \begin{bmatrix} 2 & -3 & 4 \\ 7 & -1 & 3 \\ -1 & 2 & -2 \end{bmatrix}$$

using the elimination method.

We have

$$\begin{bmatrix} 2 & -3 & 4 \\ 7 & -1 & 3 \\ -1 & 2 & -2 \end{bmatrix} \begin{bmatrix} 1 & 0 & 0 \\ 0 & 1 & 0 \\ 0 & 0 & 1 \end{bmatrix}$$

$$\begin{bmatrix} 1 & -\frac{3}{2} & 2 \\ 7 & -1 & 3 \\ -1 & 2 & -2 \end{bmatrix} \begin{bmatrix} \frac{1}{2} & 0 & 0 \\ 0 & 1 & 0 \\ 0 & 0 & 1 \end{bmatrix} \qquad (\tfrac{1}{2}\text{R1})$$

$$\begin{bmatrix} 1 & -\frac{3}{2} & 2 \\ 0 & \frac{19}{2} & -11 \\ 0 & \frac{1}{2} & 0 \end{bmatrix} \begin{bmatrix} \frac{1}{2} & 0 & 0 \\ -\frac{7}{2} & 1 & 0 \\ \frac{1}{2} & 0 & 1 \end{bmatrix} \qquad \begin{array}{l}(\text{R2} - 7\text{R1}) \\ (\text{R3} + \text{R1})\end{array}$$

$$\begin{bmatrix} 1 & -\frac{3}{2} & 2 \\ 0 & 1 & -\frac{22}{19} \\ 0 & \frac{1}{2} & 0 \end{bmatrix} \begin{bmatrix} \frac{1}{2} & 0 & 0 \\ -\frac{7}{19} & \frac{2}{19} & 0 \\ \frac{1}{2} & 0 & 1 \end{bmatrix} \qquad (\tfrac{2}{19}\text{R2})$$

$$\begin{bmatrix} 1 & -\frac{3}{2} & 2 \\ 0 & 1 & -\frac{22}{19} \\ 0 & 0 & \frac{11}{19} \end{bmatrix} \begin{bmatrix} \frac{1}{2} & 0 & 0 \\ -\frac{7}{19} & \frac{2}{19} & 0 \\ \frac{13}{19} & -\frac{1}{19} & 1 \end{bmatrix} \qquad (\text{R3} - \tfrac{1}{2}\text{R2})$$

$$\begin{bmatrix} 1 & -\frac{3}{2} & 2 \\ 0 & 1 & -\frac{22}{19} \\ 0 & 0 & 1 \end{bmatrix} \begin{bmatrix} \frac{1}{2} & 0 & 0 \\ -\frac{7}{19} & \frac{2}{19} & 0 \\ \frac{13}{11} & -\frac{1}{11} & \frac{19}{11} \end{bmatrix} \qquad (\tfrac{19}{11}\text{R3})$$

$$\begin{bmatrix} 1 & -\frac{3}{2} & 0 \\ 0 & 1 & 0 \\ 0 & 0 & 1 \end{bmatrix} \begin{bmatrix} -\frac{41}{22} & \frac{2}{11} & -\frac{38}{11} \\ 1 & 0 & 2 \\ \frac{13}{11} & -\frac{1}{11} & \frac{19}{11} \end{bmatrix} \qquad \begin{array}{l}(\text{R1} - 2\text{R3}) \\ (\text{R2} + \tfrac{22}{19}\text{R3})\end{array}$$

$$\begin{bmatrix} 1 & 0 & 0 \\ 0 & 1 & 0 \\ 0 & 0 & 1 \end{bmatrix} \begin{bmatrix} -\frac{4}{11} & \frac{2}{11} & -\frac{5}{11} \\ 1 & 0 & 2 \\ \frac{13}{11} & -\frac{1}{11} & \frac{19}{11} \end{bmatrix} \qquad (\text{R1} + \tfrac{3}{2}\text{R2})$$

Thus,

$$A^{-1} = \frac{1}{11} \begin{bmatrix} -4 & 2 & -5 \\ 11 & 0 & 22 \\ 13 & -1 & 19 \end{bmatrix}$$

Example 12.7.   Rework Example 12.6 using the adjoint method.

The determinant value is

$$|A| = \begin{vmatrix} 2 & -3 & 4 \\ 7 & -1 & 3 \\ -1 & 2 & -2 \end{vmatrix} = \begin{array}{l} (2)(-1)(-2) + (-3)(3)(-1) + (4)(7)(2) \\ = -(4)(-1)(-1) - (-3)(7)(-2) - (2)(3)(2) \\ = 4 + 9 + 56 - 4 - 42 - 12 = 11 \end{array}$$

The cofactors are

$$A_{11} = \begin{vmatrix} -1 & 3 \\ 2 & -2 \end{vmatrix} = -4 \quad A_{12} = -\begin{vmatrix} 7 & 3 \\ -1 & -2 \end{vmatrix} = 11 \quad A_{13} = \begin{vmatrix} 7 & -1 \\ -1 & 2 \end{vmatrix} = 13$$

$$A_{21} = -\begin{vmatrix} -3 & 4 \\ 2 & -2 \end{vmatrix} = 2 \quad A_{22} = \begin{vmatrix} 2 & 4 \\ -1 & -2 \end{vmatrix} = 0 \quad A_{23} = -\begin{vmatrix} 2 & -3 \\ =1 & 2 \end{vmatrix} = -1$$

$$A_{31} = \begin{vmatrix} -3 & 4 \\ -1 & 3 \end{vmatrix} = -5 \quad A_{32} = -\begin{vmatrix} 2 & 4 \\ 7 & 3 \end{vmatrix} = 22 \quad A_{33} = \begin{vmatrix} 2 & -3 \\ 7 & -1 \end{vmatrix} = 19$$

Then

$$A^{-1} = \frac{\begin{bmatrix} A_{11} & A_{21} & A_{31} \\ A_{12} & A_{22} & A_{32} \\ A_{13} & A_{23} & A_{33} \end{bmatrix}}{|A|} = \frac{1}{11}\begin{bmatrix} -4 & 2 & -5 \\ 11 & 0 & 22 \\ 13 & -1 & 19 \end{bmatrix}$$

Example 12.8. *Rework Example 12.6 using the exchange method.*

The matrix is

$$A = \begin{bmatrix} 2 & -3 & 4 \\ 7 & -1 & 3 \\ -1 & 2 & -2 \end{bmatrix}$$

One application of the rules of the exchange method to row 1, column 1, gives

$$\begin{bmatrix} \frac{1}{2} & \frac{3}{2} & -2 \\ \frac{7}{2} & -1 - (\frac{1}{2})(7)(-3) & 3 - (\frac{1}{2})(7)(4) \\ -\frac{1}{2} & 2 - (\frac{1}{2})(-1)(-3) & -2 - (\frac{1}{2})(-1)(4) \end{bmatrix} = \begin{bmatrix} \frac{1}{2} & \frac{3}{2} & -2 \\ \frac{7}{2} & \frac{19}{2} & -11 \\ -\frac{1}{2} & \frac{1}{2} & 0 \end{bmatrix}$$

A second application to row 2, column 2, gives

$$\begin{bmatrix} \frac{1}{2} - (\frac{2}{19})(\frac{3}{2})(\frac{7}{2}) & \frac{3}{19} & -2 - (\frac{2}{19})(\frac{3}{2})(-11) \\ -\frac{7}{19} & \frac{2}{19} & \frac{22}{19} \\ -\frac{1}{2} - (\frac{2}{19})(\frac{1}{2})(\frac{7}{2}) & \frac{1}{19} & 0 - (\frac{2}{19})(\frac{1}{2})(-11) \end{bmatrix} = \begin{bmatrix} -\frac{1}{19} & \frac{3}{19} & -\frac{5}{19} \\ -\frac{7}{19} & \frac{2}{19} & \frac{22}{19} \\ -\frac{13}{19} & \frac{1}{19} & \frac{11}{19} \end{bmatrix}$$

A third application to row 3, column 3, gives

$$\begin{bmatrix} -\frac{1}{19} - (\frac{19}{11})(-\frac{5}{19})(-\frac{13}{19}) & \frac{3}{19} - (\frac{19}{11})(-\frac{5}{19})(\frac{1}{19}) & -\frac{5}{11} \\ -\frac{7}{19} - (\frac{19}{11})(\frac{22}{19})(-\frac{13}{19}) & \frac{2}{19} - (\frac{19}{11})(\frac{22}{19})(\frac{1}{19}) & 2 \\ \frac{13}{11} & -\frac{1}{11} & \frac{19}{11} \end{bmatrix}$$

$$= \begin{bmatrix} -\frac{4}{11} & \frac{2}{11} & -\frac{5}{11} \\ 1 & 0 & 2 \\ \frac{13}{11} & -\frac{1}{11} & \frac{19}{11} \end{bmatrix}$$

which is the answer.

Example 12.9. *Assume that the following approximate inverse (which is accurate to one decimal) for the matrix given in Example 12.6 is available*

$$B = \begin{bmatrix} -.4 & .2 & -.5 \\ 1.0 & 0 & 2.0 \\ 1.2 & -.1 & 1.7 \end{bmatrix}$$

Develop a better approximation using the iterative method as far as R^2.

We have

$$R = I - BA$$

$$= \begin{bmatrix} 1 & 0 & 0 \\ 0 & 1 & 0 \\ 0 & 0 & 1 \end{bmatrix} - \begin{bmatrix} -.4 & .2 & -.5 \\ 1.0 & 0 & 2.0 \\ 1.2 & -.1 & 1.7 \end{bmatrix} \begin{bmatrix} 2 & -3 & 4 \\ 7 & -1 & 3 \\ -1 & 2 & -2 \end{bmatrix}$$

$$= \begin{bmatrix} 1 & 0 & 0 \\ 0 & 1 & 0 \\ 0 & 0 & 1 \end{bmatrix} - \begin{bmatrix} 1.1 & 0 & 0 \\ 0 & 1.0 & 0 \\ 0 & -.1 & 1.1 \end{bmatrix} = \begin{bmatrix} -.1 & 0 & 0 \\ 0 & 0 & 0 \\ 0 & .1 & -.1 \end{bmatrix}$$

and

$$R^2 = \begin{bmatrix} .01 & 0 & 0 \\ 0 & 0 & 0 \\ 0 & -.01 & .01 \end{bmatrix}$$

Thus,

$$I + R + R^2 = \begin{bmatrix} .91 & 0 & 0 \\ 0 & 1 & 0 \\ 0 & .09 & .91 \end{bmatrix}$$

which gives

$$(I + R + R^2)B = \begin{bmatrix} .91 & 0 & 0 \\ 0 & 1 & 0 \\ 0 & .09 & .91 \end{bmatrix} \begin{bmatrix} -.4 & .2 & -.5 \\ 1.0 & 0 & 2.0 \\ 1.2 & -.1 & 1.7 \end{bmatrix}$$

$$= \begin{bmatrix} -.364 & .182 & -.455 \\ 1 & 0 & 2 \\ 1.182 & -.091 & 1.727 \end{bmatrix}$$

It is interesting to note that this answer is accurate to three decimal places. Further refinement in the answer can be obtained by carrying higher powers of R; or alternatively, by letting this answer serve as a new B and starting the process over again.

In general, if higher powers of R are carried, then fewer iterations are required, and conversely. A significant question would be whether it is more efficient to carry higher powers of R and perform fewer iterations or carry lower powers of R and perform more iterations. Although this is a difficult question to answer definitely, certain investigations using reasonable assumptions indicate that the optimum choice is to carry powers of R through R^3.[5]

12.7. FACTORIZATION OF A MATRIX

On occasion it is required to factor a square matrix, A, into the product of two matrices, a lower triangular matrix, L, and an upper triangular matrix, U. Thus, we have

$$A = LU \qquad (12.24)^{[6]}$$

The process of *factorization* is sometimes called *triangular decomposition*. Not all matrices, A, can be factored. Also, if A can be factored, the factorization is not unique and other factorizations exist.

In one popular method of factorization the diagonal elements of the lower triangular matrix are all equal to one. The rest of the elements in both triangular matrices can then be uniquely determined by straightforward algebra. For example, with third-order matrices we have

$$\begin{bmatrix} a_{11} & a_{12} & a_{13} \\ a_{21} & a_{22} & a_{23} \\ a_{31} & a_{32} & a_{33} \end{bmatrix} = \begin{bmatrix} 1 & 0 & 0 \\ l_{21} & 1 & 0 \\ l_{31} & l_{32} & 1 \end{bmatrix} \begin{bmatrix} u_{11} & u_{12} & u_{13} \\ 0 & u_{22} & u_{23} \\ 0 & 0 & u_{33} \end{bmatrix}$$

from which we get the following nine equations in nine unknowns:

$$u_{11} = a_{11}$$
$$u_{12} = a_{12}$$
$$u_{13} = a_{13}$$
$$l_{21}u_{11} = a_{21}$$
$$l_{21}u_{12} + u_{22} = a_{22}$$
$$l_{21}u_{13} + u_{23} = a_{23}$$
$$l_{31}u_{11} = a_{31}$$
$$l_{31}u_{12} + l_{32}u_{22} = a_{32}$$
$$l_{31}u_{13} + l_{32}u_{23} + u_{33} = a_{33}$$

These equations can be solved in the order listed by straightforward substitution for the nine unknown u's and l's.

Another obvious possibility is to set the diagonal elements of the upper triangular matrix equal to one. The working procedure would be similar to the case just discussed.

---

[5]This statement is based on unpublished research in the Computer Science Department at the University of Nebraska-Lincoln.
[6]By convention, A is always factored as LU rather than UL.

A third method which finds applications in more advanced work is the *Choleski method* or the *square-root method*. In this method the diagonal elements of the lower triangular matrix are equal to the diagonal elements of the upper triangular matrix. This method can only be used for matrices which satisfy certain conditions. Since it is not needed for any later development, further details of the Choleski factorization method are not given here but are contained in Appendix G.

Example 12.10. *Factor the matrix A given in Example 12.6 using the formulas developed above.*

The matrix is

$$A = \begin{bmatrix} 2 & -3 & 4 \\ 7 & -1 & 3 \\ -1 & 2 & -2 \end{bmatrix}$$

The standard formulas give

$$u_{11} = 2$$
$$u_{12} = -3$$
$$u_{13} = 4$$
$$2l_{21} = 7, \; l_{21} = \tfrac{7}{2}$$
$$\tfrac{7}{2}(-3) + u_{22} = -1, \; u_{22} = \tfrac{19}{2}$$
$$\tfrac{7}{2}(4) + u_{23} = 3, \; u_{23} = -11$$
$$2l_{31} = -1, \; l_{31} = -\tfrac{1}{2}$$
$$-\tfrac{1}{2}(-3) + \tfrac{19}{2}l_{32} = 2, \; l_{32} = \tfrac{1}{19}$$
$$-\tfrac{1}{2}(4) + \tfrac{1}{19}(-11) + u_{33} = -2, \; u_{33} = \tfrac{11}{19}$$

so that

$$L = \begin{bmatrix} 1 & 0 & 0 \\ \tfrac{7}{2} & 1 & 0 \\ -\tfrac{1}{2} & \tfrac{1}{19} & 1 \end{bmatrix} \quad \text{and} \quad U = \begin{bmatrix} 2 & -3 & 4 \\ 0 & \tfrac{19}{2} & -11 \\ 0 & 0 & \tfrac{11}{19} \end{bmatrix}$$

EXERCISES

12.1. Introduction; 12.2. Matrix definitions

1. If

$$A = \begin{bmatrix} 2 & -1 & 7 \\ -3 & 5 & 4 \end{bmatrix}$$

find A^T.

2. Show that if A is a square matrix, then $A + A^T$ is symmetric.

3. If A and B are both found to be identity matrices, are they necessarily equal? Explain.

4. What is $(A^T)^T$?

5. Show that the transpose of an upper triangular matrix is lower triangular, and conversely.

6. (a) Are all zero matrices symmetric?
   (b) Are all identity matrices symmetric?

12.3. Elementary matrix operations

7. If

$$A = \begin{bmatrix} 3 & -1 \\ 1 & 2 \end{bmatrix} \quad \text{and} \quad B = \begin{bmatrix} 4 & 1 \\ -2 & -3 \end{bmatrix}$$

   (a) Find AB.
   (b) Find BA.

8. If

$$A = \begin{bmatrix} 1 & 2 & 3 \end{bmatrix} \quad \text{and} \quad B = \begin{bmatrix} 1 \\ 2 \\ 3 \end{bmatrix}$$

   (a) Find AB.
   (b) Find BA.

9. Given the definitions of A and B in Exercise 8, find:
   (a) $3A$.
   (b) $3B$.
   (c) $[3]A$.
   (d) $[3]B$.

10. If

$$A = \begin{bmatrix} 1 & 1 \\ 1 & 1 \end{bmatrix}$$

   find A^n.

11. If w, x, y, z are all nonzero and

$$\begin{bmatrix} a & b \\ c & d \end{bmatrix}\begin{bmatrix} w & x \\ y & z \end{bmatrix} = \begin{bmatrix} aw & bx \\ cy & dz \end{bmatrix}$$

   find a, b, c, and d.

12. Given the definitions of A and B in Exercise 7, demonstrate the validity of formula (12.11).

13. With real numbers if $AB = 0$, then either $A = 0$ or $B = 0$. Find a counterexample to show that this is not necessarily true for matrices.

14. With real numbers if $AB = AC$ and A is nonzero, then $B = C$. Find a counterexample to show that this is not necessarily true for matrices.

15. With real numbers if $AB = B$ and B is nonzero, then $A = 1$. Find a counterexample to show that this is not necessarily true for matrices. (The identity matrix fulfills the role of 1 in matrix multiplication.)

16. (a) Show that the product of two upper triangular matrices is upper triangular.
    (b) Show that the product of two lower triangular matrices is lower triangular.

17. Show that

$$\begin{bmatrix} 1 & 2 & 3 \\ 3 & 2 & 0 \\ -1 & -1 & -1 \end{bmatrix} \quad \text{and} \quad \begin{bmatrix} -2 & -1 & -6 \\ 3 & 2 & 9 \\ -1 & -1 & -4 \end{bmatrix}$$

   are commutative. What is the product?

18. If A is a diagonal matrix with diagonal elements $a_{11}, a_{22}, \ldots, a_{nn}$ and B is a diagonal matrix with diagonal elements $b_{11}, b_{22}, \ldots, b_{nn}$, find AB.

19. If

$$A = \begin{bmatrix} \lambda & 1 \\ 0 & \lambda \end{bmatrix}$$

   find A^n.

12.4. Determinants

20. (a) Is the number of interchanges necessary to rearrange 53142 to 12345 even or odd?
    (b) Is the number of interchanges necessary to rearrange 52314 to 12345 even or odd?

21. Find x if

$$\begin{vmatrix} x-3 & x+1 \\ x & x+2 \end{vmatrix} = 4$$

22. Evaluate

$$\begin{vmatrix} -4 & 1 & 1 & 1 & 1 \\ 1 & -4 & 1 & 1 & 1 \\ 1 & 1 & -4 & 1 & 1 \\ 1 & 1 & 1 & -4 & 1 \\ 1 & 1 & 1 & 1 & -4 \end{vmatrix}$$

23. It is known that

$$\begin{vmatrix} 2 & 3 & -2 & 4 \\ 3 & -2 & 1 & 2 \\ 3 & 2 & 3 & 4 \\ -2 & 4 & 0 & 5 \end{vmatrix} = -286$$

   Find the value of

$$\begin{vmatrix} 4 & -2 & 3 & 3 \\ 6 & 4 & 2 & -2 \\ -4 & 0 & 3 & 1 \\ 8 & 5 & 4 & 2 \end{vmatrix}$$

24. Verify property 7 given in Section 12.4 using the matrices given in Exercise 7.

25. Find

$$\frac{d}{dx} \begin{vmatrix} x^2 & x+1 & 3 \\ 1 & 2x-1 & x^3 \\ 0 & x & -2 \end{vmatrix}$$

26. Find

$$\begin{vmatrix} 1 & 1 & 1 \\ 1 & 1+x & 1 \\ 1 & 1 & 1+y \end{vmatrix}$$

27. Show that

$$\begin{vmatrix} 1 & a & a^2 \\ 1 & b & b^2 \\ 1 & c & c^2 \end{vmatrix} = (b-c)(c-a)(a-b)$$

12.5. Evaluation of determinants by minors

28. Show that the determinant value of an upper triangular, lower triangular, or diagonal determinant is equal to the product of the elements on the diagonal.

29. Evaluate

$$\begin{vmatrix} 3 & 2 & 1 & 4 \\ 15 & 29 & 2 & 14 \\ 16 & 19 & 3 & 17 \\ 33 & 39 & 8 & 38 \end{vmatrix}$$

30. Evaluate

$$\begin{vmatrix} 3 & -2 & 5 & -6 \\ 1 & 7 & 6 & 9 \\ 2 & -1 & 3 & -2 \\ 2 & 6 & 8 & 5 \end{vmatrix}$$

31. Find

$$\begin{vmatrix} x & x^2 & x^3 & x^4 & x^5 \\ 0 & x^2 & x^3 & x^4 & x^5 \\ 0 & 0 & x^3 & x^4 & x^5 \\ 0 & 0 & 0 & x^4 & x^5 \\ 0 & 0 & 0 & 0 & x^5 \end{vmatrix}$$

32. Show that the cofactor of each element of

$$\begin{bmatrix} -\frac{1}{3} & -\frac{2}{3} & -\frac{2}{3} \\ \frac{2}{3} & \frac{1}{3} & -\frac{2}{3} \\ \frac{2}{3} & -\frac{2}{3} & \frac{1}{3} \end{bmatrix}$$

is that element.

33. Find the value of the nth-order determinant

$$\begin{vmatrix} a+b & a & a & \cdots & a \\ a & a+b & a & \cdots & a \\ a & a & a+b & \cdots & a \\ \vdots & \vdots & \vdots & & \vdots \\ a & a & a & \cdots & a+b \end{vmatrix}$$

34. Show that the value of the first determinant given in Exercise 23 is -286.

35. What is the determinant value of the nth-order scalar matrix kI?

12.6.   The inverse of a matrix

36.   Use the elimination method to find A^{-1}, if

$$A = \begin{bmatrix} 1 & -1 & 1 \\ 2 & -4 & 3 \\ 0 & -2 & 3 \end{bmatrix}$$

37.   Rework Exercise 36 using the adjoint method.

38.   Rework Exercise 36 using the exchange method.

39.   If A is a diagonal matrix with diagonal elements $d_1, d_2, \ldots, d_n$, find A^{-1}.

40.   Show that if

$$A = \begin{bmatrix} a & b \\ c & d \end{bmatrix}$$

is nonsingular, then

$$A^{-1} = \frac{\begin{bmatrix} d & -b \\ -c & a \end{bmatrix}}{\begin{vmatrix} a & b \\ c & d \end{vmatrix}}$$

41.   Verify that formula (12.18) holds if

$$A = \begin{bmatrix} 3 & -2 \\ -4 & -1 \end{bmatrix} \quad \text{and} \quad B = \begin{bmatrix} 5 & 2 \\ 2 & 1 \end{bmatrix}$$

42.   The inverse of

$$A = \begin{bmatrix} 1 & 2 \\ 2 & 7 \end{bmatrix}$$

is to be found by using the iterative method. If a starting value of

$$B = \begin{bmatrix} 2.3 & -.7 \\ -.7 & .3 \end{bmatrix}$$

is used, generate the next approximation going only as far as R.

43.   What is $(A^{-1})^{-1}$?

12.7.   Factorization of a matrix

44.   Factor the matrix given in Exercise 36, so that the diagonal elements of the lower triangular matrix are all equal to one.

45.   Rework Exercise 44, so that the diagonal elements of the upper triangular matrix are all equal to one.

46.   Factor

$$\begin{bmatrix} 1 & 2 \\ 2 & 5 \end{bmatrix}$$

such that the diagonal of the lower triangular matrix is equal to the diagonal of the upper triangular matrix and such that no negative numbers appear in either matrix.

Miscellaneous problems

47. Show that if X is a column vector with elements $x_1, x_2, \ldots, x_n$, then the *dot product* of X with itself is given by $X^T X$, where $X^T X$ is interpreted as a real number rather than as a 1×1 matrix.

48. If the determinant value of an $n \times n$ matrix A is $|A|$, what is the determinant value of kA?

49. The sum of the diagonal elements of a square matrix A is called the *trace* of A. Find the trace of the $n \times n$ scalar matrix kI.

50. A square matrix A is called *idempotent*, if $A^2 = A$. Show that the following matrix is idempotent

$$\begin{bmatrix} 2 & -2 & -4 \\ -1 & 3 & 4 \\ 1 & -2 & -3 \end{bmatrix}$$

51. A square matrix A is called *nilpotent*, if $A^2 = 0$. Show that the following matrix is nilpotent

$$\begin{bmatrix} 1 & -3 & -4 \\ -1 & 3 & 4 \\ 1 & -3 & -4 \end{bmatrix}$$

52. A square matrix A is called *involutory*, if $A^2 = I$. Show that the following matrix is involutory

$$\begin{bmatrix} 0 & 1 & -1 \\ 4 & -3 & 4 \\ 3 & -3 & 4 \end{bmatrix}$$

53. Find the matrix K such that $KU = \Delta^3 U$ where U is the column vector with elements $u_1, u_2, \ldots, u_n$ and $\Delta^3 U$ is the column vector with elements $\Delta^3 u_1, \Delta^3 u_2, \ldots, \Delta^3 u_{n-3}$.

13

Systems of linear equations

13.1. INTRODUCTION

A SYSTEM of n simultaneous linear equations in n unknowns can be written as

$$\left. \begin{array}{l} a_{11}x_1 + a_{12}x_2 + \cdots + a_{1n}x_n = c_1 \\ a_{21}x_1 + a_{22}x_2 + \cdots + a_{2n}x_n = c_2 \\ \cdots\cdots\cdots\cdots\cdots\cdots\cdots\cdots\cdots\cdots\cdots\cdots\cdots \\ a_{n1}x_1 + a_{n2}x_2 + \cdots + a_{nn}x_n = c_n \end{array} \right\} \tag{13.1}$$

The student has undoubtedly encountered systems of linear equations in previous courses in algebra. Although the problem of solving n simultaneous linear equations in n unknowns is apparently elementary, complexities do exist in computation for large systems.

Various algorithms have been devised to solve systems of linear equations. These reflect the fact that no single algorithm is best in all situations. These various algorithms must be judged on their speed and accuracy, as all algorithms are judged. Speed is of concern in solving large systems because of the large volume of computations. Also accuracy is of importance because of the roundoff error involved in performing large volumes of computations. It is not possible to say that one algorithm which is superior to a second algorithm for one system of equations will be superior for all other systems. An algorithm is sought which will give the highest degree of accuracy for the effort expended.

Algorithms for solving systems of linear equations can be classified as direct or indirect. *Direct methods* proceed along analytical lines involving a fixed number of computational steps to the answer. *Indirect methods* are iterative in nature and successively approximate the answer to the required level of accuracy. However, the number of computational steps is not fixed and continues until the required level of accuracy is obtained.

Depending upon the circumstances, either a direct method or an indirect method can be superior to the other. Direct methods are used for typical systems in most practical situations. The student might ask why an "approximate" iterative method would ever be preferred to an "exact" direct method. The reason is that the roundoff error produced in direct methods is often more severe than that produced in indirect methods. This is because no roundoff error appears in an iterative method until the last iteration. On the other hand, one problem with iterative methods is that for rapid convergence to occur, good starting values must be available.

Systems of equations with a unique solution are called *nonsingular*.[1] These equations are characterized by the fact that the determinant value of the coefficients, $|A|$, is nonzero. If the determinant value of the coefficients is equal to zero, then the system of equations is called *singular* and no unique solution exists. Singular systems have either no solution or an infinite number of solutions.

In some instances it is useful to describe singularity or nonsingularity in terms of *homogeneous linear equations*, which are equations (13.1) in which $c_1 = c_2 = \cdots = c_n = 0$. If these homogeneous linear equations have only the *trivial solution* $x_1 = x_2 = \cdots = x_n = 0$, then the system is nonsingular regardless of the values of $c_1, c_2, \ldots, c_n$. If solutions other than the trivial solution $x_1 = x_2 = \cdots = x_n = 0$ exist, then the system is singular for any values of $c_1, c_2, \ldots, c_n$.

13.2. CRAMER'S RULE

Cramer's rule for solving systems of linear equations is a direct method based on determinants. The solution is given by

$$x_i = \frac{|A_i|}{|A|} \qquad \text{for } i = 1, 2, \ldots, n \tag{13.2}$$

where $|A|$ is the determinant of the coefficients and $|A_i|$ is the determinant of the coefficients in which the ith column of a's has been replaced by the column of c's.

---

[1]This definition of "nonsingular" is seen to be consistent with "nonsingular" matrices defined in Section 12.6.

The proof of Cramer's rule can be found in standard textbooks in linear algebra. It is a method of considerable appeal in theoretical work. In general, however, it is not a widely used algorithm in computational work, since the number of computations involved in evaluating large-order determinants exceeds the number involved in alternative algorithms.

Example 13.1. **Solve the system of equations**

$$2x_1 - 3x_2 + 4x_3 = -3$$
$$7x_1 - x_2 + 3x_3 = 2$$
$$-x_1 + 2x_2 - 2x_3 = 2$$

using Cramer's rule.

We have

$$|A| = \begin{vmatrix} 2 & -3 & 4 \\ 7 & -1 & 3 \\ -1 & 2 & -2 \end{vmatrix} = 11$$

$$|A_1| = \begin{vmatrix} -3 & -3 & 4 \\ 2 & -1 & 3 \\ 2 & 2 & -2 \end{vmatrix} = 6$$

$$|A_2| = \begin{vmatrix} 2 & -3 & 4 \\ 7 & 2 & 3 \\ -1 & 2 & -2 \end{vmatrix} = 11$$

$$|A_3| = \begin{vmatrix} 2 & -3 & -3 \\ 7 & -1 & 2 \\ -1 & 2 & 2 \end{vmatrix} = -3$$

so that the solution is

$$x_1 = \tfrac{6}{11} \qquad x_2 = 1 \qquad x_3 = -\tfrac{3}{11}$$

The student should verify this solution by substitution.

13.3. GAUSSIAN ELIMINATION

Gaussian elimination is a direct method of solving systems of linear equations which is a refinement of the approach taken in elementary algebra courses. Despite its elementary approach, it is an effective method in many situations and is widely used in practice.

The steps in Gaussian elimination convert the system (13.1) into an equivalent upper triangular system of the form given by (13.3)

$$\left.\begin{aligned} x_1 + a'_{12}x_2 + \cdots + a'_{1n}x_n &= c'_1 \\ x_2 + \cdots + a'_{2n}x_n &= c'_2 \\ \cdots\cdots\cdots\cdots \\ x_n &= c'_n \end{aligned}\right\} \qquad (13.3)$$

Note that the diagonal coefficients are equal to one.

Once the equations are in upper triangular form the system can easily be solved by a process known as *back-substitution*. In this process the last equation is solved for x_n, this value is then substituted into the next-to-last equation which is solved for x_{n-1}, and so forth.

The equations are converted to upper triangular form by a series of row operations. The coefficient of x_1 in the first row is set equal to one by division along the first row. Then multiples of row 1 are added to rows 2 through n to produce all zeros in the first column for rows 2 through n. The coefficient of x_2 in the second row is similarly set equal to one, and then multiples of row 2 are added to rows 3 through n to produce all zeros in the second column for rows 3 through n. This process is continued until the equations are in upper triangular form with diagonal coefficients equal to one. Each of the elements on the diagonal, before division in that row to make the diagonal coefficient equal to one, is called a *pivot*.

The procedure just described can be refined to improve the accuracy of the method by reducing roundoff error. In this refinement we rearrange the equations each time a new pivot is called for, so that we use the largest pivot possible (in absolute value), rather than using the equations in the original order. This insures that all multiples of the pivot row added to remaining rows are less than or equal to one in absolute value, thereby tending to minimize the effect of roundoff error. As the student recalls, multiplying by a number less than one reduces roundoff error, and conversely. (See Exercise 15 (a) in Chapter 1.) In rearranging equations, we cannot reuse an equation which has already contained a pivot, that is, any rearrangement must involve only the pivot row and lower rows.

The student should note that the rearrangement of rows is irrelevant if exact values are to be computed, since the same answers will be obtained regardless of the arrangement. However, if values are being rounded off in decimal form, as is almost always the case in computer applications to large systems of equations, then the rearrangement is an important part of the process.

Gaussian elimination provides a simple method of evaluating determinants as a by-product. From the properties of determinants discussed in Chapter 12 it is seen that the value of an upper triangular determinant is equal to the product of the elements on the diagonal. Also, it is seen that the effect of interchanging two rows in a determinant is to change the sign of the determinant value. The other operations in Gaussian elimination perform row operations which do not change the value of the determinant.

Thus, the determinant value of the coefficients, $|A|$, is equal to the product of the pivots in Gaussian elimination, modified by a sign change if the number of row interchanges is odd.

Example 13.2. Rework Example 13.1 using Gaussian elimination. Use decimal form and carry four decimal places in all computations.

We have the following:

$$2x_1 - 3x_2 + 4x_3 = -3$$
$$7x_1 - x_2 + 3x_3 = 2 \qquad \text{(original equations)}$$
$$-x_1 + 2x_2 - 2x_3 = 2$$

$$7x_1 - x_2 + 3x_3 = 2$$
$$2x_1 - 3x_2 + 4x_3 = -3 \qquad \text{(rearranging to bring the largest number in column 1, that is, 7, to be the first pivot)}$$
$$-x_1 + 2x_2 - 2x_3 = 2$$

$$1.0000x_1 - .1429x_2 + .4286x_3 = .2857 \qquad \text{(R1/7)}$$
$$2.0000x_1 - 3.0000x_2 + 4.0000x_3 = -3.0000$$
$$-1.0000x_1 + 2.0000x_2 - 2.0000x_3 = 2.0000$$

$$1.0000x_1 - .1429x_2 + .4286x_3 = .2857$$
$$- 2.7142x_2 + 3.1428x_3 = -3.5714 \qquad \text{(R2 - 2R1)}$$
$$1.8571x_2 - 1.5714x_3 = 2.2857 \qquad \text{(R3 + R1)}$$

$$1.0000x_1 - .1429x_2 + .4286x_3 = .2857$$
$$1.0000x_2 - 1.1579x_3 = 1.3158 \qquad \text{(R2/ - 2.7142)}$$
$$1.8571x_2 - 1.5714x_3 = 2.2857$$

$$1.0000x_1 - .1429x_2 + .4286x_3 = .2857$$
$$1.0000x_2 - 1.1579x_3 = 1.3158$$
$$.5789x_3 = - .1579 \qquad \text{(R3 - 1.8571R2)}$$

$$1.0000x_1 - .1429x_2 + .4286x_3 = .2857$$
$$1.0000x_2 - 1.1579x_3 = 1.3158$$
$$1.0000x_3 = - .2728 \qquad \text{(R3/.5789)}$$

Solving by back-substitution, we have

$$x_3 = -.2728 \qquad x_2 = .9999 \qquad x_1 = .5455$$

The exact values to four decimal places are

$$x_3 = -\tfrac{3}{11} = -.2727 \qquad x_2 = 1.0000 \qquad x_1 = \tfrac{6}{11} = .5455$$

so that the maximum absolute actual error in any of the answers is .0001 and all answers are correct to three decimal places.

Example 13.3. Evaluate the determinant of the coefficients in Example 13.1 using Gaussian elimination.

The exact value of $|A|$ is equal to 11 from Example 13.1. The product of the pivots in Example 13.2 is equal to $(7.0000)(-2.7142)(.5789) = -10.9988$. Since there was one row interchange, this becomes 10.9988, which is accurate to two decimal places when rounded.

The amount of computation involved in Gaussian elimination is much less than the amount of computation involved in traditional methods of evaluating determinants, particularly for large n. Gaussian elimination is generally the most efficient method of evaluating large-order determinants.

The amount of computation involved in evaluating determinants directly as discussed in Chapter 12 tends to be proportional to $n!$, which increases very rapidly with increasing n. On the other hand, the amount of computation in Gaussian elimination tends to be more nearly proportional to only n^3, which increases much more slowly. This latter statement is developed more fully in Exercises 39 and 40.

13.4. MATRIX INVERSION METHOD

The system of linear equations given by (13.1) can be written in the form

$$Ax = c \tag{13.4}$$

where x and c are column vectors

$$x = \begin{bmatrix} x_1 \\ x_2 \\ \vdots \\ x_n \end{bmatrix} \quad \text{and} \quad c = \begin{bmatrix} c_1 \\ c_2 \\ \vdots \\ c_n \end{bmatrix}$$

The solution vector x can be found by multiplying formula (13.4) from the left by A^{-1} to give

$$A^{-1}(Ax) = (A^{-1}A)x = x = A^{-1}c \tag{13.5}$$

Thus, if the inverse A^{-1} is known, the solution is immediately obtained upon multiplying the vector c from the left by A^{-1}.

The matrix inversion method is direct, if the elimination method, adjoint method, or exchange method of inversion is used. It is indirect, if the iterative method of inversion is used.

In Section 12.6 the discussion of matrix inversion by the elimination method indicates that the order in which computations are performed has an impact on the amount of roundoff error in the result.

The recommended computational steps in the elimination method of matrix inversion are the same as the computational steps in Gaussian elimination. This necessitates a third row operation in addition to the two listed in

Section 12.6; namely, interchanging rows in order to obtain the largest pivot. Thus, legitimate row operations are as follows:

1. Any row can be multiplied by a nonzero scalar.
2. Any multiple of one row can be added to any other row.
3. Any two rows can be interchanged.

It must be carefully noted that if row interchanges are made in an attempt to minimize roundoff error, then a different matrix inverse is obtained. This must be offset by performing the same row interchanges in the c-vector, so that the correct answers are obtained when formula (13.5) is applied.

Section 12.6 also indicates that the exchange method of matrix inversion can be refined to minimize roundoff error. A generalized version of the exchange method can be defined by allowing the choice of pivot to be the element in any row and column which have not been used before, rather than restricting the choice to a diagonal element.

The generalized rules of the exchange method are as follows:

1. Replace the pivot a_{ij} by its reciprocal.
2. Each other element in the same column is divided by a_{ij}.
3. Each other element in the same row is divided by a_{ij} and the sign is changed.
4. All other elements, a_{kl}, are replaced by

$$a_{kl} - \frac{a_{kj} a_{il}}{a_{ij}} \tag{13.6}$$

which is a generalized version of the *rectangular rule*.

The array is usually set up with x's and c's listed as in Table 13.1. For example, if an exchange is made between c_3 and x_2, the above rules give the result in Table 13.2. The label c_3 now appears across the top, while the label x_2 appears down the left-hand side. After two more exchanges, the c's will be across the top, the x's will be down the left-hand side, and the inversion is complete. It must be noted that if elements off the diagonal are chosen, then the order of the x's and c's is rearranged. This means that a different matrix inverse is obtained which must be offset in formula (13.5) by using the x's and

TABLE 13.1

	x_1	x_2	x_3
c_1	a_{11}	a_{12}	a_{13}
c_2	a_{21}	a_{22}	a_{23}
c_3	a_{31}	a_{32}	a_{33}

TABLE 13.2

	x_1	c_3	x_3
c_1	$a_{11} - \dfrac{a_{12}\,a_{31}}{a_{32}}$	$\dfrac{a_{12}}{a_{32}}$	$a_{13} - \dfrac{a_{12}\,a_{33}}{a_{32}}$
c_2	$a_{21} - \dfrac{a_{22}\,a_{31}}{a_{32}}$	$\dfrac{a_{22}}{a_{32}}$	$a_{23} - \dfrac{a_{22}\,a_{33}}{a_{32}}$
x_2	$-\dfrac{a_{31}}{a_{32}}$	$\dfrac{1}{a_{32}}$	$-\dfrac{a_{33}}{a_{32}}$

c's in their rearranged orders, respectively. This is illustrated in Example 13.5.

The recommended procedure to minimize roundoff error, assuming computations are in decimal form rather than exact fractions, is to successively pick as the next pivot the largest remaining element (in absolute value) which lies in both a row and a column that have not previously contained a pivot. Since all the other revised elements contain terms divided by the pivot, this procedure tends to minimize the effect of roundoff error. As the student recalls, dividing by a number larger than one reduces roundoff error, and conversely.

If several different systems of linear equations are to be solved, which are identical except for the c's, the matrix inversion method is a particularly useful algorithm. The inverse matrix A^{-1} need only be computed once and can then be applied to each c-vector.

Example 13.4.   Rework Example 13.1 using the matrix inversion method.

The inverse matrix computed in Example 12.6 is

$$A^{-1} = \frac{1}{11}\begin{bmatrix} -4 & 2 & -5 \\ 11 & 0 & 22 \\ 13 & -1 & 19 \end{bmatrix}$$

Applying formula (13.5) gives

$$x = A^{-1}c = \frac{1}{11}\begin{bmatrix} -4 & 2 & -5 \\ 11 & 0 & 22 \\ 13 & -1 & 19 \end{bmatrix}\begin{bmatrix} -3 \\ 2 \\ 2 \end{bmatrix} = \begin{bmatrix} \frac{6}{11} \\ 1 \\ -\frac{3}{11} \end{bmatrix}$$

which agrees with Example 13.1.

Example 13.5.   Apply the exchange method once using the appropriate pivot to the matrix of the coefficients given in Example 13.1.

The array is as follows:

	x_1	x_2	x_3
c_1	2	−3	4
c_2	7	−1	3
c_3	−1	2	−2

The appropriate choice for first pivot is 7 in row 2, column 1. One application of the rules of the exchange method gives the following:

	c_2	x_2	x_3
c_1	$\frac{2}{7}$	$-\frac{19}{7}$	$\frac{22}{7}$
x_1	$\frac{1}{7}$	$\frac{1}{7}$	$-\frac{3}{7}$
c_3	$-\frac{1}{7}$	$\frac{13}{7}$	$-\frac{11}{7}$

Two more applications of the rules of the exchange method will complete the inversion. The appropriate choice for second pivot is $\frac{22}{7}$ in row 1, column 3. The third pivot, of necessity, is in row 3, column 2. The rest of the inversion is contained in Exercise 11.

The student should note that the advantage of choosing pivots in this order is evident only when decimal form is carried rather than exact fractions as in this example.

The student should also note that the original array

	x_1	x_2	x_3
c_1	a_{11}	a_{12}	a_{13}
c_2	a_{21}	a_{22}	a_{23}
c_3	a_{31}	a_{32}	a_{33}

becomes

	c_2	c_3	c_1
x_3	b_{11}	b_{12}	b_{13}
x_1	b_{21}	b_{22}	b_{23}
x_2	b_{31}	b_{32}	b_{33}

after three exchanges. In solving a system of equations, if the entries in the second array constitute the matrix B, then the solution is given by

$$Bc = x$$

where

$$c = \begin{bmatrix} c_2 \\ c_3 \\ c_1 \end{bmatrix} \quad \text{and} \quad x = \begin{bmatrix} x_3 \\ x_1 \\ x_2 \end{bmatrix}$$

The rearranged order of c and x should be carefully noted.

13.5. MATRIX FACTORIZATION METHOD

The factorization, or triangular decomposition, of A into the product of a lower and an upper triangular matrix is given by formula (12.24)

$$A = LU \tag{12.24}$$

This factorization leads to an alternative direct algorithm for solving a system of linear equations.

Formula (13.4) gives

$$Ax = c \tag{13.4}$$

This can be written as

$$LUx = c$$

If Ux is set equal to y, then formula (13.4) becomes two equations

$$Ux = y \tag{13.7a}$$

and

$$Ly = c \tag{13.7b}$$

Formula (13.7b) is solved by back-substitution to give y, and then formula (13.7a) is solved by another back-substitution to give x.

Example 13.6. *Rework Example 13.1 using the matrix factorization method.*

The lower and upper triangular matrices from Example 12.10 are

$$L = \begin{bmatrix} 1 & 0 & 0 \\ \frac{7}{2} & 1 & 0 \\ -\frac{1}{2} & \frac{1}{19} & 1 \end{bmatrix} \quad \text{and} \quad U = \begin{bmatrix} 2 & -3 & 4 \\ 0 & \frac{19}{2} & -11 \\ 0 & 0 & \frac{11}{19} \end{bmatrix}$$

Now $Ly = c$ gives

$$\begin{bmatrix} 1 & 0 & 0 \\ \frac{7}{2} & 1 & 0 \\ -\frac{1}{2} & \frac{1}{19} & 1 \end{bmatrix} \begin{bmatrix} y_1 \\ y_2 \\ y_3 \end{bmatrix} = \begin{bmatrix} -3 \\ 2 \\ 2 \end{bmatrix}$$

or

$$
\begin{aligned}
y_1 &= -3 \\
\tfrac{7}{2}y_1 + y_2 &= 2 \\
-\tfrac{1}{2}y_1 + \tfrac{1}{19}y_2 + y_3 &= 2
\end{aligned}
$$

These equations solved by back-substitution[2] give

$$y_1 = -3 \qquad y_2 = \tfrac{25}{2} \qquad y_3 = -\tfrac{3}{19}$$

Continuing, $Ux = y$ gives

$$\begin{bmatrix} 2 & -3 & 4 \\ 0 & \frac{19}{2} & -11 \\ 0 & 0 & \frac{11}{19} \end{bmatrix} \begin{bmatrix} x_1 \\ x_2 \\ x_3 \end{bmatrix} = \begin{bmatrix} -3 \\ \frac{25}{2} \\ -\frac{3}{19} \end{bmatrix}$$

or

$$
\begin{aligned}
2x_1 - 3x_2 + 4x_3 &= -3 \\
\tfrac{19}{2}x_2 - 11x_3 &= \tfrac{25}{2} \\
\tfrac{11}{19}x_3 &= -\tfrac{3}{19}
\end{aligned}
$$

These equations solved by back-substitution give

$$x_3 = -\tfrac{3}{11} \qquad x_2 = 1 \qquad x_1 = \tfrac{6}{11}$$

which agrees with Example 13.1.

13.6. GAUSS-SEIDEL METHOD

The *Gauss-Seidel method* of solving systems of linear equations is a widely used iterative method. The iterative approach for systems of linear equations is similar to the iterative approach developed in Chapter 11 for one non-linear equation in one unknown in which a value of x is found such that $f(x) = 0$. In the case of systems of linear equations the situation is somewhat more complex, since it is required to find a set of values $x_1, x_2, \ldots, x_n$ which satisfy $Ax = c$.

---

[2] " Forward-substitution " might be a more appropriate phrase.

To apply the Gauss-Seidel method the equations given in (13.1) are re-written as follows:

$$x_1 = \frac{1}{a_{11}}(c_1 - a_{12}x_2 - a_{13}x_3 - \cdots - a_{1n}x_n)$$

$$x_2 = \frac{1}{a_{22}}(c_2 - a_{21}x_1 - a_{23}x_3 - \cdots - a_{2n}x_n)$$

$$\cdots\cdots\cdots\cdots\cdots\cdots\cdots\cdots\cdots\cdots\cdots\cdots\cdots$$

$$x_n = \frac{1}{a_{nn}}(c_n - a_{n1}x_1 - a_{n2}x_2 - \cdots - a_{n,n-1}x_{n-1})$$

$\qquad$ (13.8)

Next a set of starting values $x_1^0, x_2^0, \ldots, x_n^0$ is chosen. The starting values

$$x_i^0 = \frac{c_i}{a_{ii}} \qquad \text{for } i = 1, 2, \ldots, n \qquad (13.9)$$

are frequently used if more accurate starting values are not available. If a set of starting values is available that is believed to be a better approximation, then these values should be used instead of those given by formula (13.9).

$\qquad$The starting values to be used are then substituted into the first equation in (13.8) to generate x_1^1, the first approximation for x_1. This new value x_1^1 is used together with the other starting values in the second equation in (13.8) to generate x_2^1, the first approximation for x_2. This process continues through the remaining $n - 2$ equations until first approximations for all n values of x_i are obtained, which completes the first cycle. It is important to note that each new value of x_i^1 is used as soon as it becomes available in the remaining equations. This feature makes the Gauss-Seidel method particularly suited to computer applications, where updated variables are immediately stored for future reference.

$\qquad$Once the first cycle is completed, the above procedure is repeated for as many cycles as necessary to obtain the required level of accuracy. The process is stopped when further iterations do not change the answers obtained to the required number of decimal places. Thus, on the kth cycle the following equations appear:

$$x_1^{k+1} = \frac{1}{a_{11}}(c_1 - a_{12}x_2^k - a_{13}x_3^k - \cdots - a_{1n}x_n^k)$$

$$x_2^{k+1} = \frac{1}{a_{22}}(c_2 - a_{21}x_1^{k+1} - a_{23}x_3^k - \cdots - a_{2n}x_n^k)$$

$$\cdots\cdots\cdots\cdots\cdots\cdots\cdots\cdots\cdots\cdots\cdots\cdots\cdots$$

$$x_n^{k+1} = \frac{1}{a_{nn}}(c_n - a_{n1}x_1^{k+1} - a_{n2}x_2^{k+1} - \cdots - a_{n,n-1}x_{n-1}^{k+1})$$

$\qquad$ (13.10)

Unfortunately, the Gauss-Seidel method as just described does not always converge. Generally, convergence requires the presence of a matrix with a *dominant diagonal*, that is, the absolute values of a_{ii} should be "large" in some sense when compared with absolute values of a_{ij} for $i \neq j$. Note in particular that if any of the diagonal elements is equal to zero, then immediate divergence results, since division by zero is impossible.

Certain types of problems tend to produce systems of equations with a dominant diagonal. For example, systems of differential equations are often of this type. (See Exercise 30 in Chapter 10.) In other types of problems some rearrangement of the equations may be required in order to produce a matrix with a dominant diagonal. This rearrangement in many cases produces a convergent system, even though the original system may have been divergent.

It is difficult to describe analytically the conditions under which convergence of the Gauss-Seidel method does or does not occur. One useful result which is a sufficient condition for convergence is given by

$$\frac{1}{|a_{ii}|} \sum_{j \neq i} |a_{ij}| < 1 \qquad \text{for } i = 1, 2, \ldots, n \qquad (13.11)[3]$$

In other words, formula (13.11) states that if the sum of the absolute values of the nondiagonal elements in each row is less than the absolute value of the diagonal element in that same row, then convergence occurs. However, formula (13.11) is not a necessary condition for convergence, that is, convergence may occur even if formula (13.11) is not satisfied. Another sufficient condition for convergence is for the matrix of coefficients to be *positive definite*.[4]

An alternative iterative method which appears in the literature is the *Jacobi method*. This method is identical to the Gauss-Seidel method except that new values enter the process in a group at the start of each cycle rather than entering the process immediately as they become available. Thus, for the Jacobi method (13.10) becomes

$$
\left.
\begin{aligned}
x_1^{k+1} &= \frac{1}{a_{11}} \left(c_1 - a_{12} x_2^k - a_{13} x_3^k - \cdots - a_{1n} x_n^k \right) \\
x_2^{k+1} &= \frac{1}{a_{22}} \left(c_2 - a_{21} x_1^k - a_{23} x_3^k - \cdots - a_{2n} x_n^k \right) \\
&\cdots\cdots\cdots\cdots\cdots\cdots\cdots\cdots\cdots\cdots\cdots\cdots\cdots\cdots\cdots\cdots \\
x_n^{k+1} &= \frac{1}{a_{nn}} \left(c_n - a_{n1} x_1^k - a_{n2} x_2^k - \cdots - a_{n,\,n-1} x_{n-1}^k \right)
\end{aligned}
\right\} \qquad (13.12)
$$

---

[3] For a proof of formula (13.11) see S. D. Conte and Carl deBoor, *Elementary Numerical Analysis* (New York: McGraw-Hill Book Co., 1972), pp. 159–61. This reference also contains a more general theoretical result which gives necessary and sufficient conditions for convergence on pp. 161–63.

[4] Positive definite matrices are defined in Appendix G.

The Jacobi method is not as widely used as the Gauss-Seidel method, since convergence is not as rapid. In fact it can be shown that convergence of the Gauss-Seidel method is twice as fast as the Jacobi method.[5] However, examples exist in which the Jacobi method converges while the Gauss-Seidel method does not converge, and conversely. Generally, the Jacobi method is not as well suited to computer applications as is the Gauss-Seidel method.

Systems of equations with a "large" number of nondiagonal elements equal to zero are called *sparse*, while systems with a "small" number of non-diagonal zero elements are called *dense*. The Gauss-Seidel and other iterative methods are particularly effective for sparse systems, since the volume of computation is much less than for dense systems and since convergence is usually quite rapid. Direct methods, such as Gaussian elimination, tend to be more widely used for dense systems.

Example 13.7. *Rework Example 13.1 using the Gauss-Seidel method. Use decimal form and carry four decimal places in all computations. Assume starting values $x_1^0 = .6$, $x_2^0 = 1.1$, and $x_3^0 = -.2$ are given.*

The equations as given are

$$2x_1 - 3x_2 + 4x_3 = -3$$
$$7x_1 - x_2 + 3x_3 = 2$$
$$-x_1 + 2x_2 - 2x_3 = 2$$

The equations do not have a dominant diagonal. If they are rewritten as in (13.8) with no rearrangement, we have

$$x_1 = \tfrac{1}{2}(-3 + 3x_2 - 4x_3)$$
$$x_2 = -2 + 7x_1 + 3x_3$$
$$x_3 = -\tfrac{1}{2}(2 + x_1 - 2x_2)$$

The values of x_n^k computed using (13.10) for the first five cycles are given in Table 13.3. It is apparent that despite the presence of excellent starting values, divergence is occurring.

TABLE 13.3

n \ k	0	1	2	3	4	5
1	.6000	.5500	.4250	.4750	.7812	.8220
2	1.1000	1.2500	.9000	.3875	.9184	2.3374
3	−.2000	−.0250	−.3125	−.8500	−.4722	.9264

[5] For a proof of this result see Carl-Erik Fröberg, *Introduction to Numerical Analysis* (Boston: Addison-Wesley Publishing Co., Inc., 1969), pp. 95–100.

However, it is possible to obtain a convergent iteration by rearranging the equations to produce a more dominant diagonal. An appropriate rearrangement may require some judgment. In this case we shall try the equations in the following order:

$$7x_1 - x_2 + 3x_3 = 2$$
$$-x_1 + 2x_2 - 2x_3 = 2$$
$$2x_1 - 3x_2 + 4x_3 = -3$$

These equations rewritten as in (13.8) become

$$x_1 = \tfrac{1}{7}(\ 2 + x_2 - 3x_3)$$
$$x_2 = \tfrac{1}{2}(\ 2 + x_1 + 2x_3)$$
$$x_3 = \tfrac{1}{4}(-3 - 2x_1 + 3x_2)$$

The values of x_n^k computed using (13.10) for the first five cycles are given in Table 13.4. Convergence to the true answers to four decimal places, $x_1 = .5455$, $x_2 = 1.0000$, $x_3 = -.2727$, is occurring, although the convergence is

TABLE 13.4

k \ n	0	1	2	3	4	5
1	.6000	.5286	.5304	.5335	.5361	.5381
2	1.1000	1.0643	1.0491	1.0384	1.0300	1.0234
3	−.2000	−.2161	−.2284	−.2380	−.2456	−.2515

rather slow and many more cycles will be necessary to produce four-decimal accuracy. The student should note that convergence is occurring despite the fact that formula (13.11) is not satisfied. For the first equation we do have that $7 > 1 + 3$, but for the second equation $2 \not> 1 + 2$ and for the third equation $4 \not> 2 + 3$. This illustrates that formula (13.11) is not a necessary condition for convergence, even though it is sufficient.

13.7. RELAXATION

Relaxation is an iterative method of solving systems of linear equations. It is a generalization of the Gauss-Seidel method designed to accelerate convergence. In general, convergence with relaxation requires a matrix with a dominant diagonal, just as the Gauss-Seidel method does.

The first step in relaxation is to compute *residuals*. The residual for the *i*th equation is denoted by r_i. For the system given by (13.1) each element is divided by the diagonal element in that row and the residuals are defined by

$$\left. \begin{aligned} r_1 &= \frac{1}{a_{11}}\left(a_{11}x_1 + a_{12}x_2 + \cdots + a_{1n}x_n - c_1\right) \\[2mm] r_2 &= \frac{1}{a_{22}}\left(a_{21}x_1 + a_{22}x_2 + \cdots + a_{2n}x_n - c_2\right) \\[2mm] &\cdots\cdots\cdots\cdots\cdots\cdots\cdots\cdots\cdots\cdots\cdots\cdots\cdots\cdots\cdots \\[2mm] r_n &= \frac{1}{a_{nn}}\left(a_{n1}x_1 + a_{n2}x_2 + \cdots + a_{nn}x_n - c_n\right) \end{aligned} \right\} \quad (13.13\text{a})$$

which can be expressed as

$$\left. \begin{aligned} r_1 &= a'_{11}x_1 + a'_{12}x_2 + \cdots + a'_{1n}x_n - c'_1 \\ r_2 &= a'_{21}x_1 + a'_{22}x_2 + \cdots + a'_{2n}x_n - c'_2 \\ &\cdots\cdots\cdots\cdots\cdots\cdots\cdots\cdots\cdots\cdots\cdots\cdots \\ r_n &= a'_{n1}x_1 + a'_{n2}x_2 + \cdots + a'_{nn}x_n - c'_n \end{aligned} \right\} \quad (13.13\text{b})$$

where $a'_{ij} = \dfrac{a_{ij}}{a_{ii}}$ and $c'_i = \dfrac{c_i}{a_{ii}}$ for $i = 1, 2, \ldots, n$ and $j = 1, 2, \ldots, n$.

We now attempt to find a set of values of $x_1, x_2, \ldots, x_n$ such that the residuals $r_1, r_2, \ldots, r_n$ are all zero or at least as small as possible. For most systems the size of the residuals gives an indication of the size of the changes required in the values of x_i.

The above concept can be systematized in a *residual table*, which shows how $r_1, r_2, \ldots, r_n$ are affected by changes in $x_1, x_2, \ldots, x_n$. Table 13.5 is a residual table for the system given by (13.13b). Note that the values in the right half of the residual table are given by A'^T, that is, the transpose of the matrix of coefficients A' from (13.13b).

As seen above, each equation is multiplied by a factor which makes the elements on the diagonal equal to one, that is, $a'_{11} = a'_{22} = \cdots = a'_{nn} = 1$. If this is done, then it is apparent from Table 13.5 that $\Delta r_i = \Delta x_i = 1$ for $i = 1, 2, \ldots, n$ which considerably simplifies the mechanics of relaxation.

TABLE 13.5

Δx_1	Δx_2	$\cdots$	Δx_n	Δr_1	Δr_2	$\cdots$	Δr_n
$+1$				a'_{11}	a'_{21}	$\cdots$	a'_{n1}
	$+1$			a'_{12}	a'_{22}	$\cdots$	a'_{n2}
		$\cdots$		$\vdots$	$\vdots$		$\vdots$
			$+1$	a'_{1n}	a'_{2n}	$\cdots$	a'_{nn}

The acceleration in convergence using relaxation is accomplished by always working on the equation with the largest residual (in absolute value) rather than working on the equations in a fixed, predetermined order as with the Gauss-Seidel method. Thus, efficiency is increased by always focusing on the equation with maximum error (residual).

If the equation with the largest residual is the jth equation, then x_j is adjusted to reduce r_j. Note, however, that the process of adjusting x_j to reduce r_j also changes all the other residuals as well.

Relaxation ceases to be a unique method at this point and becomes a family of methods, since different adjustments to x_j can be made. Thus, some judgment enters the process. It would appear that a logical approach would be to reduce the largest residual to zero each time. However, experience has shown that more rapid convergence occurs if the largest residual is changed by slightly more than this amount, such that the residual actually changes signs but is still reduced in absolute value. This process is known as *overrelaxation*. It should be reemphasized that overrelaxation does not refer to a single well-defined method, but instead refers to a family of methods in which the skill of the practitioner affects the speed of convergence.

A completely correct solution to a nonsingular system of linear equations exists only when no roundoff error is present anywhere in the solution and when the residuals are reduced to zero. One common source of roundoff error arises from expressing coefficients (that is, the a's) in rounded decimal form. If the coefficients are not exact, then even reducing the residuals to zero does not generate a completely correct solution. Furthermore, because of roundoff error in the computations, it is usually not possible to reduce all the residuals to exactly zero.

Thus, it is necessary to develop a rule to determine when the iteration should be stopped. One possibility is to continue the iteration until the largest residual (in absolute value) is less than or equal to a certain amount. A second possibility is to continue the iteration until the sum of the absolute values of the residuals is less than or equal to a certain amount. Other possibilities also suggest themselves. Unfortunately, none of these criteria is sufficient to guarantee any given level of accuracy in the values of x in the solution vector, as will be seen in Section 13.8.

The student should note that relaxation is more suited to hand computation than it is to computer applications. Although it could be adapted to computers, the decision as to how much to relax or overrelax would have to be precisely defined. Furthermore, computer time is consumed in searching for the largest residual. To avoid these problems in computer applications the residuals are often reduced to exactly zero. This procedure results in some acceleration of the convergence if the equation with the largest residual is used each time, although not as much acceleration as overrelaxation generally produces. However, in order to reduce the time spent searching for the largest residual, the equations are often taken in a fixed, predetermined order, which reduces the acceleration further. In fact, the student should note that if this latter procedure is followed, then the relaxation method is identical with the Gauss-Seidel method. Thus, the Gauss-Seidel method can be viewed as a special case of the family of relaxation methods.

Example 13.8. Rework Example 13.7 using relaxation.

The equations arranged in dominant diagonal form are

$$7x_1 - x_2 + 3x_3 = 2$$
$$-x_1 + 2x_2 - 2x_3 = 2$$
$$2x_1 - 3x_2 + 4x_3 = -3$$

If the equations are written in residual form with diagonal elements equal to one and if decimal form is used, we have

$$r_1 = x_1 - .1429x_2 + .4286x_3 - .2857$$
$$r_2 = -.5x_1 + x_2 - x_3 - 1$$
$$r_3 = .5x_1 - .75 x_2 + x_3 + .75$$

The residual table is given in Table 13.6.

TABLE 13.6

Δx_1	Δx_2	Δx_3	Δr_1	Δr_2	Δr_3
+1			1.0000	−.5000	.5000
	+1		−.1429	1.0000	−.7500
		+1	.4286	−1.0000	1.0000

The steps involved in the iteration are summarized in Table 13.7. The following observations should help the student's understanding of Table 13.7:

1. The "successive changes to x" are chosen arbitrarily using over-relaxation (see statement 5).
2. The "cumulative values of x" start at step 0 with the starting values given in the problem. Each succeeding step contains the adjustments from "successive changes to x." The last step gives the answers as far as 10 iterations.

TABLE 13.7

	Successive changes to x			Cumulative values of x		
Step	x_1	x_2	x_3	x_1	x_2	x_3
0				.6000	1.1000	−.2000
1	−.0800			.5200	1.1000	−.2000
2		−.0500		.5200	1.0500	−.2000
3			−.0400	.5200	1.0500	−.2400
4		−.0400		.5200	1.0100	−.2400
5	+.0200			.5400	1.0100	−.2400
6			−.0300	.5400	1.0100	−.2700
7		−.0150		.5400	.9950	−.2700
8		+.0060		.5400	1.0010	−.2700
9	+.0050			.5450	1.0010	−.2700
10			−.0020	.5450	1.0010	−.2720

	Successive changes to r			Cumulative values of r		
Step	r_1	r_2	r_3	r_1	r_2	r_3
0				.0714	0	.0250
1	−.0800	+.0400	−.0400	−.0086	.0400	−.0150
2	+.0071	−.0500	+.0375	−.0015	−.0100	.0225
3	−.0171	+.0400	−.0400	−.0186	.0300	−.0175
4	+.0057	−.0400	+.0300	−.0129	−.0100	.0125
5	+.0200	−.0100	+.0100	.0071	−.0200	.0225
6	−.0129	+.0300	−.0300	−.0058	.0100	−.0075
7	+.0021	−.0150	+.0112	−.0037	−.0050	.0037
8	−.0009	+.0060	−.0045	−.0046	.0010	−.0008
9	+.0050	−.0025	+.0025	.0004	−.0015	.0017
10	−.0009	+.0020	−.0020	−.0005	.0005	−.0003

3. The "successive changes to r" are computed from Table 13.6 based on the "successive changes to x." The entries in Table 13.6 are multiplied by the change to x for the appropriate value of x.

4. The "cumulative values of r" are computed at step 0 using the three residual formulas appearing immediately before Table 13.6. Each succeeding step contains the adjustments from "successive changes to r." The last step gives the remaining residuals after 10 iterations.

5. The numbers appearing in "successive changes to x" are based on "cumulative values of r" using overrelaxation. Thus, the largest cumulative value of r (in absolute value) at step i is chosen which leads to a slightly larger change to x of the opposite sign at step $i + 1$. The amount of the adjustment is arbitrary other than that it satisfies this condition.

6. Steps 7 and 8 illustrate what happens when one overrelaxes too much on step 7. If this happens, then the same equation is chosen again for the next step and the adjustment is in the opposite direction. In no other case in " successive changes to x " is the same value of x used twice in succession.

7. Stopping the iteration at 10 steps is arbitrary, and it could be continued for more steps. However, Table 13.7 adequately illustrates the principles involved.

8. "Cumulative values of r" tend to be monotonic decreasing (in absolute value) with only a few exceptions. This is the sign of a converging iteration.

9. It is possible for a residual in one equation to equal zero and still have no true answers. For example, see r_2 in step 0.

10. The true answers to four decimal places are $x_1 = .5455$, $x_2 = 1.0000$, and $x_3 = -.2727$. The answers produced by relaxation using only 10 steps are superior to those produced in Example 13.7 using the Gauss-Seidel method. In Example 13.7, the Gauss-Seidel method is carried through five cycles which is equivalent to 15 steps, since there are three equations per cycle. Thus the more rapid convergence using relaxation is apparent.

13.8.   ILL-CONDITIONED SYSTEMS

In Section 13.7 it is implicitly assumed that small residuals are associated with small errors in the values of x in the solution vector. In most systems of linear equations this relationship holds. However, exceptions do exist and these are called *ill-conditioned systems*.

These systems are characterized by the fact that small changes in the coefficients of the equations (that is, the a's) and the constants (that is, the c's) cause correspondingly large changes in the solution (that is, the x's). Thus, ill-conditioned systems are inherently unstable, and small residuals do not necessarily mean that the solution vector is close to the true answer.

In theory a system of linear equations is either singular or nonsingular depending upon whether the determinant value of the coefficients is zero or nonzero. However, in computation involving roundoff error, the situation is not always this clear-cut. For example, a singular system with no unique solution can become a nonsingular system when roundoff error is introduced into the solution algorithm.[6] Naturally, any solution developed for this

---

[6]Conceivably the converse could also be true, that is, a nonsingular system could be turned into a singular system by roundoff error.

system is completely spurious. Ill-conditioned systems are characterized by the fact that the determinant value of the coefficients is close to zero. For this reason, ill-conditioned systems are often called "near-singular."

Instability in a system of equations is a problem for which no good solution exists. Usually the instability becomes apparent during the course of a solution algorithm. However, in some cases the instability can be well disguised, and solutions which are substantially in error are accepted as correct. Instability in large systems of equations is particularly difficult to determine, and the effect of roundoff error is more substantial in these systems than it is for smaller systems. In fact, roundoff error may become significant enough to introduce instability into an otherwise stable system.

Example 13.9.   *Demonstrate that the following system is ill-conditioned:*

$$x + 1.001y = 2.001$$
$$x + \quad y = 2$$

The trial solution $x = 2$, $y = 0$ produces residuals of .001 and 0, which appear to be acceptably small. However, the trial solution $x = 0$, $y = 2$ produces residuals of $-.001$ and 0, which are equally good. Clearly, the "true" solution which produces zero residuals is $x = 1$, $y = 1$. However, interpreting $x = 1$, $y = 1$ as the "true" solution is correct only if no roundoff error is present in the problem. For example, if the numbers 1.001 and 2.001 are rounded numbers, then it is impossible to say what the "true" solution is with any degree of accuracy at all. The student should note that the determinant value of the coefficients is equal to $-.001$, which is close to zero. Thus, the system is "near-singular."

13.9.   UNDERDETERMINED SYSTEMS

The system of linear equations given in (13.1) has an equal number of equations and unknowns. In certain situations it is necessary to analyze systems of linear equations in which the number of equations and unknowns are not equal.

If the number of equations is equal to m and the number of unknowns is equal to n, then the system can be written as

$$\left.\begin{array}{l} a_{11}x_1 + a_{12}x_2 + \cdots + a_{1n}x_n = c_1 \\ a_{21}x_1 + a_{22}x_2 + \cdots + a_{2n}x_n = c_2 \\ \cdots\cdots\cdots\cdots\cdots\cdots\cdots\cdots\cdots\cdots\cdots\cdots \\ a_{m1}x_1 + a_{m2}x_2 + \cdots + a_{mn}x_n = c_m \end{array}\right\} \qquad (13.14)$$

The following are definitions and observations about the system given in (13.14):

1. If $m = n$, then the original system considered in this chapter, that is, (13.1), is obtained.

2. If the system has at least one solution, then the system is said to be *consistent*. If no solution exists, then the system is said to be *inconsistent*.

3. A nonsingular system with an equal number of equations and unknowns is consistent and has a unique solution.

4. A singular system with an equal number of equations and unknowns may be either consistent, in which case an infinite number of solutions exist; or it may be inconsistent, in which case no solution exists.

5. Any consistent system with more than one solution is called *under-determined*. Any inconsistent system is called *overdetermined*. Thus, all systems can be categorized as:
   (a) Having a unique solution.
   (b) Underdetermined.
   (c) Overdetermined.

6. If $m < n$ and the system is consistent, then it is underdetermined. This is the typical manner in which underdetermined systems arise. However, underdetermined systems may arise with $m = n$ or even with $m > n$, if there is not enough information to develop a unique solution. For example, the system:

$$x + y = 1$$
$$2x + 2y = 2$$
$$3x + 3y = 3$$

is underdetermined even though there are more equations than unknowns.

7. Overdetermined systems typically arise when $m > n$. However, overdetermined systems may arise with $m = n$ or even with $m < n$, if there is conflicting information. For example, the system:

$$x + y + z = 1$$
$$2x + 2y + 2z = 1$$

is overdetermined even though there are fewer equations than unknowns.

Although overdetermined systems do not possess any solutions, it is possible to analyze them further. Overdetermined systems are considered further in Chapter 14. The balance of this section considers underdetermined systems.

In analyzing underdetermined systems it is convenient to write the system given by (13.14) as the following *augmented matrix*:

$$\begin{bmatrix} a_{11} & a_{12} & \cdots & a_{1n} & | & c_1 \\ a_{21} & a_{22} & \cdots & a_{2n} & | & c_2 \\ \vdots & \vdots & & \vdots & | & \vdots \\ a_{m1} & a_{m2} & \cdots & a_{mn} & | & c_m \end{bmatrix} \qquad (13.15)$$

The augmented matrix (13.15) is now manipulated with the three row operations given in Section 13.4:

1. Any row can be multiplied by a nonzero scalar.
2. Any multiple of one row can be added to any other row.
3. Any two rows can be interchanged.

The student should carefully note that when working with augmented matrices only row operations can be performed, that is, column operations are not valid.

To solve the underdetermined system given in (13.14), the augmented matrix (13.15) is converted to *canonical form*[7] by these row operations. For our purposes canonical form is defined by the following:

1. In the ith row, the first nonzero element is one unless all the elements are zero (see statement 4). Denote the column in which the one appears as j_i.
2. For the m rows we have $j_1 < j_2 < \cdots < j_m$ (assuming no rows contain all zeros), that is, the ones move downward and to the right quite similar to the diagonal in a square matrix, only in this case certain columns may be skipped if $n > m$.
3. In each column numbered j_i, for $i = 1, 2, \ldots, m$, all elements except the one in the ith row are equal to zero.
4. Any rows which contain all zeros should be interchanged to the bottom of the augmented matrix. These equations contain no new information not contained in the other equations.

The solution to the system of equations can immediately be obtained from the augmented matrix in canonical form by converting it back to equation form. This procedure is demonstrated in Example 13.10.

The student should note that many of the terms and concepts presented in this section are rather sketchy. A more complete and rigorous development can be obtained from textbooks in linear algebra.

---

[7] Many different types of canonical form have been devised of which this is one of the most common.

Example 13.10.   ***Solve the following system of equations:***

$$x_1 + x_2 - 2x_3 + x_4 + 3x_5 = 1$$
$$2x_1 - x_2 + 2x_3 + 2x_4 + 6x_5 = 2$$
$$3x_1 + 2x_2 - 4x_3 - 3x_4 - 9x_5 = 3$$

In this example we have three equations in five unknowns. Thus, the system is underdetermined, unless it is inconsistent (which, as we shall see, it is not). The solution is developed as follows:

$$\begin{bmatrix} 1 & 1 & -2 & 1 & 3 & | & 1 \\ 2 & -1 & 2 & 2 & 6 & | & 2 \\ 3 & 2 & -4 & -3 & -9 & | & 3 \end{bmatrix}$$ (original aug-mented matrix)

$$\begin{bmatrix} 1 & 1 & -2 & 1 & 3 & | & 1 \\ 0 & -3 & 6 & 0 & 0 & | & 0 \\ 0 & -1 & 2 & -6 & -18 & | & 0 \end{bmatrix}$$ (R2 − 2R1) (R3 − 3R1)

$$\begin{bmatrix} 1 & 1 & -2 & 1 & 3 & | & 1 \\ 0 & 1 & -2 & 0 & 0 & | & 0 \\ 0 & -1 & 2 & -6 & -18 & | & 0 \end{bmatrix}$$ $(-\tfrac{1}{3}R2)$

$$\begin{bmatrix} 1 & 1 & -2 & 1 & 3 & | & 1 \\ 0 & 1 & -2 & 0 & 0 & | & 0 \\ 0 & 0 & 0 & -6 & -18 & | & 0 \end{bmatrix}$$ (R3 + R2)

$$\begin{bmatrix} 1 & 1 & -2 & 1 & 3 & | & 1 \\ 0 & 1 & -2 & 0 & 0 & | & 0 \\ 0 & 0 & 0 & 1 & 3 & | & 0 \end{bmatrix}$$ $(-\tfrac{1}{6}R3)$

$$\begin{bmatrix} 1 & 0 & 0 & 0 & 0 & | & 1 \\ 0 & 1 & -2 & 0 & 0 & | & 0 \\ 0 & 0 & 0 & 1 & 3 & | & 0 \end{bmatrix}$$ (R1 − R2 − R3)

The last augmented matrix above is in canonical form. The system expressed in equation form now becomes

$$x_1 \qquad\qquad\qquad = 1$$
$$x_2 - 2x_3 \qquad\quad = 0$$
$$x_4 + 3x_5 = 0$$

If we choose $x_3 = a$ and $x_5 = b$ to be arbitrary, then the *general solution* is given by

$$x_1 = 1$$
$$x_2 = 2a$$
$$x_3 = a$$
$$x_4 = -3b$$
$$x_5 = b$$

Any real values of a and b can be selected to give a *particular solution.*[8] Obviously, since the system is underdetermined, there is no unique particular solution. However, it is interesting to note that one of the five values of x, namely x_1, does have a unique value.

13.10. EIGENVALUES AND EIGENVECTORS

Let A be a square matrix and let x be a column vector as defined in formula (13.4). If λ is a real number such that the linear system

$$Ax = \lambda x \qquad (13.16)$$

has solutions other than $x = 0$, then λ is called an *eigenvalue* (or *latent root*) of the matrix A. Any solution vector x corresponding to a given eigenvalue is called an *eigenvector* of the matrix A.

Formula (13.16) can be rewritten as

$$(A - \lambda I)x = 0 \qquad (13.17)$$

where I is the identity matrix. Eigenvalues can now be developed by setting

$$|A - \lambda I| = 0 \qquad (13.18)$$

that is, by setting the determinant value of $A - \lambda I$ equal to zero. If $|A - \lambda I| \neq 0$, then the only solution to formula (13.17) is the unique solution $x = 0$.

If the determinant given by formula (13.18) is expanded, it is seen to be a polynomial in λ. This polynomial is called the *characteristic polynomial.*[9] The roots of the characteristic polynomial are the desired eigenvalues. In general, if A is an $n \times n$ matrix, then A has n eigenvalues, since the characteristic polynomial is of degree n. However, these n roots need not be all distinct nor all real. For our purposes we shall ignore any complex roots of the characteristic polynomial.

Once the eigenvalues are obtained, then the eigenvectors associated with each eigenvalue can be found using formula (13.17). It is important to note that any eigenvector associated with an eigenvalue is not unique. For example, if x is an eigenvector associated with λ, then kx is also an eigenvector for any constant $k \neq 0$. In finding eigenvectors one value of x must be chosen arbitrarily, and then the remaining values of x which are consistent can be found.

Example 13.11.   Find the eigenvalues and eigenvectors of the matrix

$$A = \begin{bmatrix} -3 & 3 \\ -10 & 8 \end{bmatrix}$$

---

[8]The student should note that the usage of the terms "general solution" and "particular solution" is somewhat similar to the usage of these terms in Sections 9.2 and 10.1.

[9]The student should not confuse the "characteristic polynomial" given here with the "characteristic equation" given in Section 9.4.

Formula (13.18) gives

$$\begin{vmatrix} -3 - \lambda & 3 \\ -10 & 8 - \lambda \end{vmatrix} = 0$$

or

$$(-3 - \lambda)(8 - \lambda) - (3)(-10) = 0$$

or

$$\lambda^2 - 5\lambda + 6 = 0$$

The eigenvalues are the roots of this equation, that is, $\lambda = 2$ and $\lambda = 3$. Since A is a 2×2 matrix, there is a maximum of two distinct, real eigenvalues.

The eigenvectors corresponding to $\lambda = 2$ can be found from equation (13.17) written as

$$-5x_1 + 3x_2 = 0$$
$$-10x_1 + 6x_2 = 0$$

Clearly, the system is underdetermined as it must be. If a unique solution existed, it would have to be $x_1 = x_2 = 0$. However, it is the other solutions in which we are interested. One eigenvector is given by

$$x = \begin{bmatrix} 3 \\ 5 \end{bmatrix}$$

Any other vector kx where k is a nonzero constant is also an eigenvector.

Similarly, the eigenvectors corresponding to $\lambda = 3$ can be found from the system

$$-6x_1 + 3x_2 = 0$$
$$-10x_1 + 5x_2 = 0$$

Again the system is underdetermined as it must be. One eigenvector is given by

$$x = \begin{bmatrix} 1 \\ 2 \end{bmatrix}$$

All other vectors kx, where $k \neq 0$, are also eigenvectors.

EXERCISES

13.1. Introduction

1. Show that the following system does not have a unique solution:

$$x_1 + 2x_2 + x_3 = 3$$
$$2x_1 + 3x_2 + x_3 = 5$$
$$3x_1 + 5x_2 + 2x_3 = 1$$

13.2. Cramer's rule

2. Solve the following system of equations using Cramer's rule:

$$x_1 - x_2 + x_3 = 9$$
$$2x_1 - 4x_2 + 3x_3 = 28$$
$$- 2x_2 + 3x_3 = 18$$

3. In the linear system of equations

$$a_{11}x_1 + a_{12}x_2 = c_1$$
$$a_{21}x_1 + a_{22}x_2 = c_2$$

it is known that the constants are $c_1 = 8$, $c_2 = 11$, and it is known that the solution is $x_1 = 5$, $x_2 = -2$. Find the matrix of coefficients, A, if it is known that $|A| = 1$ and that the two diagonal elements in the matrix are equal.

4. Attempt to solve the singular system

$$x + 2y = 3$$
$$2x + 4y = 5$$

using Cramer's rule. Where does the method fail?

13.3. Gaussian elimination

5. Rework Exercise 2 using Gaussian elimination.

6. The following system of equations is to be solved using Gaussian elimination with rearrangement to minimize roundoff error:

$$3x_1 + 4x_2 + 4x_3 = 11$$
$$4x_1 + 7x_2 + 4x_3 = 15$$
$$5x_1 + 4x_2 + 3x_3 = 12$$

If three decimal places are carried in all computations, what do the equations become when expressed in upper triangular form?

7. Solve the upper triangular system developed in Exercise 6 by back-substitution. Carry the answers to three decimal places.

8. Using the result developed in Exercise 6, find

$$\begin{vmatrix} 3 & 4 & 4 \\ 4 & 7 & 4 \\ 5 & 4 & 3 \end{vmatrix}$$

What is the exact value of the determinant?

9. Rework Exercise 4 using Gaussian elimination. Where does the method fail?

13.4.   Matrix inversion method

10.   Rework Exercise 2 using the matrix inversion method. (The inverse matrix was obtained in Exercise 36 in Chapter 12.)

11.   (a)   Continue Example 13.5 by making a second exchange between c_1 and x_3.
   (b)   Continue Example 13.5 by making a third exchange between c_3 and x_2.

12.   (a)   Show that one application of the exchange method to row 1, column 1 of the following matrix leaves the matrix unchanged:

$$\begin{bmatrix} 1 & 0 & 0 \\ 1 & 1 & 0 \\ 1 & 1 & 1 \end{bmatrix}$$

   (b)   If a second application of the exchange method to row 2, column 2 is made, find the resulting matrix.
   (c)   Finish the inversion by applying the exchange method to row 3, column 3.

13.   Rework Exercise 4 using the matrix inversion method. Where does the method fail?

13.5.   Matrix factorization method

14.   Rework Exercise 2 using the matrix factorization method. (A matrix factorization was obtained in Exercise 44 in Chapter 12.)

15.   Solve the system of equations

$$x_1 + 2x_2 = 5$$
$$2x_1 + 5x_2 = 11$$

by the matrix factorization method. (A matrix factorization was obtained in Exercise 46 in Chapter 12.)

16.   (a)   Show that the system

$$2x_1 + 2x_2 + x_3 = 5$$
$$x_1 + x_2 + x_3 = 3$$
$$3x_1 + 2x_2 + x_3 = 6$$

   has a unique solution.
   (b)   Show that the system given in (a) cannot be solved by the matrix factorization method in which $A = LU$.

17.   Rework Exercise 4 using the matrix factorization method.
   (a)   Can the singular matrix be factored?
   (b)   Where does the method fail?

13.6.   Gauss-Seidel method

18.   Rework Exercise 2 using the Gauss-Seidel method. Carry exact fractions.
   (a)   What are the starting values given by formula (13.9)?
   (b)   Find the approximate answer after three cycles using the above starting values.
   (c)   Is the iteration converging?
   (d)   Is formula (13.11) satisfied?

19. The Gauss-Seidel method is to be used to find the solution for the following system of equations:

$$12x_1 \qquad + 4x_3 = 5$$
$$3x_1 + 7x_2 \qquad = 3$$
$$7x_2 + 12x_3 = 5$$

If starting values of $x_1^0 = x_2^0 = x_3^0 = .3$ are used, find the three iterated values on the first cycle to three decimal places.

20. For the system of equations given in Exercise 19 it is known that the value of x_3 on the fourth iteration is equal to .260. Find the value of x_3 on the third iteration. (These values would not be produced by the starting values given in Exercise 19.)

21. Rework Exercise 4 using the Gauss-Seidel method. Where does the method fail?

13.7. Relaxation

22. If the system of equations in Exercise 2 is to be solved using relaxation with the starting values found in Exercise 18 (a), find the residuals given by formula (13.13b).

23. Continuing Exercise 22, use overrelaxation until all three residuals equal zero. Carry exact fractions in all computations.

24. Rework Example 13.8 through 15 steps by reducing the residuals to exactly zero each time and by cycling through the equations in a fixed order. This process is equivalent to the Gauss-Seidel method and should produce answers consistent with Example 13.7 (except for minor rounding discrepancies).

25. Rework Example 13.8 through 10 steps by reducing the residuals to exactly zero each time but by always working on the equation with the largest residual. The rate of convergence of this approach is superior to the Gauss-Seidel method (Exercise 24), but is inferior to using overrelaxation (Example 13.8).

26. Compare the three approaches developed in Exercise 24, Exercise 25, and Example 13.8 with the true answer. Use the answers after 10 steps for each method.

27. Use overrelaxation to solve the following system:

$$x_1 + \tfrac{1}{3}x_2 + \tfrac{1}{2}x_3 = \quad 1$$
$$\tfrac{1}{2}x_1 + \quad x_2 + \tfrac{1}{4}x_3 = -2$$
$$\tfrac{1}{2}x_1 + \tfrac{1}{6}x_2 + \quad x_3 = \quad 2$$

Starting values of $x_1^0 = x_2^0 = x_3^0 = 0$ are given. Exact fractions are to be used in all computations. The following table is also given for successive changes to x:

Step	x_1	x_2	x_3
1		-2	
2			$+3$
3		-1	
4	$+1$		
5			-1

13.8. Ill-conditioned systems

28. How can the instability in the system of equations given in Example 13.9 be interpreted geometrically?

29. What is the true answer to Example 13.9 if 1.001 is really a rounded value of 1.0005 and if 2.001 is really a rounded value of 2.0015?

30. (a) Solve the system:

$$x - \quad y = 1$$
$$x - 1.001y = 0$$

(b) Solve the system:

$$x - \quad y = 1$$
$$x - .999y = 0$$

13.9. Underdetermined systems

31. Solve the following system of equations:

$$x_1 + 2x_2 - 3x_3 - 4x_4 = 6$$
$$x_1 + 3x_2 + x_3 - 2x_4 = 4$$
$$2x_1 + 5x_2 - 2x_3 - 5x_4 = 10$$

32. Solve the following system of equations:

$$x_1 + x_2 + x_3 = 4$$
$$2x_1 + 5x_2 - 2x_3 = 3$$

33. Show that the following system of equations is inconsistent:

$$x_1 + x_2 + 2x_3 + x_4 = 5$$
$$2x_1 + 3x_2 - x_3 - 2x_4 = 2$$
$$4x_1 + 5x_2 + 3x_3 \quad = 7$$

What is the canonical augmented matrix which shows the inconsistency?

13.10. Eigenvalues and eigenvectors

34. (a) Find the one eigenvalue of the matrix

$$\begin{bmatrix} 0 & 1 & 0 \\ 0 & 0 & 1 \\ 1 & -3 & 3 \end{bmatrix}$$

(b) Find an eigenvector associated with this eigenvalue.

35. Find the eigenvalues of the matrix

$$\begin{bmatrix} 2 & -2 & 3 \\ 1 & 1 & 1 \\ 1 & 3 & -1 \end{bmatrix}$$

36. If the matrix A has the two eigenvalues 1, 2 with associated eigenvectors $\begin{bmatrix} 1 \\ 1 \end{bmatrix}$, $\begin{bmatrix} 1 \\ 2 \end{bmatrix}$, respectively, find A.

Miscellaneous problems

37. In certain cases the Gauss-Seidel method can be applied to nonlinear equations. If starting values of $x_0 = .7$, $y_0 = .4$, and $z_0 = 1.1$ are used, find the approximate answers to three decimal places for the following system of equations after two cycles:

$$x - .1y^2 + .05z^2 = .7$$
$$y + .3x^2 - .1xz = .5$$
$$z + .4y^2 + .1xy = 1.2$$

38. The value of a 100×100 determinant is to be evaluated using Gaussian elimination. If each pivot is subject to a maximum relative roundoff error of .1 %, then find:
   (a) The approximate maximum relative roundoff error in the value of the determinant.
   (b) The approximate estimated relative roundoff error in the value of the determinant.

39. (a) Show that Gaussian elimination (not counting back-substitution) involves

$$1 + 2 + \cdots + n = \frac{n^2 + n}{2}$$

   divisions (include the c-vector in this count).
   (b) Show that Gaussian elimination (not counting back-substitution) involves

$$1 \cdot 2 + 2 \cdot 3 + \cdots + (n-1)(n) = \frac{n^3 - n}{3}$$

   multiplications.
   (c) Show that Gaussian elimination (not counting back-substitution) involves

$$1 \cdot 2 + 2 \cdot 3 + \cdots + (n-1)(n) = \frac{n^3 - n}{3}$$

   additions (or subtractions).

40. (a) Show that back-substitution, in which all diagonal elements equal one, involves

$$1 + 2 + \cdots + (n-1) = \frac{n^2 - n}{2}$$

   multiplications.
   (b) Show that back-substitution, in which all diagonal elements equal one, involves

$$1 + 2 + \cdots + (n-1) = \frac{n^2 - n}{2}$$

   additions (or subtractions).

14

Linear programming

14.1. INTRODUCTION

LINEAR PROGRAMMING is concerned with maximizing or minimizing a linear function subject to a set of linear constraints. This chapter develops two approaches which can be used to solve linear programming problems:

1. A geometrical solution.
2. The simplex method.

These two approaches are described in Sections 14.2 and 14.3, respectively.[1]

Linear programming has many applications in economics and other social sciences. For example, consider a manufacturing firm that wishes to maximize profits subject to constraints arising from the availability of scarce resources and the capacity of its productive facilities. Linear programming has proven to be an effective technique to find optimal solutions to problems of this type.

Linear programming is a relatively recent mathematical development, having its origins in World War II. The development of linear programming at that time was stimulated by the need to find optimal allocations of scarce

---

[1] A thorough mathematical development of linear programming requires a background in vector spaces and matrices beyond the scope of Chapter 12. The student is referred to W. Allen Spivey, *Linear Programming: An Introduction* (New York: The Macmillan Co., 1963), for proofs of certain results not proved in this chapter.

resources with which to conduct the war effort. In the postwar period the techniques of linear programming have been widely adopted by business and government planners to improve the quality of management decision making.

The general linear programming problem can be formalized as follows:

maximize the linear function

$$f = c_1 x_1 + c_2 x_2 + \cdots + c_n x_n \tag{14.1a}$$

subject to the m constraints

$$\left.\begin{array}{l} a_{11} x_1 + a_{12} x_2 + \cdots + a_{1n} x_n \le b_1 \\ a_{21} x_1 + a_{22} x_2 + \cdots + a_{2n} x_n \le b_2 \\ \cdots\cdots\cdots\cdots\cdots\cdots\cdots\cdots\cdots\cdots\cdots \\ a_{m1} x_1 + a_{m2} x_2 + \cdots + a_{mn} x_n \le b_m \end{array}\right\} \tag{14.2a}$$

and to the n constraints

$$x_i \ge 0 \qquad \text{for } i = 1, 2, \ldots, n \tag{14.3a}$$

It is assumed that $b_i \ge 0$ for $i = 1, 2, \ldots, m$.

The general problem can be expressed in matrix notation as follows:

maximize

$$f = c^T x \tag{14.1b}$$

subject to

$$Ax \le b \tag{14.2b}$$

and

$$x \ge 0 \tag{14.3b}$$

where A is the matrix of coefficients in (14.2a) and b, c, x are column vectors.

The formulation of the linear programming problem given above is actually more general than it may appear to be. The following generalizations should be noted:

1. If the problem is to minimize rather than maximize f, we can replace f with $-f$ and maximize $-f$. The answer obtained is then the negative of the required answer.

2. If any of the constraints in (14.2a) involve " $\ge$ " or " $=$ " instead of " $\le$," linear programming can still be used. These extensions are considered in subsequent sections.

3. If any of the constraints in (14.2a) involve negative b_i's, these constraints should be multiplied by -1 and the sense of the inequality reversed to produce positive b_i's. The resulting inequalities can be handled as described in generalization 2 above.

Any point $(x_1, x_2, \ldots, x_n)$ which satisfies the $n + m$ constraints is called a *feasible point*. A feasible point for which f is maximized is called a *solution point* or an *optimal feasible point*.

In certain instances there may be no feasible points in which case no solution points exist. This happens if the constraints are inconsistent. For example, the constraints

$$2x_1 + x_2 \leq 3$$
$$4x_1 + 2x_2 \geq 7$$

are inconsistent, so that no feasible points exist.

In certain other instances there may be exactly one feasible point in which case the solution is immediate. This may happen if the constraints are equalities. For example, the constraints

$$2x_1 + x_2 = 4$$
$$3x_1 - x_2 = 1$$

immediately give the unique solution $x_1 = 1$, $x_2 = 2$.

However, the above illustrations are not typical, and in general an infinite number of feasible points exist. Thus, it becomes necessary to examine this infinite set of feasible points to find any which maximize f. There is no guarantee that any such solution point exists or, even if it does, that it is unique. The possibility of nonexistent or nonunique solutions is considered further in subsequent sections.

An important theorem in linear programming is that any solution which exists occurs at an *extreme feasible point*, which is defined as a feasible point at which at least n of the $n + m$ constraints become equalities. This theorem is of great significance in finding solutions to linear programming problems, since it dramatically reduces the number of feasible points which could be solution points. In fact, the set of extreme feasible points is a finite set, since the number of constraints is finite. The concept of an extreme feasible point and the fact that the number of such points is finite will be clarified by the example considered in Section 14.2.

14.2. A GEOMETRICAL SOLUTION

If $n = 2$, then any solutions to a linear programming problem can readily be determined geometrically. The geometrical solution is not only an efficient algorithm for problems in which $n = 2$, but it also offers a valuable insight into the nature of problems for which $n > 2$.

The set of feasible points is derived by applying the $n + m$ constraints. Each constraint specifies a half-plane within which the point must lie in order to satisfy that constraint. The set of feasible points is the intersection of these $n + m$ half-planes. If the intersection is empty, then the constraints are called *inconsistent* and no solution exists.

It is not difficult to show that the set of feasible points is a *convex set*. A set of points is called convex if for any two points in the set, p_1 and p_2, the point $kp_1 + (1 - k)p_2$, where $0 \leq k \leq 1$, is also in the set. This definition taken verbally states that all straight lines connecting any two points in the set are also contained in the set.

Figure 14.1 illustrates a convex set in the plane, while Figure 14.2 illustrates a set which is not convex.

FIGURE 14.1   **FIGURE 14.2**

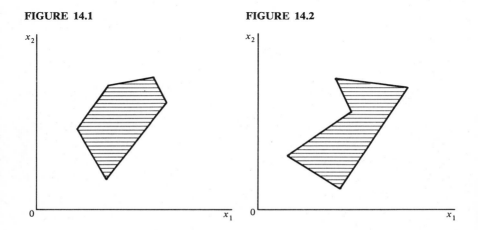

The boundary of the set of feasible points is composed of those points for which the constraint inequalities become equalities. Since none of the inequalities is a strict inequality, the boundary of the set is included in the set.[2]

As discussed in Section 14.1, any solutions of a linear programming problem occur at extreme feasible points where at least n of the $n + m$ constraints become equalities. We are considering the case $n = 2$, so that the extreme feasible points occur at the intersections of two of the constraints taken as equalities. In Figure 14.1 there are five such extreme feasible points at the five "corners" of the polygon. We have mentioned the fact that any solution point is an extreme feasible point and how important this result is in facilitating the search for solution points. Although an infinite number of

---

[2]In mathematical terms a set including its boundary points is called *closed* and one which does not include its boundary points is called *open*. An open set would arise if the constraint inequalities were strict inequalities. In linear programming problems the complications of working with open sets are avoided by not using strict inequalities. Nothing is thereby lost in applying linear programming to practical problems.

feasible points exist, only a finite number of these, namely five, are extreme feasible points and thus capable of being solution points.

Another theorem in linear programming states that if the set of feasible points is bounded, as in Figure 14.1, then both a maximum and minimum for any linear function f exist.[3] If the set of feasible points is not bounded, then either a maximum or minimum or both may not exist. Figure 14.3 illustrates a nonbounded convex set of feasible points which may lack a solution depending on f.

The geometrical solution to a linear programming problem is obtaining by visualizing the linear function f to be maximized or minimized as a family of parallel lines. Figure 14.4 illustrates this for the linear function $f = x_1 + 2x_2$. The family of lines is moved across the convex set of feasible points until an

FIGURE 14.3

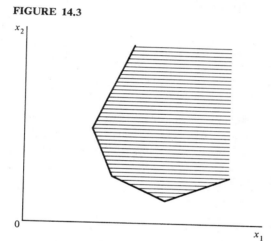

FIGURE 14.4

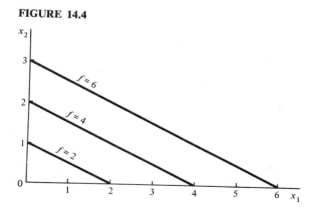

---

[3]This result is obtained from the well-known *Weierstrass theorem* in analysis.

extreme feasible point at which *f* is maximized or minimized is located. Such an extreme feasible point is a solution point.

It is now apparent why a convex set of feasible points which is not bounded, as in Figure 14.3, may or may not have a solution. The determining factor of whether or not a solution exists is the slope of *f* in relation to the boundary of the set of feasible points.

It is also apparent why a solution may or may not be unique. For example, if *f* is parallel to one of the sides of the set of feasible points, then both end points along that side are extreme feasible points which have the same value of *f* and may be solution points. If this is the case, then all boundary points lying on the line connecting the two extreme feasible points also have the same value of *f* and may be solution points.

Example 14.1. **Maximize**

$$f = x_1 + 2x_2$$

subject to

$$
\begin{aligned}
x_1 + x_2 &\geq 1 & x_1 &\geq 0 \\
-x_1 + x_2 &\leq 3 & x_2 &\geq 0 \\
2x_1 + 3x_2 &\leq 24 \\
x_1 &\leq 6
\end{aligned}
$$

Figure 14.5 is the geometrical representation of this example. The solid lines are the boundary of the set of feasible points, while the dashed lines represent the family of lines given by *f*. In this example there are six extreme

FIGURE 14.5

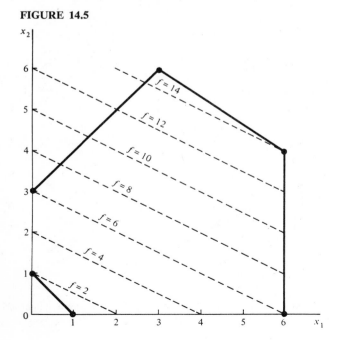

feasible points $(0, 1)$, $(0, 3)$, $(3, 6)$, $(6, 4)$, $(6, 0)$, $(1, 0)$. The solution point is the extreme feasible point at which f is maximized. It is clear that the solution point is $(3, 6)$ and the maximum value of f at this point is 15.

Example 14.2.   Rework Example 14.1 if the problem is to minimize $f = x_2 - x_1$ subject to the same constraints.

In this example the family of lines has slope 1 and the value of f decreases as the lines move downward to the right. Thus, f is minimized at the extreme feasible point $(6, 0)$ and the minimum value of f at this point is -6.

14.3.   THE SIMPLEX METHOD

Although the geometrical solution to linear programming problems discussed in Section 14.2 is an efficient algorithm for problems in which $n = 2$, it becomes much more difficult to apply if $n = 3$ and impossible to apply if $n \geq 4$.

The *simplex method* is an analytical solution to linear programming problems which can be applied for any positive integer n and which is readily capable of being programmed for computers. The concept behind the simplex method is to start at one extreme feasible point and compute f. The method then moves to another extreme feasible point at the end of a line segment from the original point along the boundary of the convex set of feasible points such that a larger value of f is obtained. This process continues until f can no longer be increased and a solution point results or until it is seen that no solution exists.

In order to apply the simplex method the form of the general linear programming problem given in Section 14.1 is modified by the introduction of *slack variables* $x_{n+1}, x_{n+2}, \ldots, x_{n+m}$. These slack variables are added to the left-hand side of the inequalities in (14.2a) in order to produce equalities. Thus the set of constraints given by (14.2a) is rewritten as

$$\left.\begin{array}{l} a_{11}x_1 + a_{12}x_2 + \cdots + a_{1n}x_n + x_{n+1} = b_1 \\ a_{21}x_1 + a_{22}x_2 + \cdots + a_{2n}x_n + x_{n+2} = b_2 \\ \cdots\cdots\cdots\cdots\cdots\cdots\cdots\cdots\cdots\cdots\cdots\cdots\cdots\cdots \\ a_{m1}x_1 + a_{m2}x_2 + \cdots + a_{mn}x_n + x_{n+m} = b_m \end{array}\right\} \quad (14.4)$$

Like the original variables $x_1, x_2, \ldots, x_n$, the slack variables are also non-negative, that is,

$$x_i \geq 0 \qquad \text{for } i = n + 1, n + 2, \ldots, n + m \qquad (14.5)$$

It should be carefully noted that regardless of any positive values which the m slack variables may assume in the solution process, no contribution is made to the value of f given by formula (14.1a).

The above formulation of a linear programming problem can be extended to include constraints which involve " $\geq$ " or " $=$ " instead of " $\leq$." If we have a constraint of the form

$$a_{i1}x_1 + a_{i2}x_2 + \cdots + a_{in}x_n \geq b_i$$

then we introduce a *surplus variable*, instead of a slack variable, and the constraint becomes

$$a_{i1}x_1 + a_{i2}x_2 + \cdots + a_{in}x_n - x_{n+i} = b_i$$

If we have a constraint of the form

$$a_{i1}x_1 + a_{i2}x_2 + \cdots + a_{in}x_n = b_i$$

then we introduce neither a slack nor surplus variable, and leave the equality unchanged. The simplex method automatically handles the equality as any other equality in the solution process.

In Section 14.1 an extreme feasible point is defined as a feasible point at which at least n of the $n + m$ constraints become equalities. With the introduction of slack (surplus) variables it is possible to rephrase this definition to state that an extreme feasible point is a feasible point at which at least n of the $n + m$ variables $x_1, \ldots, x_{n+m}$ are equal to zero.

It is apparent that (14.4) is a system of m linear equations in $n + m$ unknowns. The matrix of coefficients is

$$A = \begin{bmatrix} a_{11} & a_{12} & \cdots & a_{1n} & 1 & 0 & \cdots & 0 \\ a_{21} & a_{22} & \cdots & a_{2n} & 0 & 1 & \cdots & 0 \\ \vdots & \vdots & & \vdots & \vdots & \vdots & & \vdots \\ a_{m1} & a_{m2} & \cdots & a_{mn} & 0 & 0 & \cdots & 1 \end{bmatrix} \tag{14.6}$$

where the last m columns apply to the slack variables.[4]

We can now define $n + m + 1$ column vectors, P_i, as follows:

$$P_0 = \begin{bmatrix} b_1 \\ b_2 \\ \vdots \\ b_m \end{bmatrix}, P_1 = \begin{bmatrix} a_{11} \\ a_{21} \\ \vdots \\ a_{m1} \end{bmatrix}, \ldots, P_n = \begin{bmatrix} a_{1n} \\ a_{2n} \\ \vdots \\ a_{mn} \end{bmatrix}, P_{n+1} = \begin{bmatrix} 1 \\ 0 \\ \vdots \\ 0 \end{bmatrix}, \ldots, P_{n+m} = \begin{bmatrix} 0 \\ 0 \\ \vdots \\ 1 \end{bmatrix}$$

$$\tag{14.7}$$

The system of equations given by (14.4) can be written as

$$x_1 P_1 + x_2 P_2 + \cdots + x_{n+m} P_{n+m} = P_0 \tag{14.8}$$

---

[4]If a surplus variable is required in any constraint, then -1 appears in that row instead of 1. If any constraint is an equality, then the proper entry is 0 instead of ± 1. In this latter case, the resulting column of zeros can be eliminated from the array, since it does not affect the solution.

The simplex method requires that an extreme feasible point be known to use as a starting point. The easiest method of finding an extreme feasible point is to try the origin which often works (but not in Example 14.1). The origin should be used as the starting point, if possible. If the origin cannot be used, then the *artificial basis* method is generally used to determine an extreme feasible point.[5] The artificial basis method is described and illustrated in Example 14.5.

In general, the origin can be used as a starting point whenever the problem follows the standard format in which all constraints in (14.2a) are " $\leq$," so that a slack variable is required for each constraint, and in which $b_i \geq 0$ for $i = 1, 2, \ldots, m$. Situations in which the origin cannot be used as a starting point arise when at least one of the constraints in (14.2a) is " $\geq$ " or " $=$," or when $b_i \leq 0$ for some i (which reverses the sense of the inequality upon multiplying by -1 to make $b_i \geq 0$).

We now note that the P_i's are $m \times 1$ vectors. We can select a subset of m of the $n + m$ vectors $P_1, P_2, \ldots, P_{n+m}$ to serve as a *basis* to start the simplex method. A basis is defined as a set of m linearly independent vectors which can be linearly combined to generate all the $n + m$ vectors in the set. If the origin can be used as the starting point, then the vectors $P_{n+1}, P_{n+2}, \ldots, P_{n+m}$ are chosen as the initial basis for the problem.

The simplex method starts by constructing a *simplex tableau* as illustrated in Table 14.1. The following observations can be made about the initial simplex tableau given in Table 14.1:

1. The element in the ith row of the column vector P_j is denoted by a_{ij} for $i = 1, 2, \ldots, m$ and for $j = 0, 1, 2, \ldots, n + m$. These elements comprise the main body of the table and are given by (14.7). The student should note that the elements of $P_0, P_{n+1}, P_{n+2}, \ldots, P_{n+m}$ as given in (14.7) have been relabeled in Table 14.1. In particular, note that for the initial simplex tableau

$$a_{i0} = b_i \qquad \text{for } i = 1, 2, \ldots, m$$
$$a_{i, n+i} = 1 \qquad \text{for } i = 1, 2, \ldots, m$$
$$\underset{i \neq k}{a_{i, n+k}} = 0 \qquad \text{for } i = 1, 2, \ldots, m$$
$$k = 1, 2, \ldots, m$$

2. The initial basis vectors $P_{n+1}, P_{n+2}, \ldots, P_{n+m}$ are listed vertically in the tableau.

---

[5] In some cases it is possible to find an extreme feasible point when the origin cannot be used by geometrical means, trial and error inspection, etc. However, such ad hoc methods cannot be relied upon for all problems, particularly large and complex problems. Thus, reliance on the general artificial basis method, even in simple problems, is preferred.

TABLE 14.1

c_i		c_0	c_1	$\cdots$	c_n	c_{n+1}	$\cdots$	c_{n+m}
	Vectors / Basis	P_0	P_1	$\cdots$	P_n	P_{n+1}	$\cdots$	P_{n+m}
c_{n+1}	P_{n+1}	a_{10}	a_{11}	$\cdots$	a_{1n}	$a_{1,n+1}$	$\cdots$	$a_{1,n+m}$
c_{n+2}	P_{n+2}	a_{20}	a_{21}	$\cdots$	a_{2n}	$a_{2,n+1}$	$\cdots$	$a_{2,n+m}$
$\vdots$	$\vdots$	$\vdots$	$\vdots$	$\vdots$	$\vdots$	$\vdots$	$\vdots$	$\vdots$
c_{n+m}	P_{n+m}	a_{m0}	a_{m1}	$\cdots$	a_{mn}	$a_{m,n+1}$	$\cdots$	$a_{m,n+m}$
z_i		z_0	z_1	$\cdots$	z_n	z_{n+1}	$\cdots$	z_{n+m}
$z_i - c_i$		$z_0 - c_0$	$z_1 - c_1$	$\cdots$	$z_n - c_n$	z_{n+1} $- c_{n+1}$	$\cdots$	z_{n+m} $- c_{n+m}$

3. The coefficients of the linear function to be maximized appear across the top of the tableau. Note that $c_0 = c_{n+1} = c_{n+2} = \cdots = c_{n+m} = 0$, since only $c_1, c_2, \ldots, c_n$ appear in formula (14.1a). These values of c_i at the top of the tableau remain fixed throughout the entire solution process.

4. The values of c_i for the basis vectors are also listed in the first column of the tableau. For the initial tableau these values are all zero. However, we shall see later that in applying the simplex method the basis vectors change from step to step. Whenever a basis vector is changed, the corresponding c_i value is moved into the first column of the tableau.

5. The next-to-last line contains values of z_i. The value of z_i in the P_i column is computed as the dot product of the vector of c's at the left of the tableau and the vector P_i.

6. The last line contains values of z_i, as just computed, minus values of c_i, as given at the top of the tableau.

Once the initial simplex tableau has been completed, the following test is made:

1. If all $z_i - c_i \geq 0$, the extreme feasible point used as the starting point is a solution point and the simplex method is ended.

2. If $z_i - c_i < 0$ in some column, then one of the following is the situation:
   (a) If no a_{ij} in that column is positive, then no solution exists.
   (b) If $a_{ij} > 0$ for some element in that column, then another application of the simplex method is required.

If another application of the simplex method is required, then one of the basis vectors is exchanged for one of the nonbasis vectors. It should be noted that P_0 will never be used as a basis vector. Denote the "replacing" vector by P_s and the "replaced" vector by P_t. These two vectors are determined as follows:

1. The replacing vector, P_s, is the nonbasis vector with the most negative $z_i - c_i$ value. In the case of ties any of the vectors which are tied can be used. In order to have a fixed rule to follow, the vector with the lowest subscript i is generally chosen.
2. The replaced vector, P_t, is determined by finding the element a_{is} in the column vector, P_s, for which $\dfrac{a_{i0}}{a_{is}}$ is minimized with the conditions:
   (a) Only positive values of a_{is} are considered.
   (b) The subscript i ranges over the basis vectors appearing at this step of the simplex method.

   The basis vector appearing in row i at this step in the simplex method is the replaced vector. In the case of ties any of the vectors which are tied can be used and again it is conventional to use the basis vector with the lowest subscript i.[6] The element in the row and column thus determined is called the *pivot*.

A new simplex tableau is now constructed as follows:

1. The replacing vector, P_s, along with its associated c value, is substituted for the replaced vector, P_t, and its associated c value at the left of the tableau.
2. The elements in the pivot row are all divided by the pivot.
3. The elements in the pivot column are set equal to zero, except for the pivot itself which is equal to one.
4. The remaining elements a_{kl} are replaced by

$$a_{kl} - \frac{a_{kj}\,a_{il}}{a_{ij}} \tag{13.6}$$

which is the *rectangular rule* given in Section 13.4.
5. Revised values of z_i are computed on the next-to-last line as before.
6. Revised values of $z_i - c_i$ are computed on the last line as before.

The test described above on $z_i - c_i$ is again made. The entire simplex method continues in this fashion until the test indicates that a solution point

---

[6]In this case a tie can be interpreted as indicating *degeneracy*. Although degeneracy causes certain theoretical complications, fortunately the simplex method still proceeds to a solution point, if one exists, as long as some definite rule for resolving ties is followed.

has been obtained or that no solution exists. If a solution exists, the entries in the P_0 column give the values of x at the solution point taking into account the basis in existence at that step, while the maximum value of f appears in the last line in the P_0 column.

Example 14.3. *Maximize*

$$f = x_1 + 2x_2$$

subject to

$$
\begin{aligned}
-x_1 + x_2 &\le 3 & x_1 &\ge 0 \\
2x_1 + 3x_2 &\le 24 & x_2 &\ge 0 \\
x_1 &\le 6
\end{aligned}
$$

This example is the same as Example 14.1 without the constraint $x_1 + x_2 \ge 1$. Removing this constraint simplifies the problem by allowing us to use the origin as the starting point.

If slack variables x_3, x_4, x_5 are introduced, the constraints assume the form

$$
\begin{aligned}
-x_1 + x_2 + x_3 & & &= 3 \\
2x_1 + 3x_2 & & + x_4 &= 24 \\
x_1 & & + x_5 &= 6
\end{aligned}
$$

The initial simplex tableau is given in Table 14.2a.

TABLE 14.2a

c_i	Basis ╲ Vectors		0	1	2	0	0	0
			P_0	P_1	P_2	P_3	P_4	P_5
0	P_3		3	-1	1	1	0	0
0	P_4		24	2	3	0	1	0
0	P_5		6	1	0	0	0	1
z_i			0	0	0	0	0	0
$z_i - c_i$			0	-1	-2	0	0	0

Since $z_i - c_i < 0$ in both columns P_1 and P_2 and since positive elements appear in both columns, another application of the simplex method is indicated. The exchange vectors are determined as follows:

1. Since the most negative value of $z_i - c_i$ is -2, P_2 is the replacing vector.
2. Since $\frac{3}{1} < \frac{24}{3}$, P_3 is the replaced vector.

The pivot in the P_3 row and P_2 column is 1.

A second simplex tableau is now constructed and appears in Table 14.2b.

TABLE 14.2b

c_i			0	1	2	0	0	0
	Basis	Vectors	P_0	P_1	P_2	P_3	P_4	P_5
2	P_2		3	-1	1	1	0	0
0	P_4		15	5	0	-3	1	0
0	P_5		6	1	0	0	0	1
z_i			6	-2	2	2	0	0
$z_i - c_i$			6	-3	0	2	0	0

Since $z_i - c_i < 0$ in column P_1 and since positive elements appear in this column, another application of the simplex method is indicated. The exchange vectors are determined as follows:

1. Since the only negative value of $z_i - c_i$ is -3, P_1 is the replacing vector.
2. Since $\frac{15}{5} < \frac{6}{1}$, P_4 is the replaced vector.

The pivot in the P_4 row and P_1 column is 5.

A third simplex tableau is now constructed and appears in Table 14.2c.

Since $z_i - c_i \geq 0$ in all columns, the simplex method is ended. If we refer to Figure 14.5 (ignoring the line $x_1 + x_2 = 1$), it is apparent that the simplex method has moved around the boundary of the set of feasible points to the solution point.

In Table 14.2b the entries in the P_0 column indicate that $x_1 = 0$ (since P_1 is not a basis vector), $x_2 = 3$ (the entry on the P_2 line), and $f = 6$ (the entry on the $z_i - c_i$ line). Thus, the first step has taken us from the origin to the point $(0, 3)$ at which $f = 6$.

Similarly, in Table 14.2c the entries in the P_0 column indicate that $x_1 = 3$, $x_2 = 6$, and $f = 15$. Thus, the second step has taken us from the point $(0, 3)$ to the point $(3, 6)$ at which $f = 15$.

TABLE 14.2c

c_i			0	1	2	0	0	0
	Basis	Vectors	P_0	P_1	P_2	P_3	P_4	P_5
2	P_2		6	0	1	.4	.2	0
1	P_1		3	1	0	$-.6$	.2	0
0	P_5		3	0	0	.6	$-.2$	1
z_i			15	1	2	.2	.6	0
$z_i - c_i$			15	0	0	.2	.6	0

Example 14.4. *Rework Example 14.3 if the problem is to minimize* $f = x_2 - x_1$ *subject to the same constraints.*

This example is the same as Example 14.2 without the constraint $x_1 + x_2 \geq 1$. Removing this constraint simplifies the problem by allowing us to use the origin as the starting point.

Rather than minimizing $f = x_2 - x_1$ we maximize $-f = x_1 - x_2$. The initial tableau is the same as Table 14.2a with the revised values of c_i and is given in Table 14.3a.

TABLE 14.3a

c_i			0	1	-1	0	0	0
	Basis	Vectors	P_0	P_1	P_2	P_3	P_4	P_5
0	P_3		3	-1	1	1	0	0
0	P_4		24	2	3	0	1	0
0	P_5		6	1	0	0	0	1
z_i			0	0	0	0	0	0
$z_i - c_i$			0	-1	1	0	0	0

Since $z_i - c_i < 0$ in column P_1 and since positive elements appear in this column, another application of the simplex method is indicated. The exchange vectors are determined as follows:

1. Since the only negative value of $z_i - c_i$ is -1, P_1 is the replacing vector.
2. Since $\frac{6}{1} < \frac{24}{2}$, P_5 is the replaced vector. The pivot in the P_5 row and P_1 column is 1.

A second simplex tableau is now constructed and appears in Table 14.3b.

TABLE 14.3b

c_i			0	1	-1	0	0	0
	Basis	*Vectors*	P_0	P_1	P_2	P_3	P_4	P_5
0	P_3		9	0	1	1	0	1
0	P_4		12	0	3	0	1	-2
1	P_1		6	1	0	0	0	1
z_i			6	1	0	0	0	1
$z_i - c_i$			6	0	1	0	0	1

Since $z_i - c_i \geq 0$ in all columns, the simplex method is ended. The solution point is $x_1 = 6$, $x_2 = 0$, and the maximum value of $-f = 6$, so that the minimum value of $f = -6$. This result confirms the answer obtained in Example 14.2.

Example 14.5. **Rework Example 14.1 using the simplex method.**

In this example the constraint $x_1 + x_2 \geq 1$ is reinstated, so that the origin cannot be used as a starting point. The artificial basis method is now described and used to find a starting point.

If a surplus variable x_3 and slack variables x_4, x_5, x_6 are introduced, the problem becomes

maximize
$$f = x_1 + 2x_2$$

subject to
$$
\begin{aligned}
x_1 + x_2 - x_3 && = 1 \\
-x_1 + x_2 \quad\; + x_4 && = 3 \\
2x_1 + 3x_2 \quad\quad\; + x_5 && = 24 \\
x_1 \quad\quad\quad\quad\quad + x_6 &&= 6
\end{aligned}
$$

and $x_i \geq 0$ for $i = 1, 2, \ldots, 6$

The solutions to Examples 14.3 and 14.4 have a starting matrix at the first step which contains the identity matrix as a submatrix. Clearly, the matrix in this example

$$\begin{bmatrix} 1 & 1 & -1 & 0 & 0 & 0 \\ -1 & 1 & 0 & 1 & 0 & 0 \\ 2 & 3 & 0 & 0 & 1 & 0 \\ 1 & 0 & 0 & 0 & 0 & 1 \end{bmatrix}$$

does not contain the identity matrix as a submatrix. Whenever the identity matrix does not appear as a submatrix, then the artificial basis method is used to determine a starting value.

In the artificial basis method new variables x_i are introduced which produce the identity matrix as a submatrix. These new variables are called *artificial variables*. However, since we want these artificial variables to be zero at a solution point, we introduce them into f with a negative coefficient of such magnitude that in order to maximize f at a solution point, all the artificial variables are equal to zero.

Thus, if a surplus variable x_3, an artificial variable x_4, and slack variables x_5, x_6, x_7 are introduced, the problem becomes

maximize
$$f = x_1 + 2x_2 - Mx_4$$

subject to
$$\begin{aligned} x_1 + x_2 - x_3 + x_4 &= 1 \\ -x_1 + x_2 + x_5 &= 3 \\ 2x_1 + 3x_2 + x_6 &= 24 \\ x_1 + x_7 &= 6 \end{aligned}$$

and $x_i \geq 0$ for $i = 1, 2, \ldots, 7$, where M is an arbitrarily large positive number

The initial simplex tableau is given in Table 14.4a.

Since $z_i - c_i < 0$ in both columns P_1 and P_2 and since positive elements appear in both columns, another application of the simplex method is indicated. The exchange vectors are determined as follows:

1.   Since the most negative value of $z_i - c_i$ is $-M - 2$, P_2 is the replacing vector.
2.   Since $\frac{1}{1} < \frac{3}{1} < \frac{24}{3}$, P_4 is the replaced vector.

The pivot in the P_4 row and P_2 column is 1.

A second simplex tableau is now constructed and appears in Table 14.4b.

TABLE 14.4a

c_i		0	1	2	0	$-M$	0	0	0
Basis	Vectors	P_0	P_1	P_2	P_3	P_4	P_5	P_6	P_7
$-M$	P_4	1	1	1	-1	1	0	0	0
0	P_5	3	-1	1	0	0	1	0	0
0	P_6	24	2	3	0	0	0	1	0
0	P_7	6	1	0	0	0	0	0	1
z_i		$-M$	$-M$	$-M$	M	$-M$	0	0	0
$z_i - c_i$		$-M$	$-M-1$	$-M-2$	M	0	0	0	0

TABLE 14.4b

c_i		0	1	2	0	$-M$	0	0	0
Basis	Vectors	P_0	P_1	P_2	P_3	P_4	P_5	P_6	P_7
2	P_2	1	1	1	-1	1	0	0	0
0	P_5	2	-2	0	1	-1	1	0	0
0	P_6	21	-1	0	3	-3	0	1	0
0	P_7	6	1	0	0	0	0	0	1
z_i		2	2	2	-2	2	0	0	0
$z_i - c_i$		2	1	0	-2	$M+2$	0	0	0

It is apparent that the artificial basis method has moved us to the point $x_1 = 0$ and $x_2 = 1$ where $f = 2$. If we refer to Figure 14.5, we see that this is indeed an extreme feasible point. The regular simplex method can now be applied as before. It is also apparent that P_4 will never again appear as a basis vector because of the arbitrarily large value of M.

The rest of this example is left as Exercise 6.

14.4. DUALITY

The *duality theorem* of linear programming states that corresponding to the standard problem

$$\text{maximize}$$
$$f = c^T x \tag{14.1b}$$

$$\text{subject to}$$
$$Ax \leq b \tag{14.2b}$$
$$\text{and } x \geq 0 \tag{14.3b}$$

there exists a dual problem

$$\text{minimize}$$
$$g = y^T b \tag{14.9b}$$

$$\text{subject to}$$
$$y^T A \geq c^T \tag{14.10b}$$
$$\text{and } y \geq 0 \tag{14.11b}$$

If either problem has a solution, then the other problem has the same solution, that is, the maximum of f is equal to the minimum of g. Thus, if we solve one problem, we automatically solve its dual.

It is instructive to expand formulas (14.9b), (14.10b), (14.11b) and compare the form of the dual problem with the original problem given by formulas (14.1a), (14.2a), (14.3a). We have the following matrix definitions:

$$A = \begin{bmatrix} a_{11} & a_{12} & \cdots & a_{1n} \\ a_{21} & a_{22} & \cdots & a_{2n} \\ \vdots & \vdots & & \vdots \\ a_{m1} & a_{m2} & \cdots & a_{mn} \end{bmatrix}, \ b = \begin{bmatrix} b_1 \\ b_2 \\ \vdots \\ b_m \end{bmatrix}, \ c = \begin{bmatrix} c_1 \\ c_2 \\ \vdots \\ c_n \end{bmatrix}, \ y = \begin{bmatrix} y_1 \\ y_2 \\ \vdots \\ y_m \end{bmatrix}$$

Thus, the dual problem becomes

$$\text{minimize}$$
$$g = b_1 y_1 + b_2 y_2 + \cdots + b_m y_m \tag{14.9a}$$

$$\text{subject to}$$

$$\left. \begin{array}{c} a_{11} y_1 + a_{21} y_2 + \cdots + a_{m1} y_m \geq c_1 \\ a_{12} y_1 + a_{22} y_2 + \cdots + a_{m2} y_m \geq c_2 \\ \cdots\cdots\cdots\cdots\cdots\cdots\cdots\cdots\cdots\cdots \\ a_{1n} y_1 + a_{2n} y_2 + \cdots + a_{mn} y_m \geq c_n \end{array} \right\} \tag{14.10a}$$

$$\text{and } y_i \geq 0 \text{ for } i = 1, 2, \ldots, m \tag{14.11a}$$

In comparing the two problems, the following observations are significant:

1. One problem is a maximization problem and the other a minimization problem.
2. The constraint inequalities are " $\leq$ " in the maximization problem and " $\geq$ " in the minimization problem.
3. The coefficients of the linear function in the maximization problem become the right-hand entries for the constraint inequalities in the minimization problem, and conversely.
4. The matrix of coefficients A in the maximization problem becomes A^T in the minimization problem.
5. The variables x_i and y_i are all constrained to be nonnegative.

Example 14.6. State the dual of Example 14.3.

The original problem is

$$\text{maximize}$$
$$f = x_1 + 2x_2$$
$$\text{subject to}$$
$$\begin{aligned} -x_1 + x_2 &\leq 3 \\ 2x_1 + 3x_2 &\leq 24 \\ x_1 &\leq 6 \end{aligned}$$
$$\text{and } x_i \geq 0 \text{ for } i = 1, 2$$

The dual problem is

$$\text{minimize}$$
$$g = 3y_1 + 24y_2 + 6y_3$$
$$\text{subject to}$$
$$\begin{aligned} -y_1 + 2y_2 + y_3 &\geq 1 \\ y_1 + 3y_2 &\geq 2 \end{aligned}$$
$$\text{and } y_i \geq 0 \text{ for } i = 1, 2, 3$$

14.5. OVERDETERMINED SYSTEMS

In Section 13.9 systems of linear equations are categorized as follows:

1. Systems with a unique solution.
2. Systems with an infinite number of solutions, called *underdetermined*.
3. Systems with no solutions, called *overdetermined*.

In Chapter 13 the first two categories are considered. It remains to consider the third category.

Overdetermined systems have no solutions because the given equations are inconsistent. Typically, overdetermined systems arise when there are more equations than unknowns. The general system can be expressed as

$$\left.\begin{array}{c} a_{11}x_1 + a_{12}x_2 + \cdots + a_{1n}x_n = c_1 \\ a_{21}x_1 + a_{22}x_2 + \cdots + a_{2n}x_n = c_2 \\ \cdots\cdots\cdots\cdots\cdots\cdots\cdots\cdots \\ a_{m1}x_1 + a_{m2}x_2 + \cdots + a_{mn}x_n = c_m \end{array}\right\} \quad (14.12)$$

in which m is generally greater than n.

Since no exact solution exists, we consider the *residuals* defined by

$$r_i = a_{i1}x_1 + a_{i2}x_2 + \cdots + a_{in}x_n - c_i \qquad (14.13)[7]$$

for $i = 1, 2, \ldots, m$. We wish to minimize these residuals in some sense. Two approaches to do this have been developed and are widely used in practice. These are the following:

1. The *least-squares method*, in which $S = \sum_{i=1}^{m} r_i^2$ is minimized.

2. The *minimax method* in which max $(|r_1|, \ldots, |r_m|)$ is minimized.[8]

Both methods are initially discussed for three equations in two unknowns, and then the generalization to m equations in n unknowns is discussed.

Considering first the least-squares method, the equations become

$$r_1 = a_{11}x_1 + a_{12}x_2 - c_1$$
$$r_2 = a_{21}x_1 + a_{22}x_2 - c_2$$
$$r_3 = a_{31}x_1 + a_{32}x_2 - c_3$$

so that the sum of squares of residuals, S, is given by

$$\begin{aligned} S &= r_1^2 + r_2^2 + r_3^2 \\ &= (a_{11}^2 + a_{21}^2 + a_{31}^2)x_1^2 + (a_{12}^2 + a_{22}^2 + a_{32}^2)x_2^2 \\ &\quad + 2(a_{11}a_{12} + a_{21}a_{22} + a_{31}a_{32})x_1x_2 - 2(a_{11}c_1 + a_{21}c_2 + a_{31}c_3)x_1 \\ &\quad - 2(a_{12}c_1 + a_{22}c_2 + a_{32}c_3)x_2 + (c_1^2 + c_2^2 + c_3^2) \end{aligned}$$

---

[7]The student should note that the definition of the residuals given by formula (14.13) is not completely consistent with the definition given by (13.13b) in that elements are not divided by a diagonal element as in (13.13b).

[8]Least-squares and minimax polynomial approximation will be discussed in Sections 15.3 and 15.4, respectively. The student will note many similarities between this section and the latter two sections.

We now seek to find x_1 and x_2 such that S is minimized. This is accomplished by taking partial derivatives of S with respect to x_1 and x_2 and setting the derivatives equal to zero, which gives

$$\frac{\partial S}{\partial x_1} = 2(a_{11}^2 + a_{21}^2 + a_{31}^2)x_1 + 2(a_{11}a_{12} + a_{21}a_{22} + a_{31}a_{32})x_2$$
$$- 2(a_{11}c_1 + a_{21}c_2 + a_{31}c_3) = 0$$

$$\frac{\partial S}{\partial x_2} = 2(a_{12}^2 + a_{22}^2 + a_{32}^2)x_2 + 2(a_{11}a_{12} + a_{21}a_{22} + a_{31}a_{32})x_1$$
$$- 2(a_{12}c_1 + a_{22}c_2 + a_{32}c_3) = 0$$

These equations can be expressed as two equations in two unknowns in the following form:

$$\left. \begin{array}{l} (a_1, a_1)x_1 + (a_1, a_2)x_2 = (a_1, c) \\ (a_2, a_1)x_1 + (a_2, a_2)x_2 = (a_2, c) \end{array} \right\} \tag{14.14}$$

in which

$$(a_1, a_1) = a_{11}^2 + a_{21}^2 + a_{31}^2$$
$$(a_2, a_2) = a_{12}^2 + a_{22}^2 + a_{32}^2$$
$$(a_1, a_2) = (a_2, a_1) = a_{11}a_{12} + a_{21}a_{22} + a_{31}a_{32}$$
$$(a_1, c) = a_{11}c_1 + a_{21}c_2 + a_{31}c_3$$
$$(a_2, c) = a_{12}c_1 + a_{22}c_2 + a_{32}c_3$$

By a similar derivation the general result for m equations in n unknowns, can be determined by solving the following n equations in n unknowns:

$$\left. \begin{array}{l} (a_1, a_1)x_1 + (a_1, a_2)x_2 + \cdots + (a_1, a_n)x_n = (a_1, c) \\ (a_2, a_1)x_1 + (a_2, a_2)x_2 + \cdots + (a_2, a_n)x_n = (a_2, c) \\ \cdots\cdots\cdots\cdots\cdots\cdots\cdots\cdots\cdots\cdots\cdots\cdots\cdots\cdots\cdots\cdots\cdots \\ (a_n, a_1)x_1 + (a_n, a_2)x_2 + \cdots + (a_n, a_n)x_n = (a_n, c) \end{array} \right\} \tag{14.15}$$

where

$$(a_i, a_j) = a_{1i}a_{1j} + \cdots + a_{mi}a_{mj} \tag{14.16}$$

and

$$(a_i, c) = a_{1i}c_1 + \cdots + a_{mi}c_m \tag{14.17}$$

for $i = 1, 2, \ldots, n$ and $j = 1, 2, \ldots, n$. Formula (14.16) is the dot product of two column vectors from the matrix of coefficients; while formula (14.17) is the dot product of two column vectors, one from the matrix of coefficients and the other the vector of constants, c.

Secondly, we consider the minimax method. We seek to minimize r where $r = \max(|r_1|, |r_2|, |r_3|)$. We have

$$|r_1| \le r \qquad |r_2| \le r \qquad |r_3| \le r$$

which gives the six inequalities

$$r_1 \le r \qquad -r_1 \le r$$
$$r_2 \le r \qquad -r_2 \le r$$
$$r_3 \le r \qquad -r_3 \le r$$

The residual equations now become

$$a_{11}x_1 + a_{12}x_2 - c_1 \le r \qquad -a_{11}x_1 - a_{12}x_2 + c_1 \le r$$
$$a_{21}x_1 + a_{22}x_2 - c_2 \le r \qquad -a_{21}x_1 - a_{22}x_2 + c_2 \le r$$
$$a_{31}x_1 + a_{32}x_2 - c_3 \le r \qquad -a_{31}x_1 - a_{32}x_2 + c_3 \le r$$

If we divide through by r and set

$$y_1 = \frac{x_1}{r} \qquad y_2 = \frac{x_2}{r} \qquad y_3 = \frac{1}{r}$$

the six equations become

$$a_{11}y_1 + a_{12}y_2 - c_1 y_3 \le 1 \qquad -a_{11}y_1 - a_{12}y_2 + c_1 y_3 \le 1$$
$$a_{21}y_1 + a_{22}y_2 - c_2 y_3 \le 1 \qquad -a_{21}y_1 - a_{22}y_2 + c_2 y_3 \le 1$$
$$a_{31}y_1 + a_{32}y_2 - c_3 y_3 \le 1 \qquad -a_{31}y_1 - a_{32}y_2 + c_3 y_3 \le 1$$

Since we want to minimize r, we maximize $y_3 = 1/r$. Thus, the problem can be viewed as a linear programming problem in which $f = y_3$ is maximized subject to the above six constraints. It is also necessary for $y_i \ge 0$ for $i = 1$, 2, 3. From the above definitions of y_i, this requires $x_i \ge 0$ for $i = 1, 2$. If x_1 and x_2 are not both positive, then either a linear transformation must be made or the simplex algorithm must be modified to accommodate negative values of x_i.

The generalization of the minimax method to m equations in n unknowns is immediate.

Example 14.7.   Solve the system of linear equations

$$x_1 - x_2 = 2$$
$$x_1 + x_2 = 4$$
$$2x_1 + x_2 = 8$$

using the least-squares method.

We have

$$
\begin{aligned}
(a_1, a_1) &= 1^2 &&+ 1^2 &&+ 2^2 &&= 6 \\
(a_2, a_2) &= (-1)^2 &&+ 1^2 &&+ 1^2 &&= 3 \\
(a_1, a_2) = (a_2, a_1) &= (1)(-1) &&+ (1)(1) &&+ (2)(1) &&= 2 \\
(a_1, c) &= (1)(2) &&+ (1)(4) &&+ (2)(8) &&= 22 \\
(a_2, c) &= (-1)(2) &&+ (1)(4) &&+ (1)(8) &&= 10
\end{aligned}
$$

so that the system given by (14.14) becomes

$$
\begin{aligned}
6x_1 + 2x_2 &= 22 \\
2x_1 + 3x_2 &= 10
\end{aligned}
$$

The solution is

$$
x_1 = \tfrac{23}{7} \quad \text{and} \quad x_2 = \tfrac{8}{7}
$$

The residuals are

$$
\begin{aligned}
r_1 &= \tfrac{23}{7} - \tfrac{8}{7} - 2 = \tfrac{1}{7} \\
r_2 &= \tfrac{23}{7} + \tfrac{8}{7} - 4 = \tfrac{3}{7} \\
r_3 &= \tfrac{46}{7} + \tfrac{8}{7} - 8 = -\tfrac{2}{7}
\end{aligned}
$$

The minimum sum of squares is $\tfrac{2}{7}$, and the maximum absolute residual is $\tfrac{3}{7}$.

Example 14.8.   Rework Example 14.7 using the minimax method.

The linear programming formulation of the problem is

maximize
$$
f = y_3
$$

subject to

$$
\begin{aligned}
y_1 - y_2 - 2y_3 &\le 1 \\
y_1 + y_2 - 4y_3 &\le 1 \\
2y_1 + y_2 - 8y_3 &\le 1 \\
-y_1 + y_2 + 2y_3 &\le 1 \\
-y_1 - y_2 + 4y_3 &\le 1 \\
-2y_1 - y_2 + 8y_3 &\le 1
\end{aligned}
\qquad
\begin{aligned}
y_1 &\ge 0 \\
y_2 &\ge 0 \\
y_3 &\ge 0
\end{aligned}
$$

The solution to this linear programming problem using the simplex method is

$$
y_1 = 10 \qquad y_2 = 3 \qquad y_3 = 3 \qquad f = 3
$$

Thus, $r = 1/y_3 = \tfrac{1}{3}$, giving

$$
x_1 = \tfrac{10}{3} \quad \text{and} \quad x_2 = 1
$$

The residuals are

$$r_1 = \tfrac{10}{3} - 1 - 2 = \tfrac{1}{3}$$
$$r_2 = \tfrac{10}{3} + 1 - 4 = \tfrac{1}{3}$$
$$r_3 = \tfrac{20}{3} + 1 - 8 = -\tfrac{1}{3}$$

The minimum maximum absolute residual is $\tfrac{1}{3}$, and the sum of squares is $\tfrac{1}{3}$. The student should note that the minimax method gives a smaller maximum error than least-squares, while the latter method gives a smaller sum of squares than the former.

EXERCISES

14.1. Introduction; 14.2. A geometrical solution

1. Rework Example 14.1 to minimize $f = x_1 + 2x_2$. Does a unique solution point exist?

2. Rework Example 14.2 to maximize $f = x_2 - x_1$. Does a unique solution point exist?

3. (a) Maximize

$$f = x_1 - 2x_2$$

subject to

$$
\begin{aligned}
x_1 + 2x_2 &\le 4 \qquad x_1 \ge 0 \\
-x_1 + x_2 &\le 1 \qquad x_2 \ge 0 \\
x_1 + x_2 &\le 3
\end{aligned}
$$

(b) Minimize f.

4. (a) Maximize

$$f = x_1 + x_2$$

subject to

$$
\begin{aligned}
2x_1 + x_2 &\ge 2 \qquad x_1 \ge 0 \\
x_1 - 2x_2 &\le 2 \qquad x_2 \ge 0 \\
-x_1 + x_2 &\le 2
\end{aligned}
$$

(b) Minimize f.

5. Maximize

$$f = 2x_1 - x_2$$

subject to

$$
\begin{aligned}
x_1 + x_2 &= 2 \qquad x_1 \ge 0 \\
&\qquad\quad\ x_2 \ge 0
\end{aligned}
$$

14.3. The simplex method

6. Finish Example 14.5.

7. Rework Exercise 3 using the simplex method.

8. Rework Exercise 4 using the simplex method with an artificial basis.

9. Rework Exercise 5 using the simplex method with an artificial basis.

10. Maximize

$$f = Kx_1 + x_2$$

subject to

$$x_1 + x_2 \le 6 \qquad x_1 \ge 0$$
$$x_1 + 2x_2 \le 10 \qquad x_2 \ge 0$$

Express your answer as a function of K, where $K \ge 1$.

11. Maximize

$$f = 2x_1 + x_2$$

subject to

$$x_1 - x_2 \le 2 \qquad x_1 \ge 0$$
$$x_1 + x_2 \le 6 \qquad x_2 \ge 0$$
$$x_1 + 2x_2 \le K$$

Express your answer as a function of K, where $0 \le K \le 2$.

14.4. Duality

12. State the dual of Exercise 3(a).

13. Solve Exercise 12 using the simplex method with an artificial basis.

14. Solve Example 14.6 using the simplex method with an artificial basis.

14.5. Overdetermined systems

15. Find a least-squares solution to the overdetermined system

$$x_1 + x_2 = 1$$
$$x_1 + x_2 = 2$$

16. Find a minimax solution to the system given in Exercise 15.

17. A variable x is observed n times and the outcomes are denoted by A_i for $i = 1, 2, \ldots, n$. Find the least-squares solution to the overdetermined system

$$x = A_1$$
$$x = A_2$$
$$\vdots$$
$$x = A_n$$

18. Find the least-squares solution to the overdetermined system

$$x \quad = A$$
$$y = B$$
$$x - y = C$$
$$x + y = D$$

19. Determine the system of linear equations which must be solved to find the least-squares solution to the overdetermined system

$$x_1 - 2x_2 \quad + \quad x_4 = \quad 7$$
$$3x_1 \quad + \quad x_3 - 4x_4 = -2$$
$$3x_2 \quad - 5x_4 = \quad 1$$
$$-x_1 + \quad x_2 + \quad x_3 + 2x_4 = \quad 8$$
$$-3x_1 - \quad x_2 + 4x_3 \quad = -3$$
$$2x_1 + \quad x_2 - \quad x_3 - \quad x_4 = \quad 1$$

20. Determine the linear program which must be solved to find the minimax solution to the system given in Exercise 19.

Miscellaneous problems

21. A manufacturing firm can produce either product M or product N. For product M production processes 1 and 2 can be utilized, while for product N production processes 3 and 4 are available. Product M sells for $7 per unit and product N sells for $10 per unit. Each product requires inputs of labor, raw materials, and plant space, all of which are restricted in supply. The following table gives the volume of each input required to produce one unit of output using each process and also gives the total amount of each input which is available:

	Per unit of M		Per unit of N		Input available
	Process 1	Process 2	Process 3	Process 4	
Labor	1	2	2	3	25
Raw materials	7	5	10	7	100
Plant space	4	4	8	4	70

Denoting the output using processes 1, 2, 3, 4 by x_1, x_2, x_3, x_4 respectively, formulate the linear programming problem which must be solved to find the optimal production using each process in order to maximize revenue and not exceed the inputs available of any of the three resources.

22. A dietician can serve any of five foods 1, 2, 3, 4, 5 for which the prices per unit respectively are 5, 20, 8, 11, 12. Each food has been analyzed in units of calories, vitamins, and proteins as follows:

	Per unit of food				
	1	2	3	4	5
Calories	5	25	20	5	12
Vitamins	7	20	2	25	10
Proteins	2	16	9	8	12

It is required that the units of calories be between 1000 and 1500, the units of vitamins not less than 1000, and the units of proteins not less than 700. Denoting the number of units of foods 1, 2, 3, 4, 5 by x_1, x_2, x_3, x_4, x_5 respectively, formulate the linear programming problem which must be solved to find the optimal serving of each type of food in order to minimize the cost of the meal and still satisfy the constraints on calories, vitamins, and proteins.

23. A mining company is taking a certain kind of ore from two mines, A and B. The ore is divided into three quality groups, a, b, and c. Every week the company has to deliver 240 tons of a, 160 tons of b, and 440 tons of c. The weekly production quota for a must be met exactly, while the quotas for b and c may be exceeded. The cost per day for running mine A is $3000 and for running mine B is $2000. The daily production capabilities are as follows:

	Daily production capabilities (tons)	
	Mine A	Mine B
Ore a	60	20
Ore b	20	20
Ore c	40	80

Denoting the number of days of operation of mines A and B by x_1 and x_2, respectively, formulate the linear programming problem which must be solved to find the optimal number of days to operate each mine in order to minimize costs of operation and still produce the required volume of ore.

15

Other techniques of approximation

15.1. INTRODUCTION

A COMMON PROBLEM in applied mathematics is *approximation*, which is defined as the fitting of a relatively simple mathematical formula to data based upon a more complex mathematical formula or not based upon any known mathematical formula at all. The approximating curve can then be used for interpolation or extrapolation, or for other purposes, such as approximate differentiation.

The most common method of approximation used in practice is the collocation polynomial which has been extensively discussed in the earlier chapters of this book. However, a variety of approximation methods other than collocation have been devised and are often useful alternatives. Certain of these alternative approximation methods are the subject of this chapter.

Sections 15.2, 15.3, and 15.4 examine three methods of fitting polynomials other than by collocation. These are, respectively:

1. Osculatory polynomials.
2. Least-squares polynomials.
3. Minimax polynomials.

Finally, Sections 15.5 and 15.6 examine two particularly useful methods of approximation using nonpolynomials. These are, respectively:

1. Ratios of polynomials.
2. Exponential functions.

Naturally, many other types of nonpolynomials could also be used.[1] However, the two methods just mentioned are relatively simple to apply and are often successful in situations in which polynomials are inappropriate.

It is important to distinguish approximation methods based on smoothing curves from those based on reproducing curves. A *reproducing* curve passes through the given functional values without altering them in any way. A *smoothing* curve generally does not pass through the given functional values,[2] but instead replaces them with a smoother series of values.

The collocation polynomial is reproducing by definition. It will be seen in Sections 15.2, 15.3, and 15.4 that least-squares polynomials and minimax polynomials are smoothing, and that osculatory polynomials can be either reproducing or smoothing.

Any method of approximation which is smoothing is said to be a *graduation* method. Graduation is a process applied to data exhibiting irregularities in which the irregular functional values are replaced with a smoother series of values. Although this chapter does discuss certain smoothing formulas, a thorough treatment of the several sophisticated graduation methods which have been developed in practice is beyond the scope of this book.[3]

15.2. OSCULATORY POLYNOMIALS

As discussed in Section 4.8, interpolation formulas based upon the collocation polynomial are called *piecewise-polynomial interpolation formulas* when they are applied to successive intervals. In general, the collocation polynomials involving a given number of points, which are produced by the interpolation formula chosen, vary from interval to interval as the interpolation formula is applied to successive intervals. The result is a continuous curve which is "rough" at the points of juncture.

Osculatory interpolation is a method of modifying two interpolation curves used on either side of a point of juncture to produce greater smoothness at

---

[1] For example, successful approximation methods based upon trigonometric functions have been devised.

[2] Under certain conditions the curve may pass through the given functional values.

[3] The interested reader is encouraged to refer to T. N. E. Greville, "Graduation," Part 5 Study Note 53.1.73 (Chicago: Society of Actuaries, 1973); and to Morton D. Miller, *Elements of Graduation* (Chicago: Society of Actuaries [The Actuarial Society of America—American Institute of Actuaries], 1946), for an excellent treatment of several useful methods of graduation. These two references taken together provide a comprehensive treatment of the subject and comprise the text material for the Society of Actuaries examination covering graduation.

that point. As mentioned in the discussion of the Taylor series in Section 5.5, this greater smoothness is produced not only by having the two curves intersect at points of juncture but also by having one or more orders of derivatives match at points of juncture.

Although we are using the term "osculatory" to refer to the general family of formulas in which one or more orders of derivatives match at points of juncture, precise mathematical terminology requires that osculatory formulas must have at least first and second derivative matching. For this reason, some authors have called the general family of formulas *smooth-junction* interpolation formulas, formulas with only first derivative matching *tangential* interpolation formulas, and formulas with both first and second derivative matching *osculatory* interpolation formulas.

Osculatory polynomials can be either reproducing or smoothing. A reproducing formula passes through the given functional values and produces equality in certain derivatives as well. A smoothing formula also produces equality in certain derivatives but removes the requirement that the curve pass through the given functional values. Some authors have referred to reproducing formulas as *unmodified osculatory* interpolation formulas, while smoothing formulas are called *modified osculatory* interpolation formulas.

Osculatory interpolation requires higher degree polynomials to produce a given level of accuracy than ordinary piecewise-polynomial interpolation. For example, an ordinary interpolation formula based upon $n + 1$ points of collocation $x_0, x_1, \ldots, x_n$, such as the Newton or Gauss formulas, produces a collocation polynomial of degree n. However, an osculatory formula which also has matching of the first derivatives at the $n + 1$ points of collocation has a total of $2n + 2$ conditions resulting in an osculatory polynomial of degree $2n + 1$. In other words, in this case an osculatory polynomial of degree $2n + 1$ can only reproduce a polynomial of degree n. The extra degrees of accuracy are sacrificed in order to impose the first derivative matching condition. Continuing this example one step further, if second derivative matching is also required, then $3n + 3$ conditions are involved and the resulting osculatory polynomial is of degree $3n + 2$.

The *degree of exactness* of any interpolation formula is the highest degree polynomial which the formula reproduces exactly. In the examples cited above the degree of exactness is equal to n.

Consider an osculatory interpolation formula of the type described above based upon $n + 1$ points of collocation $x_0, x_1, \ldots, x_n$, which also requires matching of the first derivatives. As seen above, the osculatory polynomial $p(x)$ is of degree $2n + 1$. It is assumed that the $2n + 2$ values of $f(x_i)$ and $f'(x_i)$ for $i = 0, 1, \ldots, n$ are known.

Hermite's formula gives an expression for the osculatory polynomial $p(x)$ as

$$p(x) = \sum_{i=0}^{n} [1 - 2L_i'(x_i)(x - x_i)][L_i(x)]^2 f(x_i)$$

$$+ \sum_{i=0}^{n} (x - x_i)[L_i(x)]^2 f'(x_i) \qquad (15.1)$$

where $L_i(x)$ is the Lagrange coefficient defined by formula (5.2)

$$L_i(x) = \frac{(x - x_0) \cdots (x - x_{i-1})(x - x_{i+1}) \cdots (x - x_n)}{(x_i - x_0) \cdots (x_i - x_{i-1})(x_i - x_{i+1}) \cdots (x_i - x_n)} \qquad (5.2)$$

The proof of Hermite's formula consists of showing that $p(x_j) = f(x_j)$ and $p'(x_j) = f'(x_j)$ for $j = 0, 1, \ldots, n$. From formula (5.2) it is clear that

$$L_i(x_j) = \begin{cases} 0 & \text{if} \quad i \neq j \\ 1 & \text{if} \quad i = j \end{cases}$$

It is also clear that the coefficients of $f(x_i)$ and $f'(x_i)$ in formula (15.1) are both polynomials of degree $2n + 1$, since $L_i(x)$ is a polynomial of degree n. Thus, $p(x)$ is of degree $2n + 1$.

We first substitute $x = x_j$ in formula (15.1) where x_j is one of the points of collocation. Applying the results for $L_i(x_j)$ stated above and substituting in formula (15.1), we obtain $p(x_j) = f(x_j)$ for $j = 0, 1, \ldots, n$.

We next examine $p'(x)$. From formula (15.1)

$$p'(x) = \sum_{i=0}^{n} \{[1 - 2L_i'(x_i)(x - x_i)]2L_i'(x)L_i(x) - 2L_i'(x_i)[L_i(x)]^2\} f(x_i)$$

$$+ \sum_{i=0}^{n} \{(x - x_i)2L_i'(x)L_i(x) + [L_i(x)]^2\} f'(x_i) \qquad (15.2)$$

Again applying the results for $L_i(x_j)$ stated above and substituting in formula (15.2) we obtain $p'(x_j) = f'(x_j)$ for $j = 0, 1, \ldots, n$.

The derivation of formula (1.12) in Section 1.6 can be modified for Hermite's formula to show that the truncation error in formula (15.1) is given by

$$f(x) - p(x) = \frac{f^{(2n+2)}(\xi)[\pi(x)]^2}{(2n + 2)!} \qquad (15.3)$$

The proof of formula (15.3) is identical to that given for formula (1.12) in which $g(x)$ is defined to equal $f(x) - p(x) - M[\pi(x)]^2$ as the starting point of the derivation.

The *method of undetermined coefficients* can be used to find more general osculatory interpolation formulas of the type described above. For example, consider an osculatory interpolation formula based upon $n + 1$ points of collocation $x_0, x_1, \ldots, x_n$ which also requires matching of the first and second derivatives. As seen above, the osculatory polynomial $p(x)$ is of degree $3n + 2$, that is,

$$p(x) = a_{3n+2}x^{3n+2} + a_{3n+1}x^{3n+1} + \cdots + a_1 x + a_0 \qquad (15.4)$$

It is assumed that the $3n + 3$ values $f(x_i), f'(x_i),$ and $f''(x_i)$ for $i = 0, 1, \ldots, n$ are known.

Formula (15.4) can be differentiated two times. If $f(x_i) = p(x_i), f'(x_i) = p'(x_i),$ and $f''(x_i) = p''(x_i)$ for $i = 0, 1, \ldots, n,$ then a system of $3n + 3$ linear equations in the $3n + 3$ unknowns $a_0, a_1, \ldots, a_{3n+2}$ results. This system can be solved by any of the methods described in Chapter 13 to give the osculatory polynomial $p(x)$.

The method of undetermined coefficients can also be used to produce the same answers as Hermite's formula for the case of matching functional values and first derivatives.

The formulas for osculatory interpolation developed thus far assume that values of the derivatives are known at the points of collocation. Thus, if osculatory interpolation is to be applied to data for which these derivatives are not known, alternative approaches are required.

It is possible to derive a family of symmetrical osculatory interpolation formulas, which includes both reproducing and smoothing formulas, based on a *generalized Everett formula*. Everett's original formula is given by the following version of formula (4.10)

$$\begin{aligned}
u_{x+s} = \quad & su_{x+1} + \tfrac{1}{6}s(s^2 - 1)\delta^2 u_{x+1} + \tfrac{1}{120}s(s^2 - 1)(s^2 - 4)\delta^4 u_{x+1} + \cdots \\
& + \bar{s}u_x \quad + \tfrac{1}{6}\bar{s}(\bar{s}^2 - 1)\delta^2 u_x \quad + \tfrac{1}{120}\bar{s}(\bar{s}^2 - 1)(\bar{s}^2 - 4)\delta^4 u_x \quad + \cdots
\end{aligned}$$

$$(4.10)$$

where $\bar{s} = 1 - s$.

A generalized Everett formula can be expressed as

$$v_{x+s} = F(s)u_{x+1} + F(\bar{s})u_x \qquad (15.5)$$

where $F(s)$ is the operator

$$F(s) \equiv A(s) + B(s)\delta^2 + C(s)\delta^4 + \cdots \qquad (15.6)$$

The two letters u and v, to denote given and interpolated values, respectively, are used to facilitate the expression of smoothing formulas in which $v_x \neq u_x$.

Specific osculatory interpolation formulas are derived by imposing various conditions on $A(s), B(s), C(s), \ldots$. It is then assumed that $A(s), B(s), C(s), \ldots$ are the lowest degree polynomials which satisfy these conditions, respectively.

The first condition, which is almost always applied, is for the interpolating curves to intersect at the point of juncture, so that the interpolating curve is continuous. This condition can be stated as

$$v_{x+s|s=0} = v_{x-1+s|s=1} \tag{15.7}$$

From formulas (15.5) and (15.6) we have

$$
\begin{aligned}
v_{x-1+s|s=1} = \quad & A(1)u_x \quad + B(1)\delta^2 u_x \quad + C(1)\delta^4 u_x \quad + \cdots \\
+ \, & A(0)u_{x-1} + B(0)\delta^2 u_{x-1} + C(0)\delta^4 u_{x-1} + \cdots
\end{aligned}
$$

and

$$
\begin{aligned}
v_{x+s|s=0} = \quad & A(0)u_{x+1} + B(0)\delta^2 u_{x+1} + C(0)\delta^4 u_{x+1} + \cdots \\
+ \, & A(1)u_x \quad + B(1)\delta^2 u_x \quad + C(1)\delta^4 u_x \quad + \cdots
\end{aligned}
$$

Equating these two expressions gives

$$
\begin{aligned}
A(0)u_{x-1} + B(0)\delta^2 u_{x-1} + C(0)\delta^4 u_{x-1} + \cdots \\
= A(0)u_{x+1} + B(0)\delta^2 u_{x+1} + C(0)\delta^4 u_{x+1} + \cdots
\end{aligned}
$$

which holds in general if and only if

$$A(0) = B(0) = C(0) = \cdots = 0 \tag{15.8}$$

The second condition which may be applied is to make the interpolating formula reproducing. This happens if

$$u_x = v_{x-1+s|s=1} = v_{x+s|s=0}$$

If formula (15.8) is satisfied, then this result holds if and only if

$$A(1) = 1 \quad \text{and} \quad B(1) = C(1) = \cdots = 0 \tag{15.9}$$

If formula (15.9) is satisfied, then the formula is reproducing, while the formula is smoothing otherwise.

The quantities $A(1)$, $B(1)$, $C(1)$, ... are important in determining whether a formula is reproducing or smoothing and are sometimes called *coefficients of modification*, since they express the amount of deviation from the given functional values for a smoothing formula. For simplicity, these quantities are often written

$$
\begin{aligned}
A(1) &= K \\
B(1) &= L \\
C(1) &= M \\
&\;\;\vdots
\end{aligned}
$$

The third condition which is applied for an osculatory formula is to have first derivatives match. This condition is expressed as

$$v'_{x-1+s|s=1} = v'_{x+s|s=0}$$

Since $\bar{s} = 1 - s$,

$$\frac{d}{ds} F(\bar{s}) = -F'(\bar{s})$$

Thus, the general Everett formula gives

$$F'(1)u_x - F'(0)u_{x-1} = F'(0)u_{x+1} - F'(1)u_x$$

or

$$2F'(1)u_x = F'(0)\left(u_{x+1} + u_{x-1}\right)$$
$$= F'(0)\left(2 + \delta^2\right)u_x$$

from the expansion of $\delta^2 u_x$. Thus, the condition for first derivative matching is

$$2F'(1) \equiv F'(0)\left(2 + \delta^2\right) \tag{15.10}$$

Similarly, if second derivative matching is required, then

$$F''(1)u_x + F''(0)u_{x-1} = F''(0)u_{x+1} + F''(1)u_x$$

or

$$F''(0)\left(u_{x+1} - u_{x-1}\right) = 0$$

Thus, the condition for second derivative matching is

$$F''(0) \equiv 0 \tag{15.11}$$

Finally, the fourth condition which is generally applied involves the degree of exactness of the formula. Everett's formula has missing odd difference terms, so that any formula which is accurate through any even degree automatically is accurate through the succeeding odd degree. For this reason, an odd degree of exactness is usually chosen in derivations of osculatory interpolation formulas. For any given degree of exactness to hold, the formula must agree with Everett's unmodified formula through that degree, since Everett's unmodified formula is based on the collocation polynomial. Table 15.1 summarizes the conditions for degrees of exactness 1, 3, and 5.

TABLE 15.1

Degree of exactness	Conditions
1	$A(s) = s$
3	$A(s) = s$ $B(s) = \frac{1}{6}s(s^2 - 1)$
5	$A(s) = s$ $B(s) = \frac{1}{6}s(s^2 - 1)$ $C(s) = \frac{1}{120}s(s^2 - 1)(s^2 - 4)$

Specific osculatory interpolation formulas are derived by applying these four types of conditions. This type of derivation is illustrated in Examples 15.3 and 15.4.

Example 15.1.   *Use Hermite's formula to find the osculatory polynomial* $p(x)$ *based on the following data:*

i	x_i	$f(x_i)$	$f'(x_i)$
0	0	0	0
1	4	2	0

We apply formula (15.1) in which $n = 1$, $L_0(x) = \dfrac{x - x_1}{x_0 - x_1}$, $L_1(x) = \dfrac{x - x_0}{x_1 - x_0}$, $L_0'(x) = \dfrac{1}{x_0 - x_1}$, and $L_1'(x) = \dfrac{1}{x_1 - x_0}$. Note that only the term involving $f(x_1)$ needs to be computed, since $f(x_0) = f'(x_0) = f'(x_1) = 0$. Formula (15.1) now gives

$$p(x) = \left[1 - 2\frac{x - 4}{4 - 0}\right]\left[\frac{x - 0}{4 - 0}\right]^2 \cdot 2 = \frac{-x^3 + 6x^2}{16}$$

Example 15.2.   *Rework Example 15.1 using the method of undetermined coefficients.*

Since four pieces of information are given, we assume $p(x)$ is a third-degree polynomial. Thus,

$$p(x) = a_3 x^3 + a_2 x^2 + a_1 x + a_0$$

and

$$p'(x) = 3a_3 x^2 + 2a_2 x + a_1$$

We can now find four equations in the four unknowns a_0, a_1, a_2, a_3 as follows:

$$
\begin{aligned}
p(0) &= f(0) = & a_0 &= 0 \\
p(4) &= f(4) = 64a_3 + 16a_2 + 4a_1 + a_0 &= 2 \\
p'(0) &= f'(0) = & a_1 &= 0 \\
p'(4) &= f'(4) = 48a_3 + 8a_2 + a_1 &= 0
\end{aligned}
$$

The solution of this system of equations is

$$a_0 = 0 \qquad a_1 = 0 \qquad a_2 = \tfrac{3}{8} \qquad a_3 = -\tfrac{1}{16}$$

Thus

$$p(x) = \frac{-x^3 + 6x^2}{16}$$

which agrees with the answer obtained in Example 15.1.

Example 15.3.   Derive $A(s)$ and $B(s)$ for a family of four-point[4] osculatory interpolation formulas based upon the generalized Everett formula which are exact for first-degree polynomials and which have matching functional values and first derivatives at the points of juncture.

Since the degree of exactness is equal to one, Table 15.1 immediately gives

$$A(s) = s \qquad\qquad (15.12)$$

Conditions on $B(s)$ are applied as follows:

1. Interpolating curves intersect, apply formula (15.8)

$$B(0) = 0$$

2. Formula is general including both reproducing and smoothing formulas

$$B(1) = L$$

3. First derivatives are matching, apply formula (15.10)

$$2F'(1) \equiv F'(0)(2 + \delta^2)$$
$$2[1 + B'(1)\delta^2] \equiv [1 + B'(0)\delta^2](2 + \delta^2)$$
$$2B'(1)\delta^2 \equiv [1 + 2B'(0)]\delta^2 + B'(0)\delta^4$$

Thus two results hold

$$B'(0) = 0$$
$$2B'(1) = 1 + 2B'(0)$$

Since four conditions on $B(s)$ result, we assume that $B(s)$ is a third-degree polynomial

$$B(s) = b_3 s^3 + b_2 s^2 + b_1 s + b_0$$

Applying the four conditions produces the following system of linear equations:

$$B(0) = 0 = b_0$$
$$B(1) = L = b_3 + b_2 + b_1 + b_0$$
$$B'(0) = 0 = b_1$$
$$B'(1) = \tfrac{1}{2} = 3b_3 + 2b_2 + b_1$$

---

[4]The term "four-point" means that the generalized Everett formula is carried to second differences. If the differences are expanded, it is seen that four functional values are involved. In general, a "$2m$-point" formula will involve the generalized Everett formula to terms involving δ^{2m-2}.

The solution of this system of equations is

$$b_0 = 0$$
$$b_1 = 0$$
$$b_2 = 3L - \tfrac{1}{2}$$
$$b_3 = \tfrac{1}{2} - 2L$$

Thus, we have

$$B(s) = (\tfrac{1}{2} - 2L)s^3 + (3L - \tfrac{1}{2})s^2 \tag{15.13}$$

Several special cases of this general four-point formula are examined in Exercises 7–9.

Example 15.4. *Derive $A(s)$, $B(s)$, and $C(s)$ for a family of six-point osculatory interpolation formulas based upon the generalized Everett formula which are exact for third-degree polynomials and which have matching functional values, first derivatives, and second derivatives at the points of juncture.*

Since the degree of exactness is equal to three, Table 15.1 immediately gives

$$A(s) = s \tag{15.14}$$
$$B(s) = \tfrac{1}{6}(s^3 - s) \tag{15.15}$$

Conditions on $C(s)$ are applied as follows:

1. Interpolating curves intersect, apply formula (15.8)

$$C(0) = 0$$

2. Formula is general including both reproducing and smoothing formulas

$$C(1) = M$$

3. First derivatives are matching, apply formula (15.10)

$$2F'(1) \equiv F'(0)(2 + \delta^2)$$
$$2[1 + \tfrac{1}{3}\delta^2 + C'(1)\delta^4] \equiv [1 - \tfrac{1}{6}\delta^2 + C'(0)\delta^4](2 + \delta^2)$$
$$2C'(1)\delta^4 \equiv [2C'(0) - \tfrac{1}{6}]\delta^4 + C'(0)\delta^6$$

Thus two results hold

$$C'(0) = 0$$
$$2C'(1) = 2C'(0) - \tfrac{1}{6}$$

4. Second derivatives are matching, apply formula (15.11)

$$F''(0) \equiv 0$$

Thus

$$C''(0) = 0$$

Since five conditions on $C(s)$ result, we assume that $C(s)$ is a fourth-degree polynomial

$$C(s) = c_4 s^4 + c_3 s^3 + c_2 s^2 + c_1 s + c_0$$

Applying the five conditions produces the following system of linear equations:

$$
\begin{aligned}
C(0) &= & 0 & = & & & & & & c_0 \\
C(1) &= & M & = & c_4 + & c_3 + & c_2 + & c_1 + & c_0 \\
C'(0) &= & 0 & = & & & & c_1 \\
C'(1) &= & -\tfrac{1}{12} & = & 4c_4 + & 3c_3 + & 2c_2 + & c_1 \\
C''(0) &= & 0 & = & & & 2c_2
\end{aligned}
$$

The solution of this system of equations is

$$
\begin{aligned}
c_0 &= 0 \\
c_1 &= 0 \\
c_2 &= 0 \\
c_3 &= 4M + \tfrac{1}{12} \\
c_4 &= -3M - \tfrac{1}{12}
\end{aligned}
$$

Thus, we have

$$C(s) = -\left(3M + \tfrac{1}{12}\right)s^4 + \left(4M + \tfrac{1}{12}\right)s^3 \tag{15.16}$$

Several special cases of this general six-point formula are examined in Exercises 10–11.

15.3. LEAST-SQUARES POLYNOMIALS

Approximation by the fitting of *least-squares polynomials* is one of the oldest and most widely used methods of approximation in practice.[5] In this approximation method a given set of functional values is approximated by a polynomial of selected degree. The least-squares polynomial is that polynomial which minimizes the sum of squares of actual functional values minus approximated functional values.

Consider a given set of $n + 1$ points, $(x_0, y_0), (x_1, y_1), \ldots, (x_n, y_n)$, which is to be approximated by a least-squares polynomial of degree m. It is assumed that $m < n$. If $m = n$, the least-squares polynomial is identical to the collocation polynomial which produces a minimum sum of squares equal to zero; whereas if $m > n$, an infinite number of collocation polynomials through the given data points exist.

---

[5] In statistics courses least-squares approximation is presented as a *regression* method.

If $m < n$, then least-squares approximation is generally not reproducing but is smoothing.[6] Least-squares approximation may be superior to collocation in certain situations in which a large number of functional values are given. If a low-degree collocation polynomial is used, either in a piecewise fashion or in one piece throughout the entire range, then only a subset of the available data points can be used. Thus, relevant information about the nature of the true function may be lost. On the other hand, a high-degree collocation polynomial often oscillates about the true function and produces questionable approximations. Furthermore, as seen in Section 2.7, high-degree differences are often spurious because of roundoff error, which compounds the problems inherent in using high-degree collocation polynomials.

Let the least-squares polynomial $p(x)$ be denoted by

$$p(x) = a_m x^m + a_{m-1} x^{m-1} + \cdots + a_1 x + a_0 \qquad (15.17)$$

The *principle of least-squares* is to minimize S, where

$$S = \sum_{i=0}^{n} [y_i - p(x_i)]^2$$
$$= \sum_{i=0}^{n} [y_i - (a_m x_i^m + a_{m-1} x_i^{m-1} + \cdots + a_1 x_i + a_0)]^2 \qquad (15.18)$$

The sum S can be minimized by solving the $m + 1$ linear equations in the $m + 1$ unknowns $a_0, a_1, \ldots, a_m$, which result from setting

$$\frac{\partial S}{\partial a_0} = \frac{\partial S}{\partial a_1} = \cdots = \frac{\partial S}{\partial a_m} = 0 \qquad (15.19)$$

The $m + 1$ equations resulting from formula (15.19) are called the *normal equations* and are given by[7]

$$\left. \begin{aligned} a_0(n + 1) + a_1 \sum x \quad + \cdots + a_m \sum x^m &= \sum y \\ a_0 \sum x + a_1 \sum x^2 \quad + \cdots + a_m \sum x^{m+1} &= \sum xy \\ \cdots\cdots\cdots\cdots\cdots\cdots\cdots\cdots\cdots\cdots\cdots\cdots\cdots \\ a_0 \sum x^m + a_1 \sum x^{m+1} + \cdots + a_m \sum x^{2m} &= \sum x^m y \end{aligned} \right\} \qquad (15.20)$$

The system of normal equations can be solved by any of the methods described in Chapter 13 to give the least-squares polynomial $p(x)$. The system of normal equations has proven to be ill-conditioned for large m and should be used only for small m. Alternative methods for large m have been devised, but are beyond the scope of this book.

---

[6]Least-squares approximation is reproducing, if by chance the $n + 1$ given points happen to lie on a polynomial of degree m.

[7]The labels i and the limits of summation are omitted for convenience, since there is no chance of ambiguity.

One important and frequently used special case of least-squares approximation is the least-squares straight line. In this case $m = 1$ and the normal equations become

$$a_0(n + 1) + a_1 \sum x = \sum y$$
$$a_0 \sum x + a_1 \sum x^2 = \sum xy$$

The solution of this system of equations using Cramer's rule is

$$
\left.
\begin{aligned}
a_0 &= \frac{(\sum x^2)(\sum y) - (\sum xy)(\sum x)}{(n + 1) \sum x^2 - (\sum x)^2} \\
a_1 &= \frac{(n + 1) \sum xy - (\sum x)(\sum y)}{(n + 1) \sum x^2 - (\sum x)^2}
\end{aligned}
\right\}
\qquad (15.21)[8]
$$

It is mentioned in Section 7.1 that approximate differentiation formulas based on least-squares polynomials are superior to formulas based on collocation polynomials in certain cases. This will often be true when the given functional values are subject to random errors which produce a somewhat irregular series. In this case the slope of the collocation polynomial is generally a poor estimate of the slope of the true curve because of the irregularities in the data. However, if a least-squares polynomial is used, a better estimate of the derivative is often obtained, since the least-squares polynomial performs some preliminary smoothing before the required derivative is approximated.

As an example, we consider an approximate differentiation formula for $Df(x)$ based upon the five given values $f(x - 2h)$, $f(x - h)$, $f(x)$, $f(x + h)$,

---

[8]In statistics this linear approximation is called the *least-squares regression line* and is generally written as

$$y - \mu_y = \frac{\sigma_{xy}}{\sigma_x^2} (x - \mu_x) \qquad (15.22)$$

where μ_x is the *mean* of x, μ_y is the *mean* of y, σ_x^2 is the *variance* of x, and σ_{xy} is the *covariance* between x and y. These quantities are defined as follows:

$$\mu_x = \frac{\sum x}{n + 1}$$

$$\mu_y = \frac{\sum y}{n + 1}$$

$$\sigma_x^2 = \frac{\sum x^2}{n + 1} - \mu_x^2$$

$$\sigma_{xy} = \frac{\sum xy}{n + 1} - \mu_x \mu_y$$

Formula (15.21) is easily converted into formula (15.22) with these substitutions. The details are left as Exercise 15.

$f(x + 2h)$. A formula based upon the fourth-degree collocation polynomial through these five points is given by formula (7.11b)

$$Df(x) = \frac{8[f(x + h) - f(x - h)] - [f(x + 2h) - f(x - 2h)]}{12h} \quad \text{(7.11b)}$$

We shall now derive a similar formula based upon the second-degree least-squares polynomial.

Define a new variable t such that the five given values $x - 2h$, $x - h$, x, $x + h$, $x + 2h$ correspond to $t = -2, -1, 0, 1, 2$, respectively. Denote the change of variable by

$$y_t = f(x + ht) \qquad \text{for } t = -2, -1, 0, 1, 2$$

Let the second-degree least-squares polynomial be denoted by

$$p(t) = a_2 t^2 + a_1 t + a_0$$

The normal equations become

$$5a_0 \quad + 10a_2 = \sum_{t=-2}^{2} y_t = \quad y_{-2} + y_{-1} + y_0 + y_1 + \ y_2$$

$$10a_1 \quad = \sum_{t=-2}^{2} ty_t = -2y_{-2} - y_{-1} \qquad + y_1 + 2y_2$$

$$10a_0 \quad + 34a_2 = \sum_{t=-2}^{2} t^2 y_t = \quad 4y_{-2} + y_{-1} \qquad + y_1 + 4y_2$$

The first derivative of the least-squares polynomial now serves as the approximation for the actual first derivative. Thus, we have

$$Dy_t = Dp(t)$$

and applying the chain rule

$$hDf(x + ht) = 2a_2 t + a_1$$

If we evaluate the above equation for $t = 0$, we have

$$Df(x) = \frac{a_1}{h}$$

$$= \frac{-2y_{-2} - y_{-1} + y_1 + 2y_2}{10h}$$

$$= \frac{2[f(x + 2h) - f(x - 2h)] + [f(x + h) - f(x - h)]}{10h} \quad \text{(15.23)}$$

A similar approximate differentiation formula for $Df(x)$ based upon the three given values $f(x - h), f(x), f(x + h)$, using the second-degree collocation polynomial, is given by formula (7.10b)

$$Df(x) = \frac{f(x + h) - f(x - h)}{2h} \quad \text{(7.10b)}$$

It is interesting that this same formula can also be derived from the first-degree least-squares polynomial.

Using the same derivation as used for formula (15.23), the least-squares polynomial is given by

$$p(t) = a_1 t + a_0$$

and the normal equations are

$$3a_0 = y_{-1} + y_0 + y_1$$
$$2a_1 = -y_{-1} + y_1$$

The approximate differentiation formula is again given by

$$Df(x) = \frac{a_1}{h}$$

$$= \frac{-y_{-1} + y_1}{2h}$$

$$= \frac{f(x+h) - f(x-h)}{2h}$$

which agrees with formula (7.10b).

Example 15.5. *Fit a least-squares quadratic to the function $f(x) = x^3$ for* $x = 0, 1, 2, 3$.

Values are computed as follows:

x	x^2	x^3	x^4	x^5
0	0	0	0	0
1	1	1	1	1
2	4	8	16	32
3	9	27	81	243
6	14	36	98	276

The normal equations become

$$4a_0 + 6a_1 + 14a_2 = 36$$
$$6a_0 + 14a_1 + 36a_2 = 98$$
$$14a_0 + 36a_1 + 98a_2 = 276$$

The solution of this system of equations is

$$a_0 = .3 \qquad a_1 = -4.7 \qquad a_2 = 4.5$$

Thus, the least-squares quadratic approximation is given by

$$p(x) = 4.5x^2 - 4.7x + .3$$

A comparison of values of $f(x)$ and $p(x)$ is as follows:

x	$f(x)$	$p(x)$	$f(x) - p(x)$
0	0	.3	−.3
1	1	.1	.9
2	8	8.9	−.9
3	27	26.7	.3

The maximum absolute error at any of the four given data points is .9. The minimum sum of squares is $(-.3)^2 + (.9)^2 + (-.9)^2 + (.3)^2 = 1.8$.

Example 15.6.   Compare the results for $x = 3, 4, 5, 6, 7, 8$ resulting from formula (7.11b) and formula (15.23) applied in a piecewise fashion using the data in Series A and Series B below:

x	Series A	Series B
1	1.00	1.01
2	1.41	1.40
3	1.73	1.75
4	2.00	2.04
5	2.24	2.26
6	2.45	2.44
7	2.65	2.60
8	2.83	2.83
9	3.00	3.05
10	3.16	3.14

The values in Series A are values of $\sqrt{x}$ accurate to two decimal places. The values in Series B are the values in Series A adjusted with a superimposed series of errors varying randomly over $-.05, -.04, \ldots, +.04, +.05$. The exact value of the derivative of $\sqrt{x}$ can be computed from

$$\frac{d}{dx} x^{1/2} = \tfrac{1}{2} x^{-1/2}$$

Table 15.2 summarizes the results of this example.

TABLE 15.2

x	Exact derivative	Series A (7.11b)	Series A (15.23)	Series B (7.11b)	Series B (15.23)
3	.29	.29	.31	.32	.31
4	.25	.25	.26	.25	.26
5	.22	.22	.23	.20	.21
6	.20	.20	.21	.16	.19
7	.19	.19	.19	.19	.20
8	.18	.17	.18	.24	.18

It is seen that the collocation polynomial produces results superior to those of the least-squares polynomial for Series A which is a smooth series. However, the least-squares polynomial produces better results than the collocation polynomial for Series B, which is based on data with random fluctuations present.

15.4. MINIMAX POLYNOMIALS

Approximation by the fitting of *minimax polynomials*[9] is an alternative to least-squares polynomials. As with least-squares polynomials, a given set of $n + 1$ points, $(x_0, y_0), (x_1, y_1), \ldots, (x_n, y_n)$, is approximated by a polynomial of degree m. The minimax polynomial is that polynomial for which the maximum absolute error of actual functional values minus approximated functional values at any of the $n + 1$ data points is minimized.

As with least-squares polynomials it is assumed that $m < n$. If $m \geq n$, then the problem of fitting minimax polynomials reduces to the fitting of collocation polynomials. If $m < n$, then minimax approximation is generally not reproducing but is smoothing.[10]

Let the minimax polynomial $p(x)$ be denoted by

$$p(x) = a_m x^m + a_{m-1} x^{m-1} + \cdots + a_1 x + a_0 \qquad (15.24)$$

The minimax polynomial is that polynomial for which E is minimized, where

$$E = \max |y_i - p(x_i)| \qquad \text{for } i = 0, 1, \ldots, n \qquad (15.25)$$

---

[9]Approximation by minimax polynomials is often called *Chebyshev approximation.* Several authors discuss *Chebyshev polynomials,* which involve trigonometric functions and exhibit properties close to minimax conditions. However, since it is possible to develop true minimax polynomials without involving Chebyshev polynomials, the latter are not discussed further.

[10]Minimax approximation is reproducing, if by chance the $n + 1$ given points happen to lie on a polynomial of degree m.

The minimax polynomial is characterized by the *equal-error property* which gives the result that the values of the errors at $m + 2$ of the given data points are equal to $\pm E$ with alternating signs. The values of the errors at the other $n - m - 1$ data points are less than or equal to E in absolute value.[11]

The *exchange method*[12] is an efficient algorithm for determining the minimax polynomial based upon the *equal-error property*. The following are the steps in the exchange method:

1. Arbitrarily select any $m + 2$ of the given $n + 1$ data points.
2. Fit a polynomial of degree m to these $m + 2$ points such that the errors at these $m + 2$ points are equal to $\pm h$ with alternating signs.
3. Compute the errors at the other $n - m - 1$ given data points. Denote the maximum of these $n - m - 1$ errors in absolute value by H.
4. (a) If $H \le |h|$, then the process is ended and the polynomial determined is the minimax polynomial.
   (b) If $H > |h|$, then an exchange step is necessary. Add to the arbitrarily selected $m + 2$ points, a point at which the maximum absolute error H occurs and delete from these points one of the original points, such that the errors of the new collection of $m + 2$ points alternate signs. Then repeat steps 2, 3, 4 as many times as necessary.

Example 15.7. **Fit a minimax straight line to the function $f(x) = x^3$ for** $x = 0, 1, 2, 3$.

In this case we have a first-degree minimax polynomial and four given points, so that $n = 3$ and $m = 1$. Denote the minimax polynomial by

$$p(x) = a_1 x + a_0$$

We initiate the exchange method with the arbitrary selection of the $m + 2 = 3$ points $(1, 1)$, $(2, 8)$, $(3, 27)$, which are the last three given points lying on $f(x) = x^3$. The errors $f(x) - p(x)$ for these points are respectively given by

$$1 - a_1 - a_0 = h$$
$$8 - 2a_1 - a_0 = -h$$
$$27 - 3a_1 - a_0 = h$$

The solution of this system of equations is

$$h = 3 \qquad a_0 = -15 \qquad a_1 = 13$$

---

[11] For a proof that the equal-error property uniquely characterizes and determines the minimax polynomial see Francis Scheid, *Theory and Problems of Numerical Analysis* (New York: McGraw-Hill Book Co., 1968), pp. 268–73.
[12] The student should not confuse this exchange method with the exchange method of inverting matrices discussed in Sections 12.6 and 13.4.

Computing the error at the other $n - m - 1 = 1$ point, that is, the point $(0, 0)$, we have

$$0 - 0(13) + 15 = 15$$

Thus, $H = 15$ and since $H > |h|$, an exchange step is necessary.

We now must choose $(0, 0)$ as one of the points used and delete one of the others. Since the signs must alternate, we delete the point $(1, 1)$. The new system of equations is

$$
\begin{aligned}
- a_0 &= \quad h \\
8 - 2a_1 - a_0 &= -h \\
27 - 3a_1 - a_0 &= \quad h
\end{aligned}
$$

The solution of this system of equations is

$$h = 5 \qquad a_0 = -5 \qquad a_1 = 9$$

Computing the error at $(1, 1)$, we have

$$1 - 1(9) + 5 = -3$$

Thus, $H = 3$ and since $H < |h|$, the process is completed.

The minimax straight-line approximation to $f(x) = x^3$ is given by

$$p(x) = 9x - 5$$

A comparison of values of $f(x)$ and $p(x)$ is as follows:

x	$f(x)$	$p(x)$	$f(x) - p(x)$
0	0	-5	5
1	1	4	-3
2	8	13	-5
3	27	22	5

Example 15.8.   *Fit a minimax quadratic to the function* $f(x) = x^3$ *for* $x = 0, 1, 2, 3$.

In this case we have a second-degree minimax polynomial and four given points, so that $n = 3$ and $m = 2$. Denote the minimax polynomial by

$$p(x) = a_2 x^2 + a_1 x + a_0$$

We have $m + 2 = 4$, so that all four points must be used in the initial step of the exchange method. The errors $f(x) - p(x)$ for these points are respectively given by

$$
\begin{aligned}
- a_0 &= \quad h \\
1 - \quad a_2 - \quad a_1 - a_0 &= -h \\
8 - 4a_2 - 2a_1 - a_0 &= \quad h \\
27 - 9a_2 - 3a_1 - a_0 &= -h
\end{aligned}
$$

The solution of this system of equations is

$$h = -.75 \qquad a_0 = .75 \qquad a_1 = -5 \qquad a_2 = 4.5$$

Since all four points are used, the process is completed.

The minimax quadratic approximation to $f(x) = x^3$ is given by

$$p(x) = 4.5x^2 - 5x + .75$$

A comparison of values of $f(x)$ and $p(x)$ is as follows:

x	$f(x)$	$p(x)$	$f(x) - p(x)$
0	0	.75	−.75
1	1	.25	.75
2	8	8.75	−.75
3	27	26.25	.75

The result of this example should be compared with Example 15.5, which contains the least-squares quadratic approximation for the same function. The following conclusions can be drawn:

1. The maximum absolute error at any of the four given points in Example 15.8 is .75 compared with .9 in Example 15.5. Thus, the minimax polynomial has a smaller maximum error than the least-squares polynomial, as it must.

2. The sum of squares of the errors in Example 15.8 is $4(.75)^2 = 2.25$ compared with 1.8 in Example 15.5. Thus, the least-squares polynomial has a smaller sum of squares of the errors than the minimax polynomial, as it must.

15.5. RATIOS OF POLYNOMIALS

Approximation based on ratios of polynomials[13] is a useful generalization of polynomial approximation. In fact, the set of functions which are ratios of polynomials includes all polynomials as special cases.

Since the set of ratios of polynomials contains a wider variety of functions than the set of polynomials, the chances of successful approximation are increased. For example, one type of function which is incapable of being approximated successfully with polynomials but which can be approximated successfully with ratios of polynomials is a function with singularities as illustrated in Figure 15.1.

---

[13]Ratios of polynomials are sometimes called *rational functions*.

FIGURE 15.1

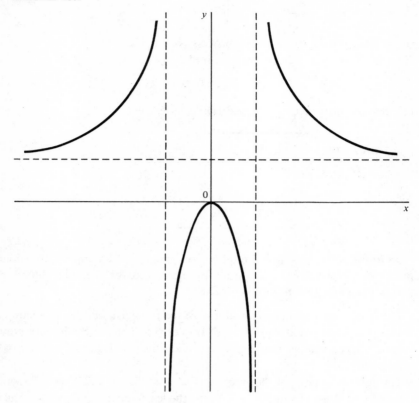

The general form of a ratio of two polynomials, $r(x)$, is given by

$$r(x) = \frac{a_m x^m + a_{m-1} x^{m-1} + \cdots + a_1 x + a_0}{b_n x^n + b_{n-1} x^{n-1} + \cdots + b_1 x + b_0} \qquad (15.26)[14]$$

where m and n are two positive integers. It is assumed that formula (15.26) is irreducible, that is, there exists no polynomial divisor common to both the numerator and denominator.

It is also assumed that no more constants are used than are required. From formula (15.26) it would appear that $m + n + 2$ constants a_i and b_i must be

---

[14]Several standard textbooks in numerical analysis discuss approximation by *continued fractions*, for example, see F. B. Hildebrand, *Introduction to Numerical Analysis* (New York: McGraw-Hill Book Co., 1956), pp. 395–412. However, the continued fractions involved can be reduced to the form given by formula (15.26).

determined. However, some of these constants may be zero. Also, one of the nonzero constants can be eliminated. For example, the ratio of quadratics

$$\frac{ax^2 + b}{cx^2 + d}$$

can be reduced to the form

$$\frac{x^2 + b'}{c'x^2 + d'}$$

for nonzero a, where

$$b' = \frac{b}{a} \qquad c' = \frac{c}{a} \qquad d' = \frac{d}{a}$$

Once the simplifications described above are made, we proceed by solving a system of linear equations based on formula (15.26) involving a number of equations equal to the number of unknown constants. We are assuming a solution by collocation,[15] so that an equal number of points of collocation and unknown constants is required.

Example 15.9.   *Fit a function of the form*

$$r(x) = \frac{ax + b}{cx + d}$$

to the function $f(x) = x^2$ for $x = 1, 2, 3$.

We let $r(x)$ be given by

$$r(x) = \frac{ax + b}{cx + 1}$$

that is, we can set $d = 1$, since one of the four constants is redundant. Three equations in three unknowns can be found by successively setting $x = 1, 2, 3$ to give

$$\frac{a + b}{c + 1} = 1$$

$$\frac{2a + b}{2c + 1} = 4$$

$$\frac{3a + b}{3c + 1} = 9$$

---

[15]It is also possible to develop least-squares and minimax methods in certain cases which parallel the approaches used for polynomials. Exercises 34 and 35 illustrate these two approaches, respectively.

These equations can be rewritten as

$$a + b - c = 1$$
$$2a + b - 8c = 4$$
$$3a + b - 27c = 9$$

The solution of this system of equations is

$$a = \tfrac{11}{6} \qquad b = -1 \qquad c = -\tfrac{1}{6}$$

Thus, we have

$$r(x) = \frac{11x - 6}{6 - x}$$

15.6. EXPONENTIAL FUNCTIONS

Approximation by exponential functions is another alternative to polynomial approximation. It is useful in many applications in which data are encountered which are more nearly exponential than polynomial in nature.

We again assume a solution by collocation,[16] so that an equal number of points of collocation and unknown constants is required.

In some cases we also assume that exponential terms are written in terms of the base, e, of the natural logarithm system. If a base other than e is involved, a simple transformation exists to convert to base e. For example, a^x can be written as e^{cx} where $c = \log_e a$.

Three tools are useful in fitting exponential functions. The first is differencing of successive terms which can be useful in determining polynomial portions of the function. The second is division of successive terms which can be useful in determining the common ratio of the exponential term. The third is taking logarithms of the exponential term, which simplifies handling the exponent. These three tools can be applied in any order an appropriate number of times depending on the nature of the exponential function being fit to the data.

Several forms of exponential functions, $e(x)$, can be used for approximations. Two of these are illustrated as examples.

The first example is a basic exponential of the form

$$e(x) = e^{p(x)} \tag{15.27}$$

where $p(x)$ is a polynomial of degree n. This case requires $n + 1$ points of collocation and can be solved by taking logarithms of the given functional

---

[16]It is also possible to develop least-squares and minimax methods in certain cases which parallel the approaches used for polynomials. Exercises 40 and 41 illustrate these two approaches, respectively.

values. Then the problem is reduced to straightforward polynomial approximation.

The second example is the sum of a polynomial and an exponential of the form

$$e(x) = p(x) + ab^x \qquad (15.28)$$

where $p(x)$ is a polynomial of degree n. This case requires $n + 3$ points of collocation and can be solved by differencing $n + 1$ times to eliminate $p(x)$ and then solving for the constants a and b. The polynomial can then be found by substitution. The solution approach is illustrated in Example 2.1.

These two examples are two of the more useful cases in practical applications and are illustrative of approaches which can be used to determine exponential functions.

Example 15.10.   Fit functions of the form

$$(1) \quad e(x) = e^{ax^2 + bx + c}$$
$$(2) \quad e(x) = a + bc^x$$

to the function $f(x) = x^2$ for $x = 1, 2, 3$.

(1)   In this case we have the following:

x	$f(x)$	$\log_e f(x)$	$\Delta \log_e f(x)$	$\Delta^2 \log_e f(x)$
1	1	0		
			1.38629	
2	4	1.38629		$-.57536$
			.81093	
3	9	2.19722		

Since $\log_e f(x) = ax^2 + bx + c$, we can use Newton's advancing difference formula to give

$$ax^2 + bx + c = \log_e f(x)$$
$$= 1.38629(x - 1) - .57536(\tfrac{1}{2})(x - 1)(x - 2)$$
$$= -.28768x^2 + 2.24933x - 1.96165$$

Thus, we have

$$e(x) = e^{-.28768x^2 + 2.24933x - 1.96165}$$

(2)   In this case we have

$$a + bc = 1$$
$$a + bc^2 = 4$$
$$a + bc^3 = 9$$

If we difference once, we have

$$bc(c - 1) = 3$$
$$bc^2(c - 1) = 5$$

If we divide, we have $c = \frac{5}{3}$. This leads directly to $b = 2.7$ and $a = -3.5$. Thus, we have

$$e(x) = -3.5 + 2.7(\tfrac{5}{3})^x$$

EXERCISES

15.1. Introduction; 15.2. Osculatory polynomials

1. Use Hermite's formula to find the osculatory polynomial based on the following data:

i	x_i	$f(x_i)$	$f'(x_i)$
0	0	0	0
1	1	1	1

2. Show that Hermite's formula applied to two points of collocation x_0 and x_1 which is evaluated at $x = \frac{1}{2}(x_0 + x_1)$ gives the formula

$$\tfrac{1}{2}[f(x_0) + f(x_1)] - \tfrac{1}{8}(x_1 - x_0)[f'(x_1) - f'(x_0)]$$

3. Use formula (15.3) to find the truncation error for the formula in Exercise 2.
4. Rework Exercise 1 using the method of undetermined coefficients.
5. Use the method of undetermined coefficients to find the osculatory polynomial based on the following data:

i	x_i	$f(x_i)$	$f'(x_i)$	$f''(x_i)$
0	0	0	0.	0
1	1	1	1	1

6. Use the method of undetermined coefficients to find the osculatory polynomial based on the following data:

i	x_i	$f(x_i)$	$f''(x_i)$
0	0	1	-2
1	1	1	4

7. Find the four-point osculatory interpolation formula from the family of formulas developed in Example 15.3 which is reproducing. This formula is called the *Karup-King formula*.

8. Find the four-point osculatory interpolation formula from the family of formulas developed in Example 15.3 for which $B(s)$ is of minimum degree.

9. Find the four-point osculatory interpolation formula from the family of formulas developed in Example 15.3 which has second derivative matching.

10. Find the six-point osculatory interpolation formula from the family of formulas developed in Example 15.4 which is reproducing.

11. Find the six-point osculatory interpolation formula from the family of formulas developed in Example 15.4 for which $C(s)$ is of minimum degree. This formula is called *Jenkins' fifth difference modified osculatory interpolation formula*.

12. *Shovelton's formula* is a six-point formula for which $A(s)$ and $B(s)$ are defined by formulas (15.14) and (15.15) and for which $C(s)$ is defined by $C(s) = \frac{1}{48}s^2(s-1)(s-5)$.
    (a) Is the formula reproducing?
    (b) Does the formula have first derivative matching?
    (c) Does the formula have second derivative matching?

13. *Henderson's formula* is a six-point formula for which $A(s)$ and $B(s)$ are defined by formulas (15.14) and (15.15) and for which $C(s)$ is defined by $C(s) = -\frac{1}{36}(s^3 - s)$.
    (a) Is the formula reproducing?
    (b) Does the formula have first derivative matching?
    (c) Does the formula have second derivative matching?

14. The following is a third difference osculatory interpolation formula based on Bessel's formula

$$v_{x+s} = \mu u_{x+\frac{1}{2}} + \left(s - \tfrac{1}{2}\right)\delta u_{x+\frac{1}{2}} + \tfrac{1}{2}s(s-1)\mu\delta^2 u_{x+\frac{1}{2}} + \tfrac{1}{2}s\left(s - \tfrac{1}{2}\right)(s-1)\delta^3 u_{x+\frac{1}{2}}$$

    (a) To what order of difference does the formula agree with Bessel's formula?
    (b) Is the formula reproducing?
    (c) Show that the formula has first derivative matching.

15.3. Least-squares polynomials

15. Derive formula (15.22) from formula (15.21).

16. (a) Fit a least-squares straight line to the following data:

x	y
-1	3
1	4
3	7
5	6

    (b) Compute the minimum sum of squares.

17. Rework Exercise 16 using a least-squares quadratic.

18. Show that $\sum_{i=0}^{n} [y_i - p(x_i)] = 0$.

19. It is known that the interval of differencing $h = 1$, and that all tabulated functional values are rounded off to three decimal places.
    (a) Find the maximum roundoff error in using formula (7.11b) to approximate $Df(x)$.
    (b) Find the maximum roundoff error in using formula (15.23) to approximate $Df(x)$.

20. A curve of the form $y = kx^2$ is to be fitted to the points $(0, 3)$, $(1, 3)$, $(2, 6)$ by the method of least-squares. Find the value of k which should be used.

21. A curve of the form $y = k/x$ is to be fitted to the points $(1, 3)$, $(2, 2)$, $(3, 1)$ by the method of least-squares. Find the value of k which should be used.

22. For a set of n observed points the following sums are known:

$$\sum x = 2n$$
$$\sum y = 8n$$
$$\sum xy = 36n$$
$$\sum x^2 = 29n$$
$$\sum y^2 = 100n$$

A line of the form $y = x + k$ is to be fitted by the method of least-squares. Find the value of k which should be used.

23. Use the second-degree least-squares polynomial to derive the following seven-point approximate differentiation formula

$$Df(x) =$$

$$\frac{3[f(x + 3h) - f(x - 3h)] + 2[f(x + 2h) - f(x - 2h)] + [f(x + h) - f(x - h)]}{28h}$$

15.4. Minimax polynomials

24. Show that if a minimax straight line is fit to the three points (x_0, y_0), (x_1, y_1), (x_2, y_2) such that the successive errors are $+h$, $-h$, $+h$, that is, such that the straight line passes through the points $(x_0, y_0 - h)$, $(x_1, y_1 + h)$, $(x_2, y_2 - h)$, then

$$h = \frac{(x_2 - x_1)y_0 - (x_2 - x_0)y_1 + (x_1 - x_0)y_2}{2(x_2 - x_0)}$$

25. (a) Use the formula for h given in Exercise 24 to verify that $h = 3$ in the first trial for the minimax straight line in Example 15.7 based on the points $(1, 1)$, $(2, 8)$, $(3, 27)$.
    (b) Use the formula for h given in Exercise 24 to verify that $h = 5$ in the second trial for the minimax straight line in Example 15.7 based on the points $(0, 0)$, $(2, 8)$, $(3, 27)$.

26. (a) Fit a minimax straight line to the following data:

x	y
-1	3
1	4
3	7
5	6

(b) Compute the minimum maximum absolute error.

27. Rework Exercise 26 using a minimax quadratic.

28. Exercises 16 and 26 are both linear approximations based upon the same data.
    (a) Find the sum of squares $\sum [y_i - p(x_i)]^2$
      (1) In Exercise 16.
      (2) In Exercise 26.
    (b) Find the maximum error max $|y_i - p(x_i)|$
      (1) In Exercise 16.
      (2) In Exercise 26.

29. Exercises 17 and 27 are both quadratic approximations based upon the same data.
    (a) Find the sum of squares $\sum [y_i - p(x_i)]^2$
      (1) In Exercise 17.
      (2) In Exercise 27.
    (b) Find the maximum error max $|y_i - p(x_i)|$
      (1) In Exercise 17.
      (2) In Exercise 27.

30. For a minimax polynomial to be unique all the values of x_i must be distinct. Show that $p_1(x) = \frac{1}{2}$ and $p_2(x) = \frac{1}{2} - x$ are both minimax straight lines for the points $(0, 0)$, $(0, 1)$, $(1, 0)$.

15.5. Ratios of polynomials

31. Fit a function of the form
$$r(x) = \frac{1}{ax + b}$$
to the functional values $f(1) = 3$ and $f(3) = 1$.

32. Fit a function of the form
$$r(x) = \frac{ax + b}{cx + d}$$
to the functional values $f(0) = \frac{1}{2}, f(1) = \frac{2}{3}, f(\infty) = 1$.

33. Fit a function of the form
$$r(x) = \frac{ax^2 + b}{cx^2 + d}$$
to the functional values $f(0) = -1, f(1) = 0, f(2) = \frac{3}{5}$.

34. Show that if a function of the form

$$r(x) = \frac{1}{a + bx}$$

is to be fit to the function $f(x) = x$ for $x = 1, 2, 3$ by the method of least-squares, the unknowns a and b can be determined by solving the following two equations in two unknowns:

$$\frac{a + b - 1}{(a + b)^3} + \frac{2a + 4b - 1}{(a + 2b)^3} + \frac{3a + 9b - 1}{(a + 3b)^3} = 0$$

$$\frac{a + b - 1}{(a + b)^3} + \frac{4a + 8b - 2}{(a + 2b)^3} + \frac{9a + 27b - 3}{(a + 3b)^3} = 0$$

35. Show that if a function of the form

$$r(x) = \frac{1}{a + bx}$$

is to be fit to the function $f(x) = x$ for $x = 1, 2, 3$ by the minimax method in which the errors alternate h, $-h$, h, the unknowns a, b, h can be determined by solving the following three equations in three unknowns:

$$a(1 - h) + b(1 - h) - 1 = 0$$
$$a(2 + h) + b(4 + 2h) - 1 = 0$$
$$a(3 - h) + b(9 - 3h) - 1 = 0$$

15.6. Exponential functions

36. Fit a function of the form

$$e(x) = a + bc^x$$

to the functional values $f(0) = 5$, $f(1) = 14$, $f(2) = 50$.

37. Fit a function of the form

$$e(x) = 2^{ax + b}$$

to the function $f(x) = x^2$ for $x = 1, 2$.

38. Fit a function of the form

$$e(x) = (ax + b)2^x$$

to the function $f(x) = x^3$ for $x = 1, 2$.

39. Fit a function of the form

$$e(x) = a^x + b^x$$

to the functional values $f(0) = 2$, $f(1) = 2.75$, $f(2) = 3.8125$.

40. Show that if a function of the form

$$e(x) = a + b^x$$

is to be fit to the function $f(x) = x$ for $x = 1, 2, 3$ by the method of least-squares, the unknowns a and b can be determined by solving the following two equations in two unknowns:

$$3a \qquad + (b + b^2 + b^3) - 6 \qquad = 0$$
$$a(b + b^2 + b^3) + (b^2 + b^4 + b^6) - (b + 2b^2 + 3b^3) = 0$$

41. Show that if a function of the form

$$e(x) = a + b^x$$

is to be fit to the function $f(x) = x$ for $x = 1, 2, 3$ by the minimax method in which errors alternate h, $-h$, h, the unknowns a, b, h can be determined by solving the following three equations in three unknowns:

$$a + b + h - 1 = 0$$
$$a + b^2 - h - 2 = 0$$
$$a + b^3 + h - 3 = 0$$

42. In certain situations "geometric" linear interpolation gives superior results to "arithmetic" linear interpolation as given by formula (3.2). *Geometric linear interpolation* is defined by

$$f(x + n) = [f(x)]^{1-n}[f(x + 1)]^n$$

assuming $f(x)$, $f(x + 1)$ are positive and $0 < n < 1$.
 (a) Show that if $n = \frac{1}{2}$, arithmetic linear interpolation gives the arithmetic mean of $f(x)$ and $f(x + 1)$.
 (b) Show that if $n = \frac{1}{2}$, geometric linear interpolation gives the geometric mean of $f(x)$ and $f(x + 1)$.
 (c) If $f(0) = 16$ and $f(1) = 81$, estimate $f(\frac{3}{4})$ using arithmetic linear interpolation.
 (d) Rework (c) using geometric linear interpolation.

16

Elements of computer programming[1]

16.1. INTRODUCTION

As HAS BEEN mentioned throughout this book, many of the techniques employed in the numerical solution to mathematical problems are now computationally practicable because of the electronic data processing equipment that has become available over the past quarter of a century.

Chapter 16 very briefly introduces the student to the most elementary concepts of electronic data processing as they apply specifically to some of the methods of the preceding chapters. It cannot be too strongly emphasized that this chapter is not intended to teach students how to write computer programs. The presentation here will be elementary to all students who have studied the rudiments of electronic data processing.

The language used is BASIC, a very elementary language designed by IBM. It is an ideal language for people who are not computer programmers; and while it is not a powerful language, it does have the advantage of simplicity and is an excellent teaching language for students who have no prior programming knowledge. Because it was developed as a simplification of FORTRAN it has many features in common with this more powerful

---

[1]This chapter is reprinted from Chapter 5 of Society of Actuaries Study Note 30-2-72 with minor editorial changes to be consistent with the format of this book.

language. A student who has mastered **BASIC** should have no difficulty learning the more advanced languages.

16.2. THE ALGORITHM

In Chapter 1 we defined the algorithm as a complete, well-defined procedure for obtaining a numerical answer to a given mathematical problem. The practical examples suggest that an algorithm can be represented as a sequence of unambiguous, ordered computational steps. This latter concept directs us to the heart of computer programming. A program is itself a set of instructions to the computer system to carry out a sequence of unambiguous, ordered procedures. The algorithm used in Example 1.2, namely the formula for solving a quadratic equation, will be used here. While it is simple, it introduces many of the fundamental steps of computer programming.

When we write the algorithm

$$x = \frac{-b \pm \sqrt{b^2 - 4ac}}{2a}$$

for example, we are implying a sequence of steps as follows:

Step 1.   Identify the coefficients of the terms x^2, x^1, and x^0 in the quadratic equation and assign them the algebraic values a, b, and c, respectively.

Step 2.   Square b and compare this result with 4 times a times c. If $b^2 \geq 4ac$ take step 3; if $b^2 < 4ac$ take step 4.

Step 3.   Take the positive square root of the discriminant $b^2 - 4ac$, and alternately add this square root to $-b$ and subtract it from $-b$, dividing the resulting sum and difference, respectively, by twice the value of a. The two results represent the required real roots.

Step 4.   Take the positive square root of $4ac - b^2$ and, alternately, add the product of i and this square root to $-b$ and subtract the product of i and this square root from $-b$, dividing the resulting sum and difference, respectively, by twice the value of a. The two results represent the required complex roots.

Note that each step must be stated in a completely unambiguous manner. It would not be sufficient to specify the steps to be taken if $b^2 > 4ac$ or if $b^2 < 4ac$, for example. We would have been unclear as to what to do if $b^2 = 4ac$ exactly; our series of steps would have been ordered, but they would not be unambiguous.

16.3. FLOWCHART DIAGRAMS

The introduction of flowcharts as a step between the statement of the algorithm and the writing of the steps in the program is helpful to the programmer in "seeing" the problem. The logical arrangement of the programming steps in flowchart form often directs the computer programmer to greater efficiencies in the program, to conceptual problems that are not apparent in the statement of the algorithm, and to errors in the program before it becomes operational (that is, in the "debugging" stage).

The flowchart diagram is a useful tool, but it is not a necessary step in the writing of a program. Many elementary programs are written without a formally documented flowchart, and this step may be omitted if the programmer has a clear concept of his algorithm. For more advanced programs, however, involving numerous steps and subroutines requiring continuous reference, the flowchart is virtually indispensable.

The more common symbols used in drawing flowchart diagrams are shown in Figure 16.1.

FIGURE 16.1

—*Terminal:* **the beginning or end of a program or a point of interruption. Denotes data being entered into or extracted from a routine, for example, read in values** a, b, c, **the coefficients of a given quadratic equation.**

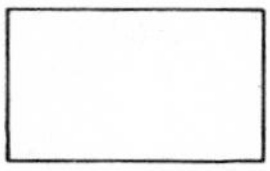

—*Process:* **computation according to a group of programmed instructions—an algorithm, for example, compute** $b^2 - 4ac$.

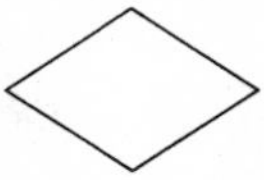

—*Decision:* **comparison of one set of data with another for equality, inequality, relative size, and routing of the program according to all possible results, for example, if** $b^2 - 4ac \geq 0$, **go to one step; if** $b^2 - 4ac < 0$, **go to another.**

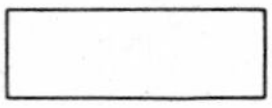

—*Counter:* **increment (or decrement) a value to measure the number of times a repetitive step has already been performed.**

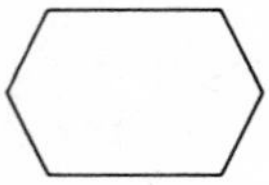

—*Subroutine:* **a predetermined group of operations, not detailed in the flowchart, designed to perform a standard set of computations that are contained in another program or which are available as a shared or "pooled" program; for example, our system might contain an already prepared program which would accept in some precisely determined format the elements of a nonsingular matrix and invert the matrix for us.**

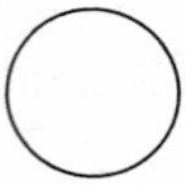

—*Connector:* **denotes entry from, or exit to, another part of the program flowchart.**

—*Direction:* **indicates step(s) to be taken next.**

Our program for computing the roots of $ax^2 + bx + c = 0$ would be diagrammed as in Figure 16.2.

FIGURE 16.2

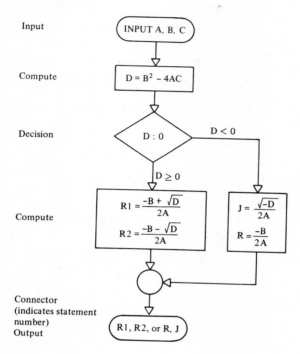

Input	INPUT A, B, C
Compute	$D = B^2 - 4AC$
Decision	$D : 0$ $D < 0$
Compute	$R1 = \dfrac{-B + \sqrt{D}}{2A}$ $R2 = \dfrac{-B - \sqrt{D}}{2A}$ $J = \dfrac{\sqrt{-D}}{2A}$ $R = \dfrac{-B}{2A}$
Connector (indicates statement number) Output	R1, R2, or R, J

16.4. PROGRAMMING LANGUAGE

The smallest element of computer program language is the *statement*. In BASIC, statements consist of input and output routines, arithmetic definitions, logic relationships, and instructions to proceed to another point in the program. The arithmetic symbols are

+	addition
−	subtraction
*	multiplication
/	division
↑(or**)	exponentiation

Whenever several arithmetic symbols are included in a single statement, and the order of operation is not specified by the use of parentheses (see below), the order assumed is first exponentiation, next multiplication and division, and finally addition and subtraction. Relational symbols are

$=$ equal to

$>$ greater than

$<$ less than

$\neq$ unequal to

$\leq$ less than or equal to

$\geq$ greater than or equal to

A partial list of BASIC functions (in effect, *canned* or predetermined subroutines) is

$\text{EXP}(X) = e^x$

$\text{SQR}(X) = \sqrt{x}$

$\text{LOG}(X) = \log_e x$

$\text{LGT}(X) = \log_{10} x$

$\text{INT}(X)$ = the integral part of x

$\text{SIN}(X)$ = sine of x radians

$\text{COS}(X)$ = cosine of x radians

$\text{TAN}(X)$ = tangent of x radians

Some examples of elementary statements in BASIC are

Input and output

INPUT A, B, C (or DATA A, B, C)

PRINT A, B, C, D, R1, R2

Arithmetic definitions

X = 2

Y = 3*X

D = (B↑2) − (4*A*C)

Parentheses are used to group operations within a multistep arithmetic statement in exactly the same manner as in algebra. If they are omitted, then the steps will be performed in the order already shown. Thus, for example, D = B↑2 − 4*A*C would give the same result as D = (B↑2) − (4*A*C), but to compute $y^{(2-x)}$ we would have to write Y↑(2 − X). The instruction Y↑2 − X would compute $y^2 - x$. The following arithmetic definitions will compute the real roots of our quadratic:

R1 = (−B + SQR(B↑2 − 4*A*C))/(2*A)

R2 = (−B − SQR(B↑2 − 4*A*C))/(2*A)

Here again the parentheses around $-B \pm SQR(B\uparrow2 - 4*A*C)$ and $2*A$ are necessary, for otherwise $-B + SQR(B\uparrow2 - 4*A*C)/2*A$ would compute

$$-b + \frac{\sqrt{b^2 - 4ac}}{2}a$$

Logical relationships—decisions

IF $D < 0$ THEN (statement number) NNNNN

In this step, D is compared with zero; if $D < 0$ then program *branches* (that is, proceeds directly) to the statement indicated; if $D \geq 0$ then it goes to the next numbered statement.

Procedural statements

GOTO (statement number) NNNNN
END (this is the last step in a BASIC program)

16.5. LOOPS

A *loop* is a series of statements within a program that defines a repetitive process. The facility to loop is vital to the iterative approach used in so many numerical analysis techniques.

In BASIC the simple loop commences with a FOR statement and ends with a NEXT statement, and is often referred to as a FOR NEXT loop. If part of our program consisted of a subroutine to calculate $f(x) = x^2$ for the first *n* numbers, for example, the loop to perform this operation would appear as

FOR $X = 1$ TO N
$F(X) = X\uparrow2$
NEXT X

This subroutine would automatically transfer from the third step back to the first step, until the Nth value had been computed (that is, there was no more NEXT X). Unless otherwise stated, the incrementation of X would be in steps of 1. To proceed at other increments we modify the first statement to: FOR $X = 1$ TO N STEP H. (This means perform the calculation at $X = 1$, $X = 1 + H$, $X = 1 + 2H, \ldots$) Thus to perform the process for odd numbers only we would write: FOR $X = 1$ TO N STEP 2.

Loops can also be *nested* one inside another, but in such a *nesting* it is absolutely critical that the inside loop be completed before the program returns to the outer loop. A BASIC program with overlapping loops cannot be properly executed.

The following nested loop is valid:

```
FOR  N = 1  TO  M
FOR  X = 1  TO  N
F(X) = X↑2
NEXT  X
NEXT  N
```

The next is invalid and cannot be executed properly:

```
FOR  N = 1  TO  M
FOR  X = 1  TO  N
F(X) = X↑2
NEXT  N
NEXT  X
```

We now have sufficient rudiments of programming language to write the entire program for the quadratic solution. Additional statement language will be introduced in Section 16.6 as it is needed.

16.6. WRITING THE BASIC PROGRAM

The BASIC program consists of a sequence of numbered statements that will be executed in consecutively numbered order. We may start with any number from 1, continuing in order through 99999. The statements need not be numbered in increments of 1; in fact, in actual practice, units of 10 or more are often used to separate the numbering system so as to allow for the insertion of additional statements to modify the original program.

Example 16.1.   Write a BASIC program to compute the roots of a quadratic equation.

Our first statement will be arbitrarily numbered 10; this has no particular significance. The program is given in Table 16.1.

The only reason we ask for the discriminant to be printed in the complex root case is so that we can distinguish the results R, J from R1, R2. If three values are printed, the last must be negative (we would not have reached statement 60 unless $D < 0$) so that if only two are printed, then these must be real roots.

We can avoid this awkward approach by introducing the PRINT USING command. This is PRINT USING followed by a numbered statement at which the print format is located, and the computed values to be printed are indicated in the order required.

TABLE 16.1

Statement number	Statement	Explanation
10	INPUT A, B, C	Input: Prepares computer to accept values of a, b, and c which will be requested each time the program is run.
20	D = (B↑2 − 4*A*C)	Computes discriminant $b^2 - 4ac$.
30	IF D ≥ 0 THEN 80	Decision: If discriminant nonnegative, program branches to subroutine commencing at statement 80.
40	J = SQR(4*A*C − B↑2)/(2*A)	Program reaches this step only if $D < 0$.
	alternatively J = ((4*A*C − B↑2)↑.5)/(2*A)	Computes $\dfrac{\sqrt{4ac - b^2}}{2a}$, the coefficient of $\pm i$ in the complex root.
50	R = −B/(2*A)	Computes the real part of complex roots.
60	PRINT R, J, D	Prints real part of complex roots, coefficient of $\pm i$, and the discriminant.
70	GOTO 110	Transfers control to end program, after complex roots printed.
80	R1 = (−B + SQR(D))/(2*A)	Computes real roots.
90	R2 = (−B − SQR(D))/(2*A)	
100	PRINT R1, R2	
110	END	

Our modification would be

60 PRINT USING 65, R, J
65 :COMPLEX ROOTS ARE # # . # # # # # PLUS AND MINUS # # . # # # # # I

The notation # indicates the positions where digits will occur in the answer. Note the advantage of incrementing the statement numbers in other than units of one; we have readily inserted another statement modifying the original program. The printout will follow the format of statement 65, placing the value of R in the first blank space (denoted by #'s in decimal format) and the value of J in the second. The modification to handle real roots would be

100 PRINT USING 105, R1, R2
105 :REAL ROOTS ARE # # . # # # # # AND # # . # # # # #

Our program completed, we can instruct the computer to name our program (Quadroot, say) by writing: NAME QUADROOT, and to save this program by writing SAVE. We then write the command RUN and the computer prints a question mark(?), asking, in effect, for the values of a, b, and c of the quadratic to be solved. These are then introduced and the results are printed. Two sample runs are shown in Appendix H, one for an equation with real and another run for an equation with complex roots.

Example 16.2.   Design a flowchart diagram and write a BASIC program to compute $\int_0^1 e^{x^2}\,dx$ to five-decimal accuracy using Romberg's method.

This example introduces programming techniques that are slightly more advanced than those used in QUADROOT, particularly the iteration loop which is fundamental to many numerical analysis programs. To prevent further complication, however, the program has been restricted to computing $\int_0^1 e^{x^2}\,dx$; in practice a more general program could be written to solve $\int_a^b f(x)\,dx$ for any function that could be integrated over the interval $a \le x \le b$.

Computational steps are as follows:

1. Compute $\frac{1}{2}(e^0 + e^1)$; call this result I_1.
2. Compute $I_2 = \frac{1}{4}[e^0 + 2e^{(1/2)^2} + e^1]$.
3. Compare I_2 and I_1; if the difference is less than acceptable error (.000005 for five-decimal accuracy) then print result; if difference exceeds .000005 then compute the next value of I_j, where

$$I_j = \frac{e^0 + 2\sum_{k=1}^{2^{j-1}-1} e^{\left(\frac{k}{2^{j-1}}\right)^2} + e^1}{2^j} \qquad j > 1$$

For example,

$$I_3 = \tfrac{1}{8}[e^0 + 2(e^{1/16} + e^{1/4} + e^{9/16}) + e^1]$$

4. Repeat procedure, computing values I_j for $j = 4, 5, \ldots, t$ and compare at each step I_j with I_{j-1} until difference D_j, say, when $D_j = |I_{j-1} - I_j|$ is less than .000005.
5. Compare D_j and D_{j-1}; if $D_j > D_{j-1}$ then values of I_j are not converging and we introduce an abort message to prevent further unnecessary computation.
6. If $D_j > .000005$ after n iterations, end program and print result. This step ensures that we will not spend excessive computation time if D_j converges to zero too slowly.

7. As soon as j reaches t, $t < n$, such that $D_j < .000005$, print result.

The flowchart for this program is shown in Appendix H.

Before writing the program we shall introduce the DIM(dimension) statement. This is another nonexecutable instruction which the program does not use for purposes of computation, but which specifies the maximum value of the variable that the program will accept. Thus, if we expect to run a maximum of 100 iterations, say, we would dimension I(J) and D(J) for J = 100. If no dimension is stated, then in BASIC language the computer assumes a maximum of 10 and will print the error message SUBSCRIPT OUT OF BOUNDS if the program attempts to compute a value for a variable higher than 10.

The program is given in Table 16.2.

The first run of this program was made for $N = 3$; this produced the approximation 1.4906789. A second run was made for $N = 4$, giving a value of 1.4697123. For $N > 11$ the following message printed: VALUE AT TERM 11 IS 1.4626522 WITH ERROR LESS THAN .0000013. Appendix H shows an actual run of this program.

Example 16.3. Write a BASIC program to apply the Newton-Raphson method of iteration to solve the equation $x^2 - e^x = 0$.

We set $f(x) = x^2 - e^x$ and note that $f(-1) = 1 - e^{-1} = .63212$ and $f(0) = -1$. Clearly, we have the conditions for a root in the range $-1 < x < 0$. Our nth iteration is of the form

$$x_{n+1} = x_n - \frac{f(x_n)}{f'(x_n)} = x_n - \frac{x_n^2 - e^{x_n}}{2x_n - e^{x_n}}$$

A very crude first approximation would be $x_1 = -.5$ from which we obtain

$$x_2 = x_1 - \frac{f(x_1)}{f'(x_1)} = -.5 - \frac{(-.5)^2 - e^{-.5}}{2(-.5) - e^{-.5}}$$

$$= -.5 - \frac{.25 - .60653}{-1 - .60653} = .7219258$$

The student should carry this iterative process two steps further and compare this result with that obtained by the program shown in Appendix H. Note how rapidly this method converges to the root if the initial value of $-.7$ is used. It would be instructive for the student interested in this phase of numerical analysis to diagram a flowchart for this procedure. Anyone with no prior programming experience should find the *BASIC Language Reference Manual* listed in the bibliography extremely helpful for self-study.

TABLE 16.2

Statement number	Statement	Explanation
10 DIM I(100), D(100)		Sets limiting number of iterations that program will perform (cannot input more than 100 with above limits on dimension).
20 INPUT N		
30 I(1) = (1 + EXP(1))/2		Computes first two values of I(J) directly.
40 I(2) = (1 + 2*EXP(0.25) + EXP(1))/4		
45 D(2) = I(1) − I(2)		First value of D(J).
50 J = 2		First value of J.
60 J = J + 1		Updates iteration count.
70 K = 1		Lower limit of sum.
80 L = (2↑(J − 1)) − 1		Upper limit of sum.
90 S = 0		Sets value of partial sum S from prior iteration back to zero to prepare for new sum.
100 FOR K = 1 TO L		
110 E = EXP((K/(L + 1))↑2)		Computes Kth term of sum.
120 S = S + E		Computes partial sum to K terms.
130 NEXT K		
140 I(J) = (1 + EXP(1) + 2*S)/(2↑J)		Computes Jth approximation to value of integral.

```
150 D(J) = I(J − 1) − I(J)
```

Computes difference between Jth and (J − 1)th approximation. Because this routine is never performed for J = 1, the calculation I(0) is avoided. BASIC does not accept arguments below 1 in this format.

```
160 D(J) = ABS(D(J))
```

Takes absolute value of error.

```
170 IF J < N THEN 200
180 PRINT USING 190, J, I(J)
190 :VALUE AT LIMITING TERM ### IS #.#####
```

If J = N print result; otherwise test for divergence.

```
195 GOTO 260
```

Routine completed.

```
200 IF D(J) < D(J − 1) THEN 230
210 PRINT USING 220, J, I(J), D(J), D(J − 1)
220 :DIVERGENCE AT TERM ## WITH VALUE #.##### AND DIFFERENCES
    #.##### AND #.#####
```

If D(J) < D(J − 1) then iteration just completed has produced an improvement in accuracy; otherwise abort program and determine reason for non-convergence.

```
225 GOTO 260
```

Routine completed.

```
230 IF D(J) > 0.000005 THEN 60
240 PRINT USING 250, J, I(J), D(J)
250 :VALUE AT TERM ### IS #.##### WITH ERROR LESS THAN #.######
```

If I(J) is converging then we check whether value D(J) still exceeds acceptable maximum. If it does, then we return to step 60 and repeat the iteration process. If it is less than 0.000005 then computation is completed.

```
260 END
```

APPENDIX A

Statistical treatment of roundoff error

1. Section 1.5

In Section 1.5 the maximum roundoff error in adding n numbers, each of which has a maximum roundoff error of E, for $E > 0$, is seen to be

$$\text{maximum roundoff error} = En \qquad (1.5)$$

This formula proves to be quite pessimistic, particularly for large n.

The theory of statistics can be used to show that a point estimate of roundoff error is given by

$$\text{estimated roundoff error} = .4607E\sqrt{n} \qquad (1.8)$$

Furthermore, it can be shown that a 95% confidence interval for roundoff error is given by

$$0 \leq \text{roundoff error} \leq 1.1316E\sqrt{n} \qquad (1.9)$$

Formulas (1.8) and (1.9) indicate that roundoff error is more nearly proportional to $\sqrt{n}$ than to n.

Let x_i be a continuous uniform random variable with frequency function

$$f(x_i) = 1 \qquad \text{for } -.5 \leq x_i \leq .5 \qquad (A.1)$$

398

This random variable provides a model for the roundoff error involved in an integer, since the true value can be up to one half on either side of the rounded value.

It is easy to show from elementary statistics that the mean and variance of x_i are given by

$$\mu_{x_i} = 0 \tag{A.2}$$

$$\sigma^2_{x_i} = \tfrac{1}{12} \tag{A.3}$$

The roundoff error involved in the sum of n numbers is given by the random variable

$$x = x_1 + x_2 + \cdots + x_n \tag{A.4}$$

where $x_1, x_2, \ldots, x_n$ are n identically and independently distributed continuous uniform random variables as defined above. Again from elementary statistics the mean and variance of x are given by

$$\mu_x = 0 \tag{A.5}$$

$$\sigma^2_x = \frac{n}{12} \tag{A.6}$$

Since we are primarily concerned with the magnitude of roundoff error rather than its sign, we shall work with the random variable $|x|$. The use of the *normal distribution*, as justified by the *central limit theorem*, is required in developing formulas (1.8) and (1.9).

Actually the 95% confidence interval given by formula (1.9) can be obtained more directly than the estimated roundoff error given by formula (1.8). In view of the nature of the problem a one-tailed confidence interval with a lower limit of zero seems more appropriate than a two-tailed confidence interval, since small roundoff errors are favorable results.

To determine the upper limit of the confidence interval, we have

$$P(|x| < K) = .95$$

or

$$P(-K < x < K) = .95$$

Thus,

$$K = 1.95996\sigma_x = 1.95996 \sqrt{\frac{n}{12}} = .56579 \sqrt{n}$$

using values from the normal distribution tables and formula (A.6). The maximum roundoff error in rounding numbers to the nearest integer is $E = .5$, so that the above result can be rewritten as

$$(1.1316)(.5)\sqrt{n}$$

In general, if we are rounding off in such a fashion that the maximum roundoff error in any one term is E instead of .5, then the general upper limit of formula (1.9) becomes

$$1.1316E\sqrt{n}$$

The point estimate of roundoff error given by formula (1.8) is derived as the mean of $|x|$, that is,

$$\mu_{|x|} = E[|x|]^1$$

Since absolute values present a complication, let us first analyze the distribution of $w = |z|$, where z is a standard normal random variable. The frequency function of a standard normal random variable is

$$f(z) = \frac{1}{\sqrt{2\pi}} e^{-\frac{1}{2}z^2} \qquad \text{for } -\infty < z < \infty \tag{A.7}$$

Now the cumulative distribution function of w is given by

$$G(w) = P(|z| \le w) = P(-w \le z \le w) = F(w) - F(-w)$$

The frequency function of w is given by

$$g(w) = G'(w) = F'(w) - F'(-w) = 2f(w)$$

$$= \frac{2}{\sqrt{2\pi}} e^{-\frac{1}{2}w^2} \qquad \text{for } 0 \le w < \infty \tag{A.8}$$

which is obvious from symmetry in any event.

The mean of w is given by

$$\mu_w = \frac{2}{\sqrt{2\pi}} \int_0^\infty w e^{-\frac{1}{2}w^2} \, dw$$

$$= -\frac{2}{\sqrt{2\pi}} e^{-\frac{1}{2}w^2} \Big|_0^\infty$$

$$= \frac{2}{\sqrt{2\pi}}$$

$$= \sqrt{\frac{2}{\pi}} \tag{A.9}$$

The student can also show by extending this analysis further that the variance of w is given by

$$\sigma_w^2 = \frac{\pi - 2}{\pi} \tag{A.10}$$

---

[1] "E" in this context is the expectation operator from statistics.

However, the variance of w is not required for our purposes, since we have already derived the confidence interval.

From formulas (A.5) and (A.6) we know that

$$\frac{x}{\sqrt{\dfrac{n}{12}}}$$

is a standard normal random variable. Therefore, applying formula (A.9)

$$E\left[\sqrt{\frac{12}{n}}\,|x|\right] = \sqrt{\frac{2}{\pi}}$$

so that

$$E[|x|] = \sqrt{\frac{n}{12}}\sqrt{\frac{2}{\pi}} = \sqrt{\frac{n}{6\pi}} = .230329\sqrt{n}$$

The maximum roundoff error in rounding numbers to the nearest integer is $E = .5$, so that the above result can be rewritten as

$$(.4607)(.5)\sqrt{n}$$

In general, if we are rounding off in such a fashion that the maximum roundoff error in any one term is E instead of .5, then the estimated roundoff error becomes

$$.4607E\sqrt{n}$$

which is formula (1.8).

Since the central limit theorem requires a "large" n, the student might well ask how accurate formula (1.8) really is. It is possible to work out exact expected values by integration from the uniform distribution, although the details become quite complex for other than very small n. For $n = 1, 2, 3, 4$, we obtain the following exact results:

$$
\begin{aligned}
E[|x_1|] &= \tfrac{1}{4} = .2500 \\
E[|x_1 + x_2|] &= \tfrac{1}{3} = .3333 \\
E[|x_1 + x_2 + x_3|] &= \tfrac{13}{32} = .4062 \\
E[|x_1 + x_2 + x_3 + x_4|] &= \tfrac{7}{15} = .4667
\end{aligned}
$$

These results are compared with those produced by formula (1.8) in Table A.1. Thus, formula (1.8) produces surprisingly good answers even for quite small n. For larger n the approximation can only get better. The power of the central limit theorem is indeed exhibited by this example.

TABLE A.1

n	Exact expected value	Formula (1.8) expected value	Actual difference	Relative difference
1	.2500	.2303	.0197	.0788
2	.3333	.3257	.0076	.0228
3	.4062	.3989	.0073	.0180
4	.4667	.4607	.0060	.0129

2. Section 2.7

(a) *Ratio of the variance in differences to the variance in functional values*

In Section 2.7 it is stated that the variance in the kth differences is approximately $\dfrac{(2k)!}{(k!)^2}$ times as great as the variance in the original functional values. It remains to derive this result.

Assume that the error in each functional value u_x is identically and independently distributed with a variance of σ^2. We must show that the variance of $\Delta^k u_x$ is equal to $\dfrac{(2k)!}{(k!)^2}\,\sigma^2$.

We have[2]

$$\operatorname{var}\left[\Delta^k u_x\right] = \operatorname{var}\left[\binom{k}{0}u_{x+k} - \binom{k}{1}u_{x+k-1} + \cdots + (-1)^k\binom{k}{k}u_x\right]$$

$$= \left[\binom{k}{0}^2 + \binom{k}{1}^2 + \cdots + \binom{k}{k}^2\right]\sigma^2$$

since $\sigma_{ax}^2 = a^2\sigma_x^2$, if a is a constant, and since $\sigma_{x\pm y}^2 = \sigma_x^2 + \sigma_y^2$, if x and y are independent.

Thus, we must show that

$$\sum_{i=0}^{k}\binom{k}{i}^2 = \frac{(2k)!}{(k!)^2} \tag{A.11}$$

This can be seen by looking at coefficients in the binomial expansion of

$$(1 + x)^k(1 + x)^k = (1 + x)^{2k}$$

---

[2] "Var" stands for variance.

The coefficient of x^k on the left-hand side is $\sum_{i=0}^{k} \binom{k}{i}^2$, while the coefficient of x^k on the right-hand side is $\dfrac{(2k)!}{(k!)^2}$.

(b) *Correlation of successive differences*

In Section 2.7 it is stated that the correlation coefficient between two successive kth differences arising strictly from independently distributed random errors is equal to $-\dfrac{k}{k+1}$. It remains to derive this result.

As above, assume that the error in each functional value u_x is identically and independently distributed with a variance of σ^2. The correlation coefficient between $\Delta^k u_x$ and $\Delta^k u_{x+1}$ is

$$\frac{E[(\Delta^k u_x)(\Delta^k u_{x+1})] - E[\Delta^k u_x]E[\Delta^k u_{x+1}]}{\sqrt{\operatorname{var}[\Delta^k u_x]}\sqrt{\operatorname{var}[\Delta^k u_{x+1}]}} \tag{A.12}$$

Since the kth differences are arising strictly from the presence of roundoff error, we have

$$E[\Delta^k u_x] = E[\Delta^k u_{x+1}] = 0$$

Also, from subsection (a) above, we have

$$\operatorname{var}[\Delta^k u_x] = \operatorname{var}[\Delta^k u_{x+1}] = \frac{(2k)!}{(k!)^2}\sigma^2$$

Thus, formula (A.12) becomes

$$\frac{E[(\Delta^k u_x)(\Delta^k u_{x+1})]}{\dfrac{(2k)!}{(k!)^2}\sigma^2}$$

The numerator of the above expression becomes

$$E[(\Delta^k u_x)(\Delta^k u_{x+1})]$$

$$= E\left[\left\{\binom{k}{0}u_{x+k} - \binom{k}{1}u_{x+k-1} + \binom{k}{2}u_{x+k-2} - \cdots + (-1)^k\binom{k}{k}u_x\right\}\right.$$

$$\left. \cdot \left\{\binom{k}{0}u_{x+k+1} - \binom{k}{1}u_{x+k} + \binom{k}{2}u_{x+k-1} - \cdots + (-1)^k\binom{k}{k}u_{x+1}\right\}\right]$$

$$= -\left[\binom{k}{0}\binom{k}{1} + \binom{k}{1}\binom{k}{2} + \cdots + \binom{k}{k-1}\binom{k}{k}\right]\sigma^2$$

We can sum the above expression by looking at coefficients in the binomial expansion of

$$(1 + x)^k(1 + x)^k = (1 + x)^{2k}$$

The coefficient of x^{k+1} on the left-hand side is

$$\binom{k}{0}\binom{k}{1} + \binom{k}{1}\binom{k}{2} + \cdots + \binom{k}{k-1}\binom{k}{k}$$

while the coefficient of x^k on the right-hand side is

$$\frac{(2k)!}{(k+1)!(k-1)!}$$

Thus, the correlation coefficient becomes

$$\frac{-\dfrac{(2k)!}{(k+1)!(k-1)!}\sigma^2}{\dfrac{(2k)!}{(k!)^2}\sigma^2} = -\frac{k}{k+1} \tag{A.13}$$

APPENDIX B

Transformation of
polynomial forms

It is possible to generalize the methods of synthetic division and inverse synthetic division discussed in Sections 2.4 and 2.5 to handle a wider variety of problems.

One generalization expresses a polynomial in factorials of $x - k$ rather than x. This can be accomplished by taking Table 2.3 and making two modifications:

1. Carry the first line of computations completely to the end of the coefficients rather than stopping one short of the last coefficient.
2. Use as index numbers $k, k + 1, k + 2, k + 3, k + 4$.

Example B.1 illustrates this generalization.

A second generalization expresses a polynomial in powers of $x - k$. This can be accomplished by taking Table 2.3 and making the same two modifications as above using as index numbers k, k, k, k, k. Example B.2 illustrates this generalization.

A completely general approach to the *transformation of polynomial forms* also exists. It includes all the methods of synthetic division and inverse synthetic division already discussed as special cases.

If a polynomial of the form

$$A_1(x - a_1)(x - a_2) \cdots (x - a_{n-2})(x - a_{n-1})(x - a_n)$$
$$+ A_2(x - a_1)(x - a_2) \cdots (x - a_{n-2})(x - a_{n-1})$$
$$+ A_3(x - a_1)(x - a_2) \cdots (x - a_{n-2})$$
$$+ \cdots$$
$$+ A_{n-1}(x - a_1)(x - a_2)$$
$$+ A_n(x - a_1)$$
$$+ A_{n+1}$$

is to be transformed into the form

$$B_1(x - b_1)(x - b_2) \cdots (x - b_{n-2})(x - b_{n-1})(x - b_n)$$
$$+ B_2(x - b_1)(x - b_2) \cdots (x - b_{n-2})(x - b_{n-1})$$
$$+ B_3(x - b_1)(x - b_2) \cdots (x - b_{n-2})$$
$$+ \cdots$$
$$+ B_{n-1}(x - b_1)(x - b_2)$$
$$+ B_n(x - b_1)$$
$$+ B_{n+1}$$

then perform the following steps:

1.   Write down the constants as follows:

$$
\begin{array}{ccccccc|ccccccc}
a_1 & a_2 & a_3 & \cdots & a_{n-2} & a_{n-1} & a_n & A_1 & A_2 & A_3 & \cdots & A_{n-1} & A_n & A_{n+1} \\
b_1 & b_2 & b_3 & \cdots & b_{n-2} & b_{n-1} & b_n & & & & & & &
\end{array}
$$

2.   Carry out the computations on the first line as follows:

$$
\begin{array}{ccccccc|ccccc}
a_1 & a_2 & a_3 & \cdots & a_{n-2} & a_{n-1} & a_n & A_1 & A_2 & A_3 & \cdots \\
b_1 & b_2 & b_3 & \cdots & b_{n-2} & b_{n-1} & b_n & 0 & s_1 & s_2 & \cdots \\
\hline
& & & & & & & A_1 & t_1 & t_2 & \cdots
\end{array}
$$

where

$$s_1 = A_1(b_1 - a_n)$$
$$t_1 = A_2 + s_1$$
$$s_2 = t_1(b_1 - a_{n-1})$$
$$t_2 = A_3 + s_2$$

and so forth.

3.   After completing the first line, eliminate b_1 and a_n and proceed similarly for the second line.

4. After each line, eliminate the first remaining b and the last remaining a and proceed similarly.

Examples B.3 and B.4 illustrate this completely general approach.

Example B.1. ***Transform*** $5x^4 - 7x^3 + 3x^2 + x - 4$ ***into factorial notation based on factorials of*** $x - 3$.

Table 2.3 can be slightly rewritten to accommodate factorials of $x - 0$ as follows:

0	5	−7	3	1	−4
	0	0	0	0	0
1	5	−7	3	1	−4
	0	5	−2	1	
2	5	−2	1	2	
	0	10	16		
3	5	8	17		
	0	15			
4	5	23			
	0				
	5				

The equivalence with Table 2.3 is apparent.

The table for factorials of $(x - 3)$ is given by the following:

3	5	−7	3	1	−4
	0	15	24	81	246
4	5	8	27	82	242
	0	20	112	556	
5	5	28	139	638	
	0	25	265		
6	5	53	404		
	0	30			
7	5	83			
	0				
	5				

Thus, $5x^4 - 7x^3 + 3x^2 + x - 4 = 5(x - 3)^{(4)} + 83(x - 3)^{(3)} + 404(x - 3)^{(2)} + 638(x - 3)^{(1)} + 242$.

Example B.2.   Transform $5x^4 - 7x^3 + 3x^2 + x - 4$ **into powers of** $x - 3$.

The table for powers of $x - 3$ is given by the following:

3	5	−7	3	1	−4
	0	15	24	81	246
3	5	8	27	82	242
	0	15	69	288	
3	5	23	96	370	
	0	15	114		
3	5	38	210		
	0	15			
3	5	53			
	0				
	5				

Thus, $5x^4 - 7x^3 + 3x^2 + x - 4 = 5(x - 3)^4 + 53(x - 3)^3 + 210(x - 3)^2 + 370(x - 3) + 242.$

Example B.3.   Convert $5x^{(4)} + 23x^{(3)} + 17x^{(2)} + 2x^{(1)} - 4$ **to standard polynomial form and confirm the answer in Section 2.5.**

The generalized approach produces the following table:

0	1	2	3	5	23	17	2	−4
0	0	0	0	0	−15	−16	−1	0
0	1	2	3	5	8	1	1	−4
0	0	0	0	0	−10	2	0	
0	1	2	3	5	−2	3	1	
0	0	0	0	0	−5	0		
0	1	2	3	5	−7	3		
0	0	0	0	0	0			
				5	−7			
				0				
				5				

The answer is $5x^4 - 7x^3 + 3x^2 + x - 4$ confirming the result of Section 2.5.

Example B.4.   Transform $(x + 6)(x + 4)(x - 3)(x - 5) - 2(x + 4)(x - 3)$ $(x - 5) + 3(x - 3)(x - 5) + (x - 5) + 4$ **into powers of** $x - 2$.

The generalized approach produces the following table:

$$
\begin{array}{cccc|ccccc}
5 & 3 & -4 & -6 & 1 & -2 & 3 & 1 & 4 \\
2 & 2 & 2 & 2 & 0 & 8 & 36 & -39 & 114 \\
\hline
5 & 3 & -4 & -6 & 1 & 6 & 39 & -38 & 118 \\
2 & 2 & 2 & 2 & 0 & 6 & -12 & -81 & \\
\hline
5 & 3 & -4 & -6 & 1 & 12 & 27 & -119 & \\
2 & 2 & 2 & 2 & 0 & -1 & -33 & & \\
\hline
5 & 3 & -4 & -6 & 1 & 11 & -6 & & \\
2 & 2 & 2 & 2 & 0 & -3 & & & \\
\hline
 & & & & 1 & 8 & & & \\
 & & & & 0 & & & & \\
\hline
 & & & & 1 & & & &
\end{array}
$$

Thus the polynomial transforms into $(x-2)^4 + 8(x-2)^3 - 6(x-2)^2 - 119(x-2) + 118$.

APPENDIX C

Further topics involving central difference formulas

1. Everett's second formula (Steffensen's formula)

The path taken by *Everett's second formula* (*Steffensen's formula*) is given in Figure 4.6.[1] The conventional form of Everett's second formula is given by

$$u_x = u_0$$

$$+ \binom{1+x}{2}\delta u_{1/2} + \binom{2+x}{4}\delta^3 u_{1/2} + \binom{3+x}{6}\delta^5 u_{1/2} + \cdots$$

$$- \binom{1-x}{2}\delta u_{-1/2} - \binom{2-x}{4}\delta^3 u_{-1/2} - \binom{3-x}{6}\delta^5 u_{-1/2} - \cdots$$

$$\text{(C.1)}$$

The similarity with Everett's formula is apparent, that is, there are two parallel paths through the difference table, only in this case they pass through the odd differences. Also the formula starts at u_0.

Everett's second formula to fifth differences can be derived from

---

[1]The formula is commonly referred to as "Everett's second formula." However, the best historical evidence indicates that Steffensen was the originator of the formula and not Everett. Steffensen referred to the formula as "Everett's second formula" because of its form.

Stirling's formula to sixth differences by expressing the even differences in terms of odd differences of degree one less.[2] We have from Stirling's formula

$$u_x = u_0 + x \cdot \tfrac{1}{2}(\delta u_{1/2} + \delta u_{-1/2}) + \frac{x^2}{2}\delta^2 u_0 + \frac{x(x^2-1)}{6}$$

$$\cdot \tfrac{1}{2}(\delta^3 u_{1/2} + \delta^3 u_{-1/2}) + \frac{x^2(x^2-1)}{24}\delta^4 u_0 + \frac{x(x^2-1)(x^2-4)}{120}$$

$$\cdot \tfrac{1}{2}(\delta^5 u_{1/2} + \delta^5 u_{-1/2}) + \frac{x^2(x^2-1)(x^2-4)}{720}\delta^6 u_0 + \cdots \qquad (4.8b)$$

Now make the following substitutions:

$$\delta^2 u_0 = \delta u_{1/2} - \delta u_{-1/2}$$
$$\delta^4 u_0 = \delta^3 u_{1/2} - \delta^3 u_{-1/2}$$
$$\delta^6 u_0 = \delta^5 u_{1/2} - \delta^5 u_{-1/2}$$

and simplify to obtain Everett's second formula.

2. Computations with Everett's formula

As discussed in Section 4.8 Everett's formula has the property that each computation can be used twice in the construction of a large table. This property reduces the volume of computations by one half, which was of considerable importance before the availability of computers. The property can best be illustrated by example.

Example C.1. *Interpolate u_x for $x = 1.2, 1.4, 1.6, 1.8, 2.2, 2.4, 2.6, 2.8$ using the data given in Example 4.3.*

Assuming fourth differences are constant, the following are extracts from Example 4.3:

x	u_x	$\delta^2 u_x$	$\delta^4 u_x$
1	1	14	24
2	16	50	24
3	81	110	24

---

[2]Note that Everett's second formula to fifth differences is accurate for sixth-degree polynomials, since even difference terms vanish.

Applying Everett's formula gives

$$u_{1.2} = (.2)(16) - (.032)(50) + (.0032)(24) + (.8)(1) - (.048)(14)$$
$$+ (.0112)(24)$$
$$= 2.0736$$

$$u_{1.4} = (.4)(16) - (.056)(50) + (.0092)(24) + (.6)(1) - (.064)(14)$$
$$+ (.0132)(24)$$
$$= 3.8416$$

$$u_{1.6} = (.6)(16) - (.064)(50) + (.0132)(24) + (.4)(1) - (.056)(14)$$
$$+ (.0092)(24)$$
$$= 6.5536$$

$$u_{1.8} = (.8)(16) - (.048)(50) + (.0112)(24) + (.2)(1) - (.032)(14)$$
$$+ (.0032)(24)$$
$$= 10.4976$$

$$u_{2.2} = (.2)(81) - (.032)(110) + (.0032)(24) + (.8)(16) - (.048)(50)$$
$$+ (.0112)(24)$$
$$= 23.4256$$

$$u_{2.4} = (.4)(81) - (.056)(110) + (.0092)(24) + (.6)(16) - (.064)(50)$$
$$+ (.0132)(24)$$
$$= 33.1776$$

$$u_{2.6} = (.6)(81) - (.064)(110) + (.0132)(24) + (.4)(16) - (.056)(50)$$
$$+ (.0092)(24)$$
$$= 45.6976$$

$$u_{2.8} = (.8)(81) - (.048)(110) + (.0112)(24) + (.2)(16) - (.032)(50)$$
$$+ (.0032)(24)$$
$$= 61.4656$$

The duplication of terms is apparent, since the first three terms for $u_{1.2}$, $u_{1.4}$, $u_{1.6}$, $u_{1.8}$ are identical to the last three terms for $u_{2.8}$, $u_{2.6}$, $u_{2.4}$, $u_{2.2}$, respectively. Similarly, the first three terms for $u_{2.2}$, $u_{2.4}$, $u_{2.6}$, $u_{2.8}$ are identical to the last three terms for $u_{3.8}$, $u_{3.6}$, $u_{3.4}$, $u_{3.2}$, if the computations are carried on through the table.

3. Throwback

Throwback is a practical technique introduced by Comrie for computing *modified differences* of a lower order which approximately reproduce the actual differences of a higher order. Thus the amount of computation is substantially reduced, while a large portion of the accuracy of using higher-order terms is retained.

We shall illustrate the technique of throwback based on Everett's formula with modified second differences to approximate the actual fourth differences. This is the most common situation in which throwback is used, although it is sometimes used with Bessel's formula or with modified differences of other than second order.

The second and fourth difference terms from the first portion of Everett's formula can be written as follows

$$\frac{x(x^2 - 1)}{3!} \delta^2 u_1 + \frac{x(x^2 - 1)(x^2 - 4)}{5!} \delta^4 u_1$$

$$= \frac{x(x^2 - 1)}{6} \left[\delta^2 u_1 + \frac{x^2 - 4}{20} \delta^4 u_1 \right]$$

and a similar manipulation can be performed for $\bar{x}$ in the second portion of the formula.

The key to the success of throwback lies in the fact that $\frac{x^2 - 4}{20}$ is relatively constant over the range of interpolation. For $0 \le x \le 1$ it is easily seen that $\frac{x^2 - 4}{20}$ ranges monotonically over $-.2$ to $-.15$. A good approximation, independent of x, can be obtained by giving $\frac{x^2 - 4}{20}$ some constant value between $-.2$ and $-.15$.

Modified second differences are then defined by

$$\delta_m^2 \equiv \delta^2 - C\delta^4 \tag{C.2}$$

where $.15 < C < .2$. The error involved in using modified second differences is

$$\frac{x(x^2 - 1)}{6} \left(\frac{x^2 - 4}{20} + C \right) \delta^4 u_1 + \frac{\bar{x}(\bar{x}^2 - 1)}{6} \left(\frac{\bar{x}^2 - 4}{20} + C \right) \delta^4 u_0$$

If we now assume that fourth differences are constant so that $\delta^4 u_0 = \delta^4 u_1 = \delta^4 u$ and if we substitute $\bar{x} = 1 - x$, we have

$$\frac{x(x - 1)(x + 1)}{6} \left[\frac{(x - 2)(x + 2)}{20} + C \right] \delta^4 u$$

$$- \frac{(x - 1)x(x - 2)}{6} \left[\frac{(x - 3)(x + 1)}{20} + C \right] \delta^4 u$$

$$= \frac{(x + 1)x(x - 1)(x - 2)[(x + 2) - (x - 3)]}{120} \delta^4 u$$

$$+ \frac{x(x - 1)[(x + 1) - (x - 2)]}{6} C\delta^4 u$$

$$= \frac{(x + 1)x(x - 1)(x - 2)}{24} \delta^4 u + \frac{x(x - 1)}{2} C\delta^4 u$$

$$= \frac{x(x - 1)}{24} [(x + 1)(x - 2) + 12C]\delta^4 u = E(x)$$

The error function just derived is denoted by $E(x)$. It is clear that $E(0) = E(1) = 0$. Furthermore, $E(x)$ is a fourth-degree polynomial and by inspection the other two roots lie in the interval $0 < x < 1$. Also, $E(x)$ is symmetrical about $x = .5$. Thus, the graph of $E(x)$ is given by Figure C.1. $E(x)$ has a maximum at $x = .5$ and two minima, one between 0 and .5 and another of equal value (from symmetry) between .5 and 1.

FIGURE C.1

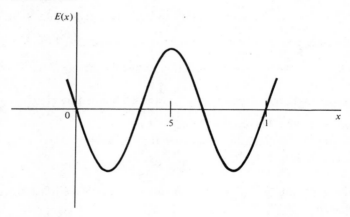

The value of C is usually chosen to minimize the maximum error. This happens when the three extrema are equal in absolute value. The value of C can then be shown to be

$$C = \frac{3 + \sqrt{2}}{24} = .1839 \qquad (C.3)$$

Thus, modified second differences are given by

$$\delta_m^2 \equiv \delta^2 - .1839\delta^4 \qquad (C.4)$$

The above analysis can still be used even if $\delta^4 u_0 \neq \delta^4 u_1$. In this case an upper bound for $E(x)$ is obtained by setting

$$|\delta^4 u| = \max \{|\delta^4 u_0|, |\delta^4 u_1|\}$$

Example C.2.   Rework Example 4.7:

(1)   Using Everett's formula to second differences.

(2)   Using Everett's formula to modified second differences with throwback.

The following are extracts from Example 4.3:

x	u_x	$\delta^2 u_x$	$\delta^4 u_x$
2	16	50	24
3	81	110	24

Example 4.7 uses all these values to produce the correct answer of 39.0625.

(1)   If we carry Everett's formula to second differences, we obtain third-degree accuracy. The result is

$$u_{2.5} = (.5)(81) + \frac{(.5)(.5^2 - 1)}{6}(110)$$

$$+ (.5)(16) + \frac{(.5)(.5^2 - 1)}{6}(50)$$

$$= 38.5$$

The relative error in this answer is

$$\frac{39.0625 - 38.5}{39.0625} = .0144$$

(2)   We first compute modified second differences

$$\delta_m^2 u_0 = 50 - (.1839)(24) = 45.5864$$
$$\delta_m^2 u_1 = 110 - (.1839)(24) = 105.5864$$

Now applying Everett's formula with these modified second differences, we obtain

$$u_{2.5} = (.5)(81) + \frac{(.5)(.5^2 - 1)}{6}(105.5864)$$

$$+ (.5)(16) + \frac{(.5)(.5^2 - 1)}{6}(45.5864)$$

$$= 39.0517$$

The relative error in this answer is

$$\frac{39.0625 - 39.0517}{39.0625} = .0003$$

The substantial improvement using throwback is apparent. Furthermore, this is the maximum error that can arise from throwback, since an extremum for $E(x)$ occurs at $x = .5$ (see Figure C.1).

APPENDIX D

Summation n

THE OPERATOR $[n]$, read "summation n," is briefly mentioned in Section 6.3. It forms the basis of the largely antiquated *summation method of graduation*.

The operator $[n]$ when applied to a functional value at one point forms the sum of n functional values spaced one unit apart with the original functional value being centrally located. Two significantly different cases arise depending upon whether n is even or odd, as the following examples illustrate:

$$[3]u_0 = u_{-1} + u_0 + u_1$$
$$[4]u_0 = u_{-3/2} + u_{-1/2} + u_{1/2} + u_{3/2}$$

In the case of an odd n, all functional values are at integral points and u_0 is included. In the case of an even n, all functional values are at half-integral points and u_0 is not included.

A general operator formula valid for either odd or even n is given by

$$[n] \equiv E^{-\frac{n-1}{2}} + E^{-\frac{n-3}{2}} + \cdots + E^{\frac{n-3}{2}} + E^{\frac{n-1}{2}} \tag{D.1}$$

If this expression is summed as a geometric progression, the following condensed formula results

$$[n] \equiv \frac{E^{n/2} - E^{-n/2}}{E^{1/2} - E^{-1/2}} \tag{D.2}$$

417

A formula for [n] in terms of differences can be derived from Stirling's formula in a fashion very similar to the derivation of formula (6.11b). We have as far as third differences

$$[n] \equiv \sum_{-\frac{n-1}{2}}^{\frac{n-1}{2}} \left[1 + x\mu\delta + \frac{x^2}{2!}\delta^2 + \frac{x(x^2-1)}{3!}\mu\delta^3 + \cdots \right]$$

The odd difference terms immediately drop out (from symmetry) leaving

$$[n] \equiv \sum_{-\frac{n-1}{2}}^{\frac{n-1}{2}} [1 + \tfrac{1}{2}\{x^{(2)} + x^{(1)}\}\delta^2 + \cdots]$$

$$\equiv \left[x + \tfrac{1}{2}\{\tfrac{1}{3}x^{(3)} + \tfrac{1}{2}x^{(2)}\}\delta^2 + \cdots \right]_{-\frac{n-1}{2}}^{\frac{n+1}{2}}$$

$$\equiv n + \frac{n^3 - n}{24}\delta^2 + \cdots \qquad (D.3)$$

upon simplification. This formula, accurate to third differences, is commonly used in problems involving summation n. It is interesting to note that the expression $\dfrac{n^3 - n}{24}$ is always an integer if n is an odd integer.

The summation method of graduation often applies these operators successively. For example, if [3] is applied twice, we have

$$[3]^2 u_0 = [3](u_{-1} + u_0 + u_1)$$
$$= [(u_{-2} + u_{-1} + u_0) + (u_{-1} + u_0 + u_1) + (u_0 + u_1 + u_2)]$$
$$= u_0 + (u_{-1} + u_0 + u_1) + (u_{-2} + u_{-1} + u_0 + u_1 + u_2)$$
$$= ([1] + [3] + [5])u_0$$

Thus, the product of the two summation operators expands into the sum of individual summation operators.

A general rule can be given for expressing [p][q] in terms of individual summation operators for any positive integers p and q. The rule is to take the larger of p and q and consider it as the central value and then arrange a number of terms around this value with spacing of two, where the number of terms is equal to the smaller of p and q. The above example of $[3]^2$ satisfies this general rule. Other examples are

$$[5][3] \equiv [3] + [5] + [7]$$
$$[5][2] \equiv [4] + [6]$$
$$[4][2] \equiv [3] + [5]$$

The student should verify these examples of the general rule from first principles as in the derivation of the expression for $[3]^2$.

The summation method of graduation uses averaging as a means of smoothing a rough series of data. The average of n terms in a series is given by the operator $[n]/n$. To third differences we have

$$\frac{[n]}{n} \equiv 1 + \frac{n^2 - 1}{24} \delta^2 \tag{D.4}$$

Thus, this averaging process does not leave a third-degree polynomial unchanged.

In the summation method of graduation the averaging operation is usually performed more than once. For example, if performed three times

$$\frac{[p][q][r]}{pqr}$$

involves an error in a third-degree polynomial of

$$\frac{p^2 + q^2 + r^2 - 3}{24} \delta^3$$

One common summation formula used in practice sets $p = q = r = 5$. This produces an error of $3\delta^2$. The functional values to which the operator is applied are usually adjusted before the graduation, so that a third-degree polynomial will be reproduced. If each u_x is replaced by $-u_{x-2} + u_{x-1} + u_x + u_{x+1} - u_{x+2}$, then a third-degree polynomial will be reproduced.

To see this, we have

$$\frac{[5]^3}{125}\left[-u_{x-2} + u_{x-1} + u_x + u_{x+1} - u_{x+2}\right]$$

$$= \frac{(5 + 5\delta^2)^3}{125}\left[-E^{-2} + E^{-1} + E^0 + E^1 - E^2\right]u_x$$

$$= (1 + \delta^2)^3[1 + (2 + \delta^2) - (2 + 4\delta^2)]u_x$$

$$= (1 + \delta^2)^3(1 - 3\delta^2)u_x$$

$$= (1 + 3\delta^2)(1 - 3\delta^2)u_x$$

$$= u_x$$

to third differences. The graduation formula for u_x given by

$$\frac{[5]^3}{125}(-u_{x-2} + u_{x-1} + u_x + u_{x+1} - u_{x+2}) \tag{D.5}$$

involves 17 terms when expanded and is called *Higham's formula*.

APPENDIX E

Truncation errors for approximate differentiation and approximate integration

1. Section 7.5

The derivation of truncation error in approximate differentiation given by

$$f'(x) - p'(x) = \frac{f^{(n+1)}(\xi)\pi'(x)}{(n+1)!} \tag{7.14}$$

in Section 7.5 ignores the fact that ξ is a function of x. However, it is possible to show that under typical conditions formula (7.14) is still a valid representation for the truncation error in approximate differentiation.

The truncation error of the collocation polynomial is given by

$$f(x) - p(x) = \frac{f^{(n+1)}(\xi)\pi(x)}{(n+1)!} \tag{1.12}$$

If it is remembered that ξ is a function of x and if both sides are differentiated with respect to x, then we have

$$f'(x) - p'(x) = \frac{f^{(n+1)}(\xi)\pi'(x)}{(n+1)!} + \frac{\pi(x)}{(n+1)!} \cdot \frac{d}{dx} f^{(n+1)}(\xi) \tag{E.1}$$

Now if x is one of the points of collocation $x_0, x_1, x_2, \ldots, x_n$, then $\pi(x) = 0$ and the second term in formula (E.1) vanishes. Thus, formula (7.14) is valid whenever the approximate differentiation formula is being applied at one of the points of collocation. This is the case for the examples and exercises in Section 7.5.

2. Section 8.3

(a) *Theoretical development of truncation errors*

The derivation of truncation error in approximate integration given by

$$\int f(x)\, dx - \int p(x)\, dx = \frac{f^{(n+1)}(\xi)}{(n+1)!} \int \pi(x)\, dx \qquad (8.14)$$

in Section 8.3 ignores the fact that ξ is a function of x. However, it is possible to show that this approach does produce valid truncation error expressions such as those given by formulas (8.15) through (8.17) for the Newton-Cotes formulas.

(1) *Trapezoidal rule*

The *mean value theorem for integrals* states

$$\int_a^b f(t)g(t)\, dt = g(\xi) \int_a^b f(t)\, dt \qquad (E.2)$$

where $a \leq \xi \leq b$, provided $f(t)$ does not undergo a sign change over the interval $a \leq t \leq b$. Formula (8.15) is obtained by a direct application of formula (E.2), since $\pi(x)$ does not undergo a sign change for two points of collocation.

(2) *Simpson's rule*

Unfortunately the use of formula (E.2), which works so efficiently to derive the truncation error for the trapezoidal rule, cannot be directly applied for higher-order Newton-Cotes formulas, since $\pi(x)$ undergoes a sign change over the range of collocation. Thus, some manipulation is required to obtain an expression to which formula (E.2) can be applied.

Assume that the points of collocation are $-h$, 0, h. Then the actual error, $F(h)$, is given by

$$F(h) = \int_{-h}^{h} f(x)\, dx - \frac{h}{3}[f(-h) + 4f(0) + f(h)]$$

The first three derivatives of $F(h)$ are given by

$$F^{(1)}(h) = [f(h) + f(-h)] - \frac{1}{3}[f(-h) + 4f(0) + f(h)]$$

$$- \frac{h}{3}[f^{(1)}(h) - f^{(1)}(-h)]$$

$$= \frac{2}{3}[f(h) + f(-h)] - \frac{4}{3}f(0) - \frac{h}{3}[f^{(1)}(h) - f^{(1)}(-h)]$$

$$F^{(2)}(h) = \frac{2}{3}[f^{(1)}(h) - f^{(1)}(-h)] - \frac{1}{3}[f^{(1)}(h) - f^{(1)}(-h)]$$

$$- \frac{h}{3}[f^{(2)}(h) + f^{(2)}(-h)]$$

$$= \frac{1}{3}[f^{(1)}(h) - f^{(1)}(-h)] - \frac{h}{3}[f^{(2)}(h) + f^{(2)}(-h)]$$

$$F^{(3)}(h) = \frac{1}{3}[f^{(2)}(h) + f^{(2)}(-h)] - \frac{1}{3}[f^{(2)}(h) + f^{(2)}(-h)]$$

$$- \frac{h}{3}[f^{(3)}(h) - f^{(3)}(-h)]$$

$$= - \frac{h}{3}[f^{(3)}(h) - f^{(3)}(-h)] \qquad (E.3)$$

Note that $F(0) = F^{(1)}(0) = F^{(2)}(0) = F^{(3)}(0) = 0$.
The mean value theorem[1] gives

$$\frac{f^{(3)}(h) - f^{(3)}(-h)}{2h} = f^{(4)}(\theta h)$$

or

$$F^{(3)}(h) = -\tfrac{2}{3}h^2 f^{(4)}(\theta h)$$

where θ depends on h and $-1 \le \theta \le 1$. It is assumed that the fourth derivative exists and is continuous.
We can now express $F(h)$ as

$$F(h) = -\frac{1}{3}\int_0^h (h-t)^2 t^2 f^{(4)}(\theta t)\, dt \qquad (E.4)$$

---

[1]The ordinary *mean value theorem*, not the mean value theorem for integrals, given by formula (E.2), is being used here.

The student should verify formula (E.4) by

1. Differentiating formula (E.4) three times and showing that formula (E.3) is obtained.
2. Showing that the three boundary values $F(0) = F^{(1)}(0) = F^{(2)}(0) = 0$ are satisfied by formula (E.4).

Notice that $(h - t)^2 t^2$ does not change sign for $0 \le t \le h$, so that formula (E.2) can be directly applied to formula (E.4) to give

$$-\frac{f^{(4)}(\xi)}{3} \int_0^h (h - t)^2 t^2 \, dt = -\frac{h^5}{90} f^{(4)}(\xi) \qquad (8.16)$$

(3) *Three-eighths rule*

Formula (8.17) for the three-eighths rule can be derived similarly to the derivation of formula (8.16) above. Also similar derivations exist for higher-order Newton-Cotes formulas.

(b) Composite formulas

Formulas (8.18) is obtained from formula (8.15) and gives the truncation error for the composite trapezoidal rule applied to n successive intervals.

Let the trapezoidal rule be applied n times successively over the intervals $0 \le x \le h, h \le x \le 2h, \ldots, (n - 1)h \le x \le nh$. Then from formula (8.15) the truncation error is

$$-\frac{h^3}{12} [f^{(2)}(\xi_1) + f^{(2)}(\xi_2) + \cdots + f^{(2)}(\xi_n)] \qquad (E.5)$$

for n values of ξ_i such that $(i - 1)h \le x \le ih$ for $i = 1, 2, \ldots, n$.

Now assuming that the second derivative is continuous on the interval $0 \le x \le nh$, we know that

$$f^{(2)}(\xi_1) + f^{(2)}(\xi_2) + \cdots + f^{(2)}(\xi_n) = nf^{(2)}(\xi) \qquad (E.6)$$

for some ξ, such that $0 \le \xi \le nh$, from the *intermediate value theorem*.

Thus, the truncation error given by formula (E.5) becomes

$$-\frac{nh^3}{12} f^{(2)}(\xi)$$

so that the absolute value of truncation error is

$$\frac{(b - a)h^2}{12} |f^{(2)}(\xi)| \qquad (8.18)$$

since $nh = b - a$.

The derivations of formulas (8.19) and (8.20) are similar.

APPENDIX F

Coefficients and roots of Gauss-Legendre polynomials $O_n^L(x)$

n		$i=0$	$i=1$	$i=2$
1	x_i	0.0000000000		
	H_i	2.0000000000		
2	x_i	−0.5773502692	0.5773502692	
	H_i	1.0000000000	1.0000000000	
3	x_i	−0.7745966692	0.0000000000	0.7745966692
	H_i	0.5555555556	0.8888888888	0.5555555556
4	x_i	−0.8611363116	−0.3399810436	0.3399810436
	H_i	0.3478548451	0.6521451549	0.6521451549
5	x_i	−0.9061798459	−0.5384693101	0.0000000000
	H_i	0.2369268851	0.4786286705	0.5688888888
6	x_i	−0.9324695142	−0.6612093865	−0.2386191861
	H_i	0.1713244924	0.3607615730	0.4679139346

n		$i=3$	$i=4$	$i=5$
1	x_i			
	H_i			
2	x_i			
	H_i			
3	x_i			
	H_i			
4	x_i	0.8611363116		
	H_i	0.3478548451		
5	x_i	0.5384693101	0.9061798459	
	H_i	0.4786286705	0.2369268851	
6	x_i	0.2386191861	0.6612093865	0.9324695142
	H_i	0.4679139346	0.3607615730	0.1713244924

APPENDIX G

Positive definite matrices and the Choleski factorization method

IN MORE advanced work with matrices, the *Choleski factorization method* (sometimes called the *square-root method*) is useful. It is often used to factor a matrix A which is *positive definite*. A matrix A is called *positive definite* if:

1. It is a symmetric matrix.
2. The value of $x^T A x$ is positive for any nonzero vector x.

The Choleski factorization method factors a matrix into the product of a lower triangular and an upper triangular matrix such that the diagonal elements are equal in the two matrices. When the Choleski factorization method is applied to a positive definite matrix, the diagonal elements are all positive and the upper triangular matrix is the transpose of the lower triangular matrix, that is,

$$A = LU = LL^T \qquad \text{(G.1)}$$

The elements of the lower triangular matrix L can be computed row by row in order with the following formulas:

$$l_{11} = \sqrt{a_{11}} \qquad \text{(G.2a)}$$

$$l_{ij} = \frac{1}{l_{jj}} \left(a_{ij} - \sum_{k=1}^{j-1} l_{ik} l_{jk} \right) \qquad \text{(G.2b)}$$

$$l_{ii} = \sqrt{a_{ii} - \sum_{k=1}^{i-1} l_{ik}^2} \qquad \text{(G.2c)}$$

Example G.1.   Factor the positive definite matrix A with the Choleski factorization method, if

$$A = \begin{bmatrix} 36 & -36 & 18 & 0 \\ -36 & 117 & -72 & 18 \\ 18 & -72 & 189 & -36 \\ 0 & 18 & -36 & 24 \end{bmatrix}$$

Applying the formulas given we have the following:

$$l_{11} = \sqrt{36} = 6$$

$$l_{21} = \tfrac{1}{6}(-36) = -6$$

$$l_{22} = \sqrt{117 - (-6)^2} = 9$$

$$l_{31} = \tfrac{1}{6}(18) = 3$$

$$l_{32} = \tfrac{1}{9}[-72 - (3)(-6)] = -6$$

$$l_{33} = \sqrt{189 - (3)^2 - (-6)^2} = 12$$

$$l_{41} = \tfrac{1}{6}(0) = 0$$

$$l_{42} = \tfrac{1}{9}[18 - (0)(-6)] = 2$$

$$l_{43} = \tfrac{1}{12}[-36 - (0)(3) - (2)(-6)] = -2$$

$$l_{44} = \sqrt{24 - (0)^2 - (2)^2 - (-2)^2} = 4$$

Thus, we have

$$L = \begin{bmatrix} 6 & 0 & 0 & 0 \\ -6 & 9 & 0 & 0 \\ 3 & -6 & 12 & 0 \\ 0 & 2 & -2 & 4 \end{bmatrix}$$

The student should verify that $LL^T = A$.

APPENDIX H

Illustrative computer programs for selected numerical analysis algorithms

1. Solution of quadratic equations

```
10  INPUT A,B,C
20  D = (B↑2 - 4*A*C)
30  IF D≥0 THEN 80
40  J = SQR(4*A*C - B↑2)/(2*A)
50  R = -B/(2*A)
60  PRINT USING 65,R,J
65  :COMPLEX ROOTS ARE ##.##### PLUS AND MINUS ##.##### I
70  GOTO 110
80  R1 = (-B + SQR(D))/(2*A)
90  R2 = (-B - SQR(D))/(2*A)
100 PRINT USING 105,R1,R2
105 :REAL ROOTS ARE ##.##### AND ##.#####
110 END
```

RUN

? 1,-6,8.75

REAL ROOTS ARE 3.50000 AND 2.50000

PROCESSING   0 UNITS

RUN

? 1,-1,1

COMPLEX ROOTS ARE 0.50000 PLUS AND MINUS 0.86602 I

PROCESSING   0 UNITS

2. Romberg's method to compute $\int_0^1 e^{x^2}\,dx$

```
10 DIM I(100), D(100)
20 INPUT N
30 I(1) = (1 + EXP(1))/2
40 I(2) = (1 + 2*EXP(0.25) + EXP(1))/4
45 D(2) = I(1) - I(2)
50 J = 2
60 J = J + 1
70 K = 1
80 L = (2↑(J-1))-1
90 S = 0
100 FOR K = 1 TO L
110 E = EXP((K/(L+1))↑2)
120 S = S + E
130 NEXT K
140 I(J) = (1+EXP(1)+2*S)/(2↑J)
150 D(J) = I(J-1) - I(J)
160 D(J) = ABS(D(J))
170 IF J < N THEN 200
180 PRINT USING 190, J, I(J)
190 :VALUE AT LIMITING TERM ### IS #.#######
195 GOTO 260
200 IF D(J)<D(J-1) THEN 230
210 PRINT USING 220,J,I(J),D(J),D(J-1)
220 :DIVERGENCE AT TERM ## WITH VALUE #.####### AND DIFFERENCES
    #.####### AND #.#######
225 GOTO 260
230 IF D(J)>0.000005 THEN 60
240 PRINT USING 250, J,I(J),D(J)
250 :VALUE AT TERM ### IS #.####### WITH ERROR LESS THAN #.#######
260 END
```

```
RUN

? 3
VALUE AT LIMITING TERM    3 IS 1.4906789

PROCESSING     0 UNITS


RUN

? 4
VALUE AT LIMITING TERM    4 IS 1.4697123

PROCESSING     0 UNITS


RUN

? 12
VALUE AT TERM   11 IS 1.4626522 WITH ERROR LESS THAN   .0000013

PROCESSING     4 UNITS
```

Flowchart for Romberg's method:

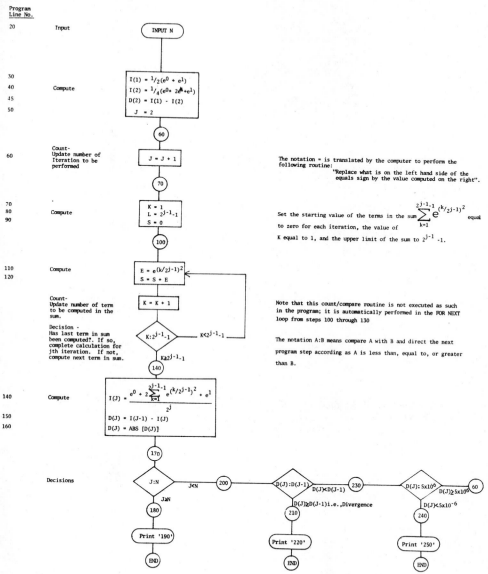

Program
Line No.

20   Input

30
40   Compute
45
50

60   Count-
      Update number of
      Iteration to be
      performed

The notation = is translated by the computer to perform the following routine:
"Replace what is on the left hand side of the equals sign by the value computed on the right".

70
80   Compute
90

Set the starting value of the terms in the sum $\sum_{k=1}^{2^{j-1}-1} e^{(k/2^{j-1})^2}$ equal to zero for each iteration, the value of K equal to 1, and the upper limit of the sum to $2^{j-1}-1$.

110   Compute
120

Count-
Update number of term
to be computed in the
sum.

Note that this count/compare routine is not executed as such in the program; it is automatically performed in the FOR NEXT loop from steps 100 through 130

Decision -
Has last term in sum
been computed?. If so,
complete calculation for
jth iteration. If not,
compute next term in sum.

The notation A:B means compare A with B and direct the next program step according as A is less than, equal to, or greater than B.

140   Compute

150
160

Decisions

3.   Solution of $x^2 - e^x = 0$, using Newton-Raphson method

```
10 INPUT X,L
15 F = (X↑2 - EXP(X))
20 D = (2*X - EXP(X))
25 A = F/D
30 E = ABS(A)
40 IF E<0.000005 THEN 80
50 X = X-A
60 N = N+1
65 IF N = L THEN 80
70 GOTO 15
80 PRINT X,N,A
90 END
```

```
RUN


? -0.5,100

-7.034674295E-01    3                    -7.042590747E-09

PROCESSING    0 UNITS


RUN


? -0.7,100

-7.034721896E-01    1                    -4.767105308E-06

PROCESSING    0 UNITS
```

Note that E–01 is read as 10^{-1}, E–09 as 10^{-9}, etc. Thus, $-7.034721896E{-}01 = -0.7034721896$.

Answers to the exercises

Chapter 1

1. (a) 63
   (b) 6
2. 307,692; the algorithm with 693 on top is superior
3. $\dfrac{\sin^2 x}{1 + \cos x}$
4. (a) 1.279
   (b) 1.274
   (d) 1.279
5. (a) .005; .0039
   (b) 0; 0
6. (a) .12
   (b) .10
7. (a) 2
   (b) 8
9. 4 decimal places
10. $-.022 \leq x \leq .022$
11. .04
12. $.12 \times 10^{18}$
13. (a) 10100101
    (b) 45
15. (a) kE
    (b) E^R

16. E
17. $1 - (1 - E^R)^n$
18. 1.5975
19. 100
20. 47,115
21. $x^3 - 2x^2 - 5x + 6$
22. $3x^2 - 2x$
23. $x(x - 1)(x - 2)$
24. $\dfrac{2\sqrt{3}}{9}$
25. $\frac{1}{12}e^\xi(2x^3 - 3x^2 + x)$
26. $\frac{1}{12}e\,|2x^3 - 3x^2 + x|$
27. $\dfrac{\sqrt{3}}{216}\,e$
28. 0
29. $\dfrac{\sqrt{42}}{12}$
33. (a) 8.971
    (b) 8.971
34. (a) 1
    (b) .894
35. (a) .333
    (b) .375
36. (a) .0228
    (b) 2.736
    (c) 2.763
37. $\dfrac{\sqrt{6}}{6} \le \xi \le \dfrac{\sqrt{2}}{2}$ or $.4082 \le \xi \le .7071$. Yes, since $\dfrac{\sqrt{42}}{12} = .5401$.

Chapter 2

3. 22
6. 186
7. 3
8. $\frac{32}{81}$
10. 129
11. $a^{bx}(a^b - 1)\left[\dfrac{(a^b - 1)^{20} - 1}{a^b - 2}\right]$
13. $\dfrac{\log_e (1 + \log_e a)}{\log_e a}$
16. $a^n \dfrac{(m + 1)!}{(m + 1 - n)!}\,[a(x + n) + b][a(x + n + 1) + b] \cdots [a(x + m) + b]$
17. 48
19. $24x + 42$
20. $\dbinom{y}{r}\Delta^{r+1}u_x$

21. -8

22. 6

23. $8x^{(4)} + 35x^{(3)} + 17x^{(2)} + 4x^{(1)} - 3$

24. $8x_2^{(4)} + 83x_2^{(3)} + 146x_2^{(2)} + 21x_2^{(1)} - 3$

25. $4! \cdot 10!$

26. $24\binom{x}{3} + 36\binom{x}{2} + 10\binom{x}{1} + 4$

27. $108x^2 + 648x + 1134$

30. (a) $\Delta - \Delta^2 + \Delta^3 - \Delta^4 + \cdots$

    (b) $\nabla + \nabla^2 + \nabla^3 + \nabla^4 + \cdots$

31. $12(x - 3)^{(-5)}$

32. $12x^2 - 24x + 14$

33. Fourth differences: $1, -3, 2, 2, -3, 1$

34. Fourth differences: $1, -5, 10, -10, 5, -1$

35. Change 79 to 77

36. $\nabla^n f(x) = f(x) - \binom{n}{1} f(x-1) + \binom{n}{2} f(x-2) - \cdots + (-1)^n f(x-n)$

40. (a) $6(x - 1)$

    (b) $6\dfrac{x+1}{(x+2)^3}$

41. -6

42. $-\frac{13}{264}$

43. $2x^2 + 2x$

44. $6h^2$

45. 30

46. $u_8 - u_4 - 4\Delta u_4 - 6\Delta^2 u_5 + 2\Delta^3 u_4$

47. u_0

49. x

50. Functional values: $1, 1, 13, 73, 241$

52. (a) (1) 0

       (2) 120

       (3) 30

53. (a) 11

54. (a) 7

Chapter 3

1. 3

2. (a) 4

    (b) 12

    (c) 6

    (d) 6

3. 1981

4. $x^3 - 2x^2 + x - 7$

6. $7x^2 - 6x$

8. 111

9. 21

10. 3
11. -16
13. .036
14. .260
15. $28x^2 - 48x$
16. $\frac{1}{3}$
17. (a) $2.5\Delta + 1.875\underset{2}{\Delta^2} + .3125\underset{2}{\Delta^3}$

    (b) $4\underset{5}{\Delta} - .12\underset{5}{\Delta^2} + .064\underset{5}{\Delta^3}$

18. $\frac{1}{4096}$
19. $\dfrac{\pi}{3} = 1.047$

20. $-.16$
21. 3.72 thousand
22. .02
23. 4 decimal places
24. $\left[\dfrac{(.01215)(\sqrt{3})}{280}\right]^{.3} = .058$

25. $u_x = u_1 + (x - 1)\nabla u_2 + \dbinom{x-1}{2}\nabla^2 u_2 + \dbinom{x}{3}\nabla^3 u_2 + \dbinom{x+1}{4}\nabla^4 u_3$

26. $C_1 = 1,\ C_2 = x - 3,\ C_3 = \dbinom{x-3}{2},\ C_4 = \dbinom{x-3}{3},\ C_5 = \dbinom{x-2}{4}$

27. 1
28. 5.28
29. $1 + 4x - 4x^2$
30. -5
31. 8
32. $\dfrac{3}{c+1}$
33. 756
34. .08
35. $(-1)^n$
36. $\frac{1}{2}x(x - 3)$

Chapter 4

1. $6x$
2. 2160
3. 12
4. $4x^2 + 1$
5. $\frac{1}{2}(x^2 + 3x + 1)$
8. $\dfrac{-\Delta}{1 + \Delta}$
9. 0
10. 1.1795

11. $\begin{aligned} A &= a + b + c + d \\ B &= -a + c + 2d \\ C &= a + d \\ D &= d \end{aligned}$

12. $f(x) = f(4) + \dfrac{x-4}{3} \underset{3}{\delta} f(5.5) + \dfrac{(x-4)(x-7)}{18} \underset{3}{\delta^2} f(4)$

        $+ \dfrac{(x-1)(x-4)(x-7)}{162} \underset{3}{\delta^3} f(5.5)$

13.   1.1749

14. $\begin{aligned} A &= a + b + c + d \\ B &= -2a - b + d \\ C &= a + d \\ D &= -a \end{aligned}$

15. $f(x) = f(3) + \dfrac{x-3}{2} \underset{2}{\delta} f(2) + \dfrac{(x-1)(x-3)}{8} \underset{2}{\delta^2} f(3)$

        $+ \dfrac{(x-1)(x-3)(x-5)}{48} \underset{2}{\delta^3} f(2)$

16.   1.1772

17. $A = 2, B = 6\frac{1}{12}, C = -1, D = -\frac{1}{12}$

18.   1.1795

21.   1.1795

25.   81

31.   (a) $\dfrac{f^{(5)}(\xi)(x^5 - 5x^3 + 4x)}{120}$

      (b) $\dfrac{K}{120} |x^5 - 5x^3 + 4x|$

      (c) $x = \pm\sqrt{\dfrac{15 + \sqrt{145}}{10}} = \pm 1.6444$

           $x = \pm\sqrt{\dfrac{15 - \sqrt{145}}{10}} = \pm.5439$

32.   (a) $\dfrac{f^{(4)}(\xi)(x^4 - 2x^3 - x^2 + 2x)}{24}$

      (b) $\dfrac{K}{24} |x^4 - 2x^3 - x^2 + 2x|$

      (c) $x = \frac{1}{2}; \; x = \dfrac{1 \pm \sqrt{5}}{2} = 1.6180 \text{ and } -.6180$

Chapter 5

1.   (b) $-\frac{7}{6}$

3.   4

4.   111

5. $n + 1$

6.   .88

9.   30

10. $\dfrac{b-e}{c-d}$

11. -2

12. 5

13. 12

14. 4

15. $\dfrac{x-3}{x-5}$

16. $m=2,\ n=-1$

17. 12

18. $(x-j)(x-k)(x-l)(x-m)Z$

19. -11

20. $2x_0 + x_1$

23. 4

25. 4

26. 11

27. $\frac{5}{8}$

28. $\frac{3}{5}$

29. $\frac{28}{39}$

30. 6

31. 6

32. (a) $u_x = u_c + (x-c)\underset{d}{\triangle}u_c + (x-c)(x-d)\underset{cd}{\triangle}{}^2u_b + (x-b)(x-c)(x-d)\underset{cde}{\triangle}{}^3u_b$
$+ (x-b)(x-c)(x-d)(x-e)\underset{bcde}{\triangle}{}^4u_a$

   (b) $u_x = u_c + (x-c)\underset{c}{\triangle}u_b + (x-b)(x-c)\underset{cd}{\triangle}{}^2u_b + (x-b)(x-c)(x-d)\underset{bcd}{\triangle}{}^3u_a$
$+ (x-a)(x-b)(x-c)(x-d)\underset{bcde}{\triangle}{}^4u_a$

   (c) $u_x = u_c + \frac{1}{2}(x-c)\left(\underset{d}{\triangle}u_c + \underset{c}{\triangle}u_b\right) + \frac{1}{2}(x-c)(2x-b-d)\underset{cd}{\triangle}{}^2u_b$
$+ \frac{1}{2}(x-b)(x-c)(x-d)\left(\underset{cde}{\triangle}{}^3u_b + \underset{bcd}{\triangle}{}^3u_a\right)$
$+ \frac{1}{2}(x-b)(x-c)(x-d)(2x-a-e)\underset{bcde}{\triangle}{}^4u_a$

   (d) $u_x = \frac{1}{2}(u_c + u_d) + \frac{1}{2}(2x-c-d)\underset{d}{\triangle}u_c + \frac{1}{2}(x-c)(x-d)\left(\underset{cd}{\triangle}{}^2u_b + \underset{de}{\triangle}{}^2u_c\right)$
$+ \frac{1}{2}(x-c)(x-d)(2x-b-e)\underset{cde}{\triangle}{}^3u_b$
$+ \frac{1}{2}(x-b)(x-c)(x-d)(x-e)\left(\underset{bcde}{\triangle}{}^4u_a + \underset{cdef}{\triangle}{}^4u_b\right)$

33. -6

34. $x^3 + x$

35. $x(x-1)\left(x-\frac{3}{2}\right)$

37. $\frac{1}{4}x^2$

38. $\frac{1}{6}(x-1)(x-4)(x-6)$

39. 4

Chapter 6

1. -193

2. 4626

3. $v_{10\frac{1}{4}} - v_{-\frac{1}{4}}$

6. 2303
8. $22(104)^{(4)}$
9. $\frac{1}{2}n(n+1)^2(n+2)$
10. $3x^2 + 3x$
11. $\dfrac{n(3n+5)}{8(3n+1)(3n+4)}$
12. $3^{120} - 1$
13. $nu_0 + \dbinom{n+1}{2}\Delta u_{-1} + \dbinom{n+2}{3}\Delta^2 u_{-1} + \dbinom{n+2}{4}\Delta^3 u_{-2} + \dbinom{n+3}{5}\Delta^4 u_{-2}$
14. 2130
15. $2(2^{n+1} - 1) - (n+1)(n+2)$
16. $6u_0 + 35\delta^2 u_0$
17. $\frac{5}{14}u_0 - \frac{1}{2}u_1 + \frac{1}{56}\sum_{x=0}^{7} u_x$
19. 715
20. $10 \cdot 4^{11}$
21. $\frac{20}{27}$
22. $\dfrac{(14n^2 - 21n + 1)n^{(7)}}{126}$
25. $\frac{1}{6}[(n+7)(n+5)(n+3) + (n+6)(n+4)(n+2) - 153]$
26. $\frac{7}{24}$
27. $15 \cdot 7!$
28. 19
29. $\dfrac{n}{n-1}$
30. $m(x-2)^{(m-1)}$
31. $\log_b (bx+c) + k$
32. $\dfrac{3^n}{n+1} - 1$
33. .693
34. 113
35. 671.46
36. .4054
37. $\frac{1}{5}n^5 + \frac{1}{2}n^4 + \frac{1}{3}n^3 - \frac{1}{30}n$
38. 1.2020569
39. 3617.35
40. 3617.46
42. $\frac{3}{2}(n+1)^{(4)} - \frac{10}{3}(n+1)^{(3)} + \frac{3}{2}(n+1)^{(2)}$
43. $\dfrac{1}{x(1-x)}$
44. $\frac{1}{8}(n+1)^{(4)} + \frac{2}{3}(n+1)^{(3)} + \frac{1}{2}(n+1)^{(2)}$
46. $\frac{315}{2}$
47. 12,210
48. $u_n - u_0$
49. 5
51. $\dfrac{1}{x+2}$

52.. $\frac{3}{2}(x^2 + x + 1)$

54. $\frac{1}{4}ax^4 - (\frac{1}{2}a - \frac{1}{3}b)x^3 + (\frac{1}{4}a - \frac{1}{2}b + \frac{1}{2}c)x^2 + (\frac{1}{6}b - \frac{1}{2}c + d)x + K$

55. $K - 2^{x-2}\dfrac{(x-1)!}{(2x-1)!}$

56. $B_0 = 1, B_1 = -\frac{1}{2}, B_2 = \frac{1}{6}, B_3 = 0, B_4 = -\frac{1}{30}$

Chapter 7

6. $x^2 + 4x + 4$
7. 1
8. $\frac{1}{6}$
9. $2\frac{1}{2}$
10. 13
11. $a = 1, b = \frac{1}{2}, c = -\frac{1}{6}$
12. $\nabla \equiv D - \dfrac{D^2}{2!} + \dfrac{D^3}{3!} - \dfrac{D^4}{4!} + \cdots$
15. $Du_x = \Delta u_x - \frac{1}{2}\Delta^2 u_{x-1} - \frac{1}{6}\Delta^3 u_{x-1} + \cdots$
16. $Du_x = \Delta u_{x-1} + \frac{1}{2}\Delta^2 u_{x-1} - \frac{1}{6}\Delta^3 u_{x-2} + \cdots$
17. (a) 4
    (b) 4
18. $\frac{33}{8}$; formula (7.4)
19. $\frac{1}{2}[f(1) - f(-1)]$
20. 4
21. 2.5
22. 12
23. $-\frac{3}{10}f(0) + \frac{1}{6}f(2) + \frac{2}{15}f(5)$
25. $\left[\dfrac{.00015}{e^{-1}}\right]^{1/3} = .075$ to the nearest .005
26. (a) $-.2560$
    (b) $-.2682$
    (c) $-.2568$
    (d) $-.2792$
27. 14
28. $[\nabla_x \nabla_y + \frac{1}{2}(\nabla_x^2 \nabla_y + \nabla_x \nabla_y^2) + \frac{1}{3}(\nabla_x^3 \nabla_y + \nabla_x \nabla_y^3) + \frac{1}{4}\nabla_x^2 \nabla_y^2]f(x, y)$
29. $\frac{1}{4}(u_{2:2} - u_{0:2} - u_{2:0} + u_{0:0})$
30. $\Delta u_x - \frac{1}{6}\Delta^2 u_x + \frac{1}{18}\Delta^3 u_x$
31. 4
32. $a = 1, b = -\frac{3}{2}$
34. 16
35. (b) $\delta \equiv D + \frac{1}{24}D^3 + \frac{1}{1920}D^5 + \cdots$
36. (b) $\mu \equiv 1 + \frac{1}{8}D^2 + \frac{1}{384}D^4 + \cdots$

Chapter 8

1. (a) 1.14202
   (b) 2.00116
2. 1.40977
3. $\dfrac{n^3}{3} + \dfrac{n^2}{2} + \dfrac{n}{6}$

5. $63\frac{1}{2}$

6. 60

7. 0

8. 156

10. 6000

11. (a) $4[mu_{-6} + nu_{-2} + pu_6 + qu_{10}]$

   (b) $\frac{1}{3}[mu_1 + nu_{1\frac{1}{3}} + pu_2 + qu_{2\frac{1}{3}}]$

12. $\dfrac{n}{6}(2n^2 + 1)$

14. $\frac{1}{24}[-f(0) + 5f(1) - 11f(2) + 23f(3) + 8f(4)]$

15. $\frac{1}{48}[17f(0) + 59f(1) + 44f(2) + 44f(3) + 59f(4) + 17f(5)]$

16. $-\frac{4}{15}$

17. $\frac{1}{48}$

18. $\sqrt{.0003} = .017$

19. $\sqrt[4]{.000375} = .139$

20. $\sqrt[4]{\dfrac{.0005}{3}} = .114$

21. $\frac{8}{75} = .107$

22. $\sqrt{.0000125} = .0035$

23. 50

24. 24

25. (a) $a = \frac{11}{6}, b = \frac{1}{12}$

   (b) Third degree

26. (a) $\frac{1}{18}[-15u_{-1} + 44u_0 + 7u_3]$

   (b) $\frac{11}{3}E$

27. $a = 2, b = \frac{2}{3}$

32. (a) .785393

33. 1.0213

35. $\frac{1}{81}$

36. (a) $I_k^1 = \dfrac{16S_k - S_{k-1}}{15}$

   (b) $I_k^m = \dfrac{16^m I_k^{m-1} - I_{k-1}^{m-1}}{16^m - 1}$

38. m^{-2n}

39. (b)

x	H
0	2

40. (b)

x	H
$-\sqrt{\frac{1}{3}}$	1
$\sqrt{\frac{1}{3}}$	1

41. (b)

x	H
$-\sqrt{\frac{3}{5}}$	$\frac{5}{9}$
0	$\frac{8}{9}$
$\sqrt{\frac{3}{5}}$	$\frac{5}{9}$

42. 13th degree
43. $\sqrt{\frac{5}{3}}$
44. $\dfrac{f^{(4)}(\xi)}{135}$
45. (a) 0
    (b) $\frac{2}{7}$
46. $-\frac{8}{5}$
47. (a) $2\displaystyle\int_0^1 e^{t^2}\,dt$

    (b) $\displaystyle\int_{-\pi/2}^{\pi/2} \cdot\, \sin^2\theta\,d\theta$ or $\displaystyle\int_{-1}^1 \sqrt{1-t^2}\,dt$
48. (a) Yes
    (b) No
49. $3\displaystyle\int_0^1 t\log_e\left(1+t^3\right)dt$
50. $\frac{3}{5}-\frac{3}{16}+\frac{3}{33}-\frac{3}{56}$
51. $\frac{9}{64}[f(0,0)+3f(0,1)+3f(0,2)+f(0,3)$
    $+\,3f(1,0)+9f(1,1)+9f(1,2)+3f(1,3)$
    $+\,3f(2,0)+9f(2,1)+9f(2,2)+3f(2,3)$
    $+\,f(3,0)+3f(3,1)+3f(3,2)+f(3,3)]$
52. $\frac{688}{3}$
53. (b) is better
54. (a) 4.398
    (b) 4.390

Chapter 9

1. (a) Second
   (b) First
2. 129
3. $\frac{3}{4}$
4. 70
5. $u_n = \dfrac{a^{n+1}-1}{a-1}$
6. $u_n = A + bn$
7. $1 + 2^x$
8. $\dfrac{x^3 + 3x^2 + 2x}{3}$
10. $\Delta f(x) = xf(x)$ and $f(0) = 1$
11. $(2x-4)3^x$
13. 2
14. $\dfrac{1}{(x-1)!}$
15. (a) $r_1^x r_2^x(r_1 - r_2)$
    (b) $-r^{2x+1}$
    (c) $R^{2x+1}\sin\theta$

16. (a) $u_x = Aa^x$
    (b) $r - a = 0$
    (c) $u_x = ca^x$
    (d) Yes

17. $c_1(\sqrt{2})^x \sin \dfrac{\pi x}{4} + c_2(\sqrt{2})^x \cos \dfrac{\pi x}{4}$

18. (a) $c_1 + c_2 x$
    (b) Yes

19. $\dfrac{1}{2\sqrt{2}}[(1 + \sqrt{2})^x - (1 - \sqrt{2})^x]$

21. $c_1(2 + \sqrt{3})^x + c_2(2 - \sqrt{3})^x$

22. $4^x(x + 2)$

23. $\frac{1}{2}[1 - (-1)^x]$

24. $2^n + \dfrac{3^n}{2} - \dfrac{3}{2}$

25. $c_1(\sqrt{2})^x \sin \dfrac{\pi x}{4} + c_2(\sqrt{2})^x \cos \dfrac{\pi x}{4} + (x^2 - 4)$

26. $2^x(c_1 + c_2 x + \frac{3}{8}x^2)$

27. $c_1 + c_2 \cdot 3^x - x \cdot 2^x$

28. $2^x(2 + \frac{5}{8}x + \frac{3}{8}x^2)$

29. $3 + 2 \cdot 3^x - x \cdot 2^x$

30. $\frac{1}{27}[-4(-\frac{1}{2})^x + 2x(-\frac{1}{2})^x + (3x^2 - 8x + 4)]$

31. $c_1 + c_2 \cdot 2^x + c_3 \cdot 3^x$

32. $-.7(-2)^x + .5(2)^x + 1.2(3)^x$

33. $n!k^{n-1}u_1$

34. e^k

35. $F(x) = \dfrac{A}{1 + Ax}$

36. Do not multiply by k

Chapter 10

1. $\frac{1}{3}(x^3 - 3x + 7)$

2. $1 + x^2 + \dfrac{x^4}{2} + \dfrac{x^6}{6}$

3.

x	y
0	1.0000
.1	1.0101
.2	1.0408
.3	1.0942

4. Same as 3

5. $y(x_0) = y_0$

$$y(x_1) = y_0 + \dfrac{h}{12}[5y'(x_0) + 8y'(x_1) - y'(x_2)]$$

$$y(x_2) = y_0 + \dfrac{h}{3}[y'(x_0) + 4y'(x_1) + y'(x_2)]$$

6.

x	y
0	1.0000
.1	1.0000
.2	1.0200
.3	1.0608

7. (a) $y' = 2xy$
   $y'' = 4x^2y + 2y$
   $y''' = 8x^3y + 12xy$
   (b) Same as 3 with roundoff error

8. (a) $.005[4\xi^2y(\xi) + 2y(\xi)]$
   (b) .0103
   (c) .0101
   (d) Yes
   (e) .0112
   (f) .0208

10. (a) $\dfrac{h^2}{2} + \left(h - \dfrac{h^2}{2}\right)x_i + \left(1 - h + \dfrac{h^2}{2}\right)y(x_i)$

11. Same as 3 with roundoff error

12. .9097

13. .9100

14. .9097

15. .9097

17. (a) 1.1734
    (b) 1.1735

18. (a) 1.1735
    (b) 1.1735

19. .3

21. .7406

22. (a) .0620
    (b) .0022
    (c) 0

23. (a) $\frac{20}{3}hE$
    (b) $2hE$

24. (a) $\frac{20}{3}hE$
    (b) $\frac{17}{12}hE$

25. (a) $2 + 3x + \frac{5}{2}x^2 + \frac{3}{2}x^3$

26. (a) (1) z
       (2) $-2y + 3z$
       (3) $-2y + 3z$
       (4) $-6y + 7z$

    (b)

x	y
0	2.000
.1	2.325
.2	2.709

27. 2.3266

28. $y' = z$
     $z' = w$
     $w' = 3x - y - 2w$

Chapter 11

1. (a)   One in interval $-3 < x < -2$
            One in interval $0 < x < 1$
            One in interval $1 < x < 2$
   (b)   One in interval $-1 < x < 0$
            Two in interval $0 < x < 1$
2. (a)   2
   (b) $x^2 + 3x + 1$
   (c) $-2.618, -.382$
3. (a)   1
   (b)   2
   (c)   None
   (d)   An infinite number
   (e)   3
4. (a)   Convergence, first-order
   (b)   Divergence
   (c)   Convergence, second-order
6.   3
8. (a) (1) $x_{n+1} = \left(x_n^2 + x_n + 1\right)^{1/3}$
       (2) $x_{n+1} = 1 + \dfrac{1}{x_n} + \dfrac{1}{x_n^2}$
   (b)   1.84
   (c) (1) $|F'(1.84)| = .46$ for iteration in (a) (1)
       (2) $|F'(1.84)| = .62$ for iteration in (a) (2)
10. (a)   .57735
    (b)   Third iteration
11. (a)   2.25
    (b)   Second-order
12.   1.516
13.   11
14. $.5 \le x \le .75$
15.   1.516
16.   1.6
17. (a)   (1) not satisfied
    (b)   (2) and (3) not satisfied
18. (a) $F'(r) = 1 - \dfrac{f'(r)(A - r)}{B}$
    (b) $f'(r) = \dfrac{B}{A - r}$
19. (a) $-.70383$
    (b)   .00036
    (c)   .00109

20. (a)   1.167
     (b)   1.253
     (c)   1.323
     (d)   .072
     (e)   .002
21. 1.516
22. (a) $x_{n+1} = \dfrac{x_n^2 + e^{-x_n}}{x_n + 1}$

     (b) $x_{n+1} = \dfrac{(x_n - 1)e^{x_n} + 1}{e^{x_n} - 4}$

     (c)   No root exists

     (d) $x_{n+1} = \dfrac{x_n \tan^2 x_n - \tan x_n + x_n}{\tan^2 x_n}$

     (e) $x_{n+1} = \dfrac{x_n \cos x_n - \sin x_n + 1}{\cos x_n + \frac{1}{4}}$

23. (a) $x^4 - x - 10$
     (b) $x^{2/3} - 10$
     (c) $x - e^{-x}$
24. (a) $\dfrac{1}{2}\left(x_n + \dfrac{1}{ax_n}\right)$
26. $x_{n+1} = 2x_n - ax_n^2$
27. $\frac{11}{7}$ or $\frac{5}{3}$
28. $x_{n+2} = \dfrac{x_n^4 + 6x_n^2 - 4x_n + 2}{4x_n^3 - 6x_n^2 + 8x_n - 3}$
29. 1.5
30. (a)   1.517
     (b) $-.001$
31. 1
32. $\frac{1}{3}[.023x^3 - .282x^2 + 1.927x + 1.332]$
33. 2.248
34. (a)   4 complex
     (b)   1 negative
          2 complex
     (c)   1 positive
          1 negative
          2 complex
     (d)   2 positive
          2 complex
37. (a)   1.8430
     (b)   1.8393

Chapter 12

1. $\begin{bmatrix} 2 & -3 \\ -1 & 5 \\ 7 & 4 \end{bmatrix}$

3. No. The orders may be different.
4. A
6. (a) No
   (b) Yes
7. (a) $\begin{bmatrix} 14 & 6 \\ 0 & -5 \end{bmatrix}$

   (b) $\begin{bmatrix} 13 & -2 \\ -9 & -4 \end{bmatrix}$

8. (a) $[14]$

   (b) $\begin{bmatrix} 1 & 2 & 3 \\ 2 & 4 & 6 \\ 3 & 6 & 9 \end{bmatrix}$

9. (a) $[3 \quad 6 \quad 9]$

   (b) $\begin{bmatrix} 3 \\ 6 \\ 9 \end{bmatrix}$

   (c) $[3 \quad 6 \quad 9]$

   (d) Undefined

10. $\begin{bmatrix} 2^{n-1} & 2^{n-1} \\ 2^{n-1} & 2^{n-1} \end{bmatrix}$

11. All zero
17. I
18. $\begin{bmatrix} a_{11}b_{11} & 0 & \cdots & 0 \\ 0 & a_{22}b_{22} & \cdots & 0 \\ \vdots & \vdots & & \vdots \\ 0 & 0 & \cdots & a_{nn}b_{nn} \end{bmatrix}$

19. $\begin{bmatrix} \lambda^n & n\lambda^{n-1} \\ 0 & \lambda^n \end{bmatrix}$

20. (a) Odd
    (b) Even
21. -5
22. 0
23. 572
24. -70
25. $-6x^5 - 12x^2 + 4x + 5$
26. xy
29. 6
30. 0. Add row 1 to row 2 and then add row 3 to row 4.
31. x^{15}
33. $b^{n-1}(na + b)$
35. k^n
36. $\begin{bmatrix} \frac{3}{2} & -\frac{1}{4} & -\frac{1}{4} \\ \frac{3}{2} & -\frac{3}{4} & \frac{1}{4} \\ 1 & -\frac{1}{2} & \frac{1}{2} \end{bmatrix}$

39. $\begin{bmatrix} d_1^{-1} & 0 & \cdots & 0 \\ 0 & d_2^{-1} & \cdots & 0 \\ \vdots & \vdots & & \vdots \\ 0 & 0 & \cdots & d_n^{-1} \end{bmatrix}$

41. $\dfrac{1}{11}\begin{bmatrix} 9 & 4 \\ -22 & -11 \end{bmatrix}$

42. $\begin{bmatrix} 2.32 & -.68 \\ -.68 & .32 \end{bmatrix}$

43. A

44. $L = \begin{bmatrix} 1 & 0 & 0 \\ 2 & 1 & 0 \\ 0 & 1 & 1 \end{bmatrix} \qquad U = \begin{bmatrix} 1 & -1 & 1 \\ 0 & -2 & 1 \\ 0 & 0 & 2 \end{bmatrix}$

45. $L = \begin{bmatrix} 1 & 0 & 0 \\ 2 & -2 & 0 \\ 0 & -2 & 2 \end{bmatrix} \qquad U = \begin{bmatrix} 1 & -1 & 1 \\ 0 & 1 & -\frac{1}{2} \\ 0 & 0 & 1 \end{bmatrix}$

46. $L = \begin{bmatrix} 1 & 0 \\ 2 & 1 \end{bmatrix} \qquad U = \begin{bmatrix} 1 & 2 \\ 0 & 1 \end{bmatrix}$

48. $k^n |A|$

49. kn

53. $K = \begin{bmatrix} -1 & 3 & -3 & 1 & 0 & \cdots & 0 & 0 & 0 & 0 \\ 0 & -1 & 3 & -3 & 1 & \cdots & 0 & 0 & 0 & 0 \\ \multicolumn{10}{c}{\dotfill} \\ 0 & 0 & 0 & 0 & 0 & \cdots & -1 & 3 & -3 & 1 \end{bmatrix}$ K is an $n - 3 \times n$ matrix

Chapter 13

2. $x_1 = 2,\ x_2 = -3,\ x_3 = 4$

3. $A = \begin{bmatrix} 2 & 1 \\ 3 & 2 \end{bmatrix}$

4. Division by zero is undefined.

5. $x_1 = 2,\ x_2 = -3,\ x_3 = 4$

6. $1.000x_1 + .800x_2 + .600x_3 = 2.400$
$1.000x_2 + .421x_3 = 1.421$
$1.000x_3 = 1.000$

7. $x_1 = 1.000,\ x_2 = 1.000,\ x_3 = 1.000$

8. $-28.994;\ -29$

9. It is impossible to convert the equations to upper triangular form. An inconsistent result is obtained.

10. $x_1 = 2,\ x_2 = -3,\ x_3 = 4$

11. (a)

	c_2	x_2	c_1
x_3	$-\frac{1}{11}$	$\frac{19}{22}$	$\frac{7}{22}$
x_1	$\frac{2}{11}$	$-\frac{5}{22}$	$-\frac{3}{22}$
c_3	0	$\frac{1}{2}$	$-\frac{1}{2}$

	c_2	c_3	c_1
x_3	$-\frac{1}{11}$	$\frac{19}{11}$	$\frac{13}{11}$
x_1	$\frac{2}{11}$	$-\frac{5}{11}$	$-\frac{4}{11}$
x_2	0	2	1

12. (b) $\begin{bmatrix} 1 & 0 & 0 \\ -1 & 1 & 0 \\ 0 & 1 & 1 \end{bmatrix}$

    (c) $\begin{bmatrix} 1 & 0 & 0 \\ -1 & 1 & 0 \\ 0 & -1 & 1 \end{bmatrix}$

13. The matrix inverse does not exist.
14. $x_1 = 2, x_2 = -3, x_3 = 4$
15. $x_1 = 3, x_2 = 1$
17. (a) Yes
    (b) The system $Ux = y$ cannot be solved. An inconsistent result is obtained.
18. (a) $x_1^0 = 9, x_2^0 = -7, x_3^0 = 6$
    (b) $x_1^3 = 1\frac{2}{3}, x_2^3 = -3\frac{2}{3}, x_3^3 = 3\frac{5}{9}$
    (c) Yes
    (d) No
19. $x_1^1 = .317, x_2^1 = .293, x_3^1 = .246$
20. .134
21. The iteration does not converge.
22. $r_1 = 13, r_2 = -9, r_3 = 4\frac{2}{3}$
23. $x_1 = 2, x_2 = -3, x_3 = 4$
24. $x_1 = .5381, x_2 = 1.0236, x_3 = -.2514$
25. $x_1 = .5371, x_2 = 1.0270, x_3 = -.2441$

26.

Method	x_1	x_2	x_3
True answer	.5455	1.0000	$-.2727$
Gauss-Seidel (Exercise 24)	.5361	1.0384	$-.2380$
Exact relaxation (Exercise 25)	.5371	1.0270	$-.2441$
Overrelaxation (Example 13.8)	.5450	1.0010	$-.2720$

27. $x_1 = 1, x_2 = -3, x_3 = 2$
28. The two lines are nearly parallel, which means that small shifts in slope can cause large shifts in the point of intersection.
29. $x = -1, y = 3$
30. (a) $x = 1001, y = 1000$
    (b) $x = -999, y = -1000$
31. $x_1 = 10 + 11a, x_2 = -2 - 4a, x_3 = a, x_4 = 0$
32. $x_1 = \frac{1}{3}(-7a + 17), x_2 = \frac{1}{3}(4a - 5), x_3 = a$
33. $\begin{bmatrix} 1 & 0 & 7 & 5 & | & 13 \\ 0 & 1 & -5 & -4 & | & -8 \\ 0 & 0 & 0 & 0 & | & -5 \end{bmatrix}$
34. (a) 1

(b) $\begin{bmatrix} 1 \\ 1 \\ 1 \end{bmatrix}$

35. $-2, 1, 3$

36. $A = \begin{bmatrix} 0 & 1 \\ -2 & 3 \end{bmatrix}$

37. $x_2 = .660, y_2 = .441, z_2 = 1.093$

38. (a) 10%
    (b) $.46\%$

Chapter 14

1. Solution point is $(1, 0); f = 1$. Yes.
2. Solution points lie on $x_2 - x_1 = 3$ between the points $(0, 3)$ and $(3, 6); f = 3$
   No.
3. (a) Solution point is $(3, 0); f = 3$
   (b) Solution point is $\left(\frac{2}{3}, \frac{5}{3}\right); f = -\frac{8}{3}$
4. (a) No solution exists
   (b) Solution point is $(1, 0); f = 1$
5. Solution point is $(2, 0); f = 4$
6. Solution point is $(3, 6); f = 15$
7. (a) Solution point is $(3, 0); f = 3$
   (b) Solution point is $\left(\frac{2}{3}, \frac{5}{3}\right); f = -\frac{8}{3}$
8. (a) No solution exists
   (b) Solution point is $(1, 0); f = 1$
9. Solution point is $(2, 0); f = 4$
10. Solution point is $(6, 0); f = 6K$
11. Solution point is $(K, 0); f = 2K$
12. Minimize
$$g = 4y_1 + y_2 + 3y_3$$
subject to
$$\begin{array}{ll} y_1 - y_2 + y_3 \geq 1 & y_1 \geq 0 \\ 2y_1 + y_2 + y_3 \geq -2 & y_2 \geq 0 \\ & y_3 \geq 0 \end{array}$$
13. Solution point is $(0, 0, 1); g = 3$
14. Solution point is $\left(\frac{1}{5}, \frac{3}{5}, 0\right); g = 15$
15. $x_1 = \frac{3}{4}, x_2 = \frac{3}{4}$ assuming $x_1 = x_2$; solution is not unique
16. $x_1 = \frac{3}{4}, x_2 = \frac{3}{4}$ assuming $x_1 = x_2$; solution is not unique
17. $x = \dfrac{1}{n}(A_1 + A_2 + \cdots + A_n)$
18. $x = \frac{1}{3}(A + C + D), y = \frac{1}{3}(B - C + D)$
19. $$\begin{array}{rrrrr} 24x_1 + & 2x_2 - & 12x_3 - & 15x_4 = & 4 \\ 2x_1 + & 16x_2 - & 4x_3 - & 16x_4 = & 1 \\ -12x_1 - & 4x_2 + & 19x_3 - & x_4 = & -7 \\ -15x_1 - & 16x_2 - & x_3 + & 47x_4 = & 25 \end{array}$$

20. Maximize
$$f = y_5$$
subject to

$$
\begin{aligned}
y_1 - 2y_2 \quad\quad + \; y_4 - 7y_5 &\leq 1 \\
3y_1 \quad\quad + \; y_3 - 4y_4 + 2y_5 &\leq 1 \\
3y_2 \quad\quad - \; 5y_4 - \; y_5 &\leq 1 \\
-y_1 + \; y_2 + \; y_3 + 2y_4 - 8y_5 &\leq 1 \\
-3y_1 - \; y_2 + 4y_3 \quad\quad + 3y_5 &\leq 1 \\
2y_1 + \; y_2 - \; y_3 - \; y_4 - \; y_5 &\leq 1 \\
-y_1 + 2y_2 \quad\quad - \; y_4 + 7y_5 &\leq 1 \\
-3y_1 \quad\quad - \; y_3 + 4y_4 - 2y_5 &\leq 1 \\
- 3y_2 \quad\quad + 5y_4 + \; y_5 &\leq 1 \\
y_1 - \; y_2 - \; y_3 - 2y_4 + 8y_5 &\leq 1 \\
3y_1 + \; y_2 - 4y_3 \quad\quad - 3y_5 &\leq 1 \\
-2y_1 - \; y_2 + \; y_3 + \; y_4 + \; y_5 &\leq 1
\end{aligned}
\qquad
\begin{aligned}
y_1 &\geq 0 \\
y_2 &\geq 0 \\
y_3 &\geq 0 \\
y_4 &\geq 0 \\
y_5 &\geq 0
\end{aligned}
$$

21. Maximize
$$f = 7x_1 + 7x_2 + 10x_3 + 10x_4$$
subject to

$$
\begin{aligned}
x_1 + 2x_2 + \; 2x_3 + 3x_4 &\leq 25 \\
7x_1 + 5x_2 + 10x_3 + 7x_4 &\leq 100 \\
4x_1 + 4x_2 + \; 8x_3 + 4x_4 &\leq 70
\end{aligned}
\qquad
\begin{aligned}
x_1 &\geq 0 \\
x_2 &\geq 0 \\
x_3 &\geq 0 \\
x_4 &\geq 0
\end{aligned}
$$

22. Minimize
$$f = 5x_1 + 20x_2 + 8x_3 + 11x_4 + 12x_5$$
subject to

$$
\begin{aligned}
5x_1 + 25x_2 + 20x_3 + \; 5x_4 + 12x_5 &\leq 1500 \\
5x_1 + 25x_2 + 20x_3 + \; 5x_4 + 12x_5 &\geq 1000 \\
7x_1 + 20x_2 + \; 2x_3 + 25x_4 + 10x_5 &\geq 1000 \\
2x_1 + 16x_2 + \; 9x_3 + \; 8x_4 + 12x_5 &\geq \; 700
\end{aligned}
\qquad
\begin{aligned}
x_1 &\geq 0 \\
x_2 &\geq 0 \\
x_3 &\geq 0 \\
x_4 &\geq 0 \\
x_5 &\geq 0
\end{aligned}
$$

23. Minimize
$$f = 3000x_1 + 2000x_2$$
subject to

$$
\begin{aligned}
60x_1 + 20x_2 &= 240 \\
20x_1 + 20x_2 &\geq 160 \\
40x_1 + 80x_2 &\geq 440
\end{aligned}
\qquad
\begin{aligned}
x_1 &\geq 0 \\
x_2 &\geq 0
\end{aligned}
$$

Chapter 15

1. $-x^3 + 2x^2$
3. $\dfrac{f^{(4)}(\xi)(x_1 - x_0)^4}{384}$
4. $-x^3 + 2x^2$
5. $\frac{1}{2}(7x^5 - 18x^4 + 13x^3)$
6. $x^3 - x^2 + 1$
7. $B(s) = \frac{1}{2}(s^3 - s^2)$
8. $B(s) = \frac{1}{4}s^2$

9. $B(s) = \frac{1}{6}s^3$
10. $C(s) = \frac{1}{12}(-s^4 + s^3)$
11. $C(s) = -\frac{1}{36}s^3$
12. (a) Yes
    (b) Yes
    (c) No
13. (a) Yes
    (b) No
    (c) Yes
14. (a) Second
    (b) Yes
16. (a) $.6x + 3.8$
    (b) 2.8
17. (a) $-.125x^2 + 1.1x + 3.925$
    (b) 1.8
19. (a) .00075
    (b) .0003
20. $\frac{27}{17}$
21. $\frac{156}{49}$
22. 6
26. (a) $.5x + 4.5$
    (b) 1
27. (a) $-.125x^2 + 1.25x + 3.625$
    (b) .75
28. (a) (1) 2.8
        (2) 4
    (b) (1) 1.96
        (2) 1
29. (a) (1) 1.8
        (2) 2.25
    (b) (1) .81
        (2) .75
31. $\dfrac{3}{x}$
32. $\dfrac{x+1}{x+2}$
33. $\dfrac{x^2-1}{x^2+1}$
36. $2 + 3 \cdot 4^x$
37. 2^{2x-2}
38. $(\frac{3}{2}x - 1)2^x$
39. $(1.25)^x + (1.5)^x$
42. (c) 64.75
    (d) 54

Bibliography

Arden, Bruce W., and Astill, Kenneth N. *Numerical Algorithms: Origins and Applications*. Boston: Addison-Wesley Publishing Co., Inc., 1970.

Ayres, Frank Jr. *Theory and Problems of Matrices*. New York: Schaum Publishing Co., 1962.

Celia, C. W. *An Introduction to Numerical Analysis*. London, England: McGraw-Hill Book Co., 1969.

Conte, S. D., and deBoor, Carl. *Elementary Numerical Analysis*. New York: McGraw-Hill Book Co., 1972.

Freeman, Harry. *Finite Differences for Actuarial Students*. Cambridge, England: Cambridge University Press, 1960.

Fröberg, Carl-Erik. *Introduction to Numerical Analysis*. Boston: Addison-Wesley Publishing Co., Inc., 1969.

Greville, T. N. E. "Graduation" Part 5 Study Note 53.1.73. Chicago: Society of Actuaries, 1973.

Hamming, Richard W. *Introduction to Applied Numerical Analysis*. New York: McGraw-Hill Book Co., 1971.

Hildebrand, F. B. *Introduction to Numerical Analysis*. New York: McGraw-Hill Book Co., 1956.

IBM Corporation. *BASIC Language Reference Manual* (GC-28-6837-0). New York: IBM Corporation, 1970.

Jennings, W. *First Course in Numerical Methods*. New York: The Macmillan Co. 1964.

451

McCalla, Thomas R. *Introduction to Numerical Methods and FORTRAN Programming*. New York: John Wiley & Sons, Inc., 1967.

Miller, Morton D. *Elements of Graduation*. Chicago: Society of Actuaries (The Actuarial Society of America—American Institute of Actuaries), 1946.

Scheid, Francis. *Theory and Problems of Numerical Analysis*. New York: McGraw-Hill Book Co., 1968.

Spivey, W. Allen. *Linear Programming: An Introduction*. New York: The Macmillan Co., 1963.

Wilkes, M. V. *A Short Introduction to Numerical Analysis*. Cambridge, England: Cambridge University Press, 1966.

Index

This book has been set in 10 point Times New Roman leaded 3 points and 9 point Times New Roman leaded 2 points. Chapter numbers are in 30 point Times New Roman Bold; chapter titles are in 18 point Times New Roman. The size of the type page is 27 by 46 picas.